NUREG-1841

U.S. Operating Experience With Thermally Treated Alloy 690 Steam Generator Tubes

Manuscript Completed: March 2006
Date Published: August 2007

Prepared by
K.J. Karwoski, G.L. Makar, and M.G. Yoder

Division of Component Integrity
Office of Nuclear Reactor Regulation
U.S. Nuclear Regulatory Commission
Washington, DC 20555-0001

ABSTRACT

This report documents the background and performance of thermally treated Alloy 690 steam generator tubing in U.S. commercial pressurized-water reactors (PWRs). The industry has used this material extensively for replacement steam generators since 1989. As of December 31, 2004, it was being used in 30 units, or about 43 percent of the operating PWRs in the United States. Of the 577,070 thermally treated Alloy 690 tubes placed in service, only 333 tubes (0.06 percent) have been plugged after approximately 173 calendar years of operation. The majority of these tubes (65 percent) were plugged prior to placing the steam generators in service. The dominant inservice degradation mode, responsible for about 24 percent of the plugged tubes, has been wear caused by a support structure or loose part. No corrosion or cracking had been detected as of the time this report was prepared. The superior performance experienced to date with thermally treated Alloy 690 tubes compared to earlier tube materials is attributed to the alloy chemistry (principally the higher chromium content), the corrosion-resistant microstructure developed by the combination of alloy chemistry and thermal processing, and design improvements in replacement steam generators.

CONTENTS

ABSTRACT . . . . -iii-

CONTENTS . . . . -v-

LIST OF FIGURES . . . . -vii-

LIST OF TABLES . . . . -viii-

EXECUTIVE SUMMARY . . . . -xi-

1 INTRODUCTION . . . . -1-
   1.1 Safety Significance . . . . -1-
   1.2 Tube Integrity Program . . . . -2-
      1.2.1 Purpose of Inspections . . . . -2-
      1.2.2 Eddy Current Testing . . . . -3-
      1.2.3 Tube Repairs . . . . -4-
      1.2.4 Leakage Monitoring . . . . -5-
   1.3 Mill-Annealed Alloy 600 Steam Generator Operating Experience . . . . -5-
   1.4 Thermally Treated Alloy 600 Tubes . . . . -6-
   1.5 Thermally Treated Alloy 690 Tubes . . . . -7-

2 STEAM GENERATOR DESIGNS IN UNITS WITH
  THERMALLY TREATED ALLOY 690 TUBES . . . . -11-
   2.1 Introduction . . . . -11-
   2.2 General Steam Generator Description . . . . -11-
   2.3 Grouping of Steam Generator Designs . . . . -13-
      2.3.1 Westinghouse Delta and "F" Model Steam Generators . . . . -13-
      2.3.2 BWI Steam Generators . . . . -14-
      2.3.3 Other Steam Generators . . . . -14-
   2.4 Individual Unit Steam Generator Designs . . . . -15-
      2.4.1 Arkansas Nuclear One Unit 2 . . . . -15-
      2.4.2 Braidwood 1 . . . . -16-
      2.4.3 Byron 1 . . . . -17-
      2.4.4 Calvert Cliffs 1 . . . . -18-
      2.4.5 Calvert Cliffs 2 . . . . -19-
      2.4.6 Catawba 1 . . . . -19-
      2.4.7 Cook 1 . . . . -21-
      2.4.8 Cook 2 . . . . -21-
      2.4.9 Farley 1 . . . . -22-
      2.4.10 Farley 2 . . . . -23-
      2.4.11 Ginna . . . . -23-
      2.4.12 Harris . . . . -27-
      2.4.13 Indian Point 3 . . . . -28-
      2.4.14 Kewaunee . . . . -28-
      2.4.15 McGuire 1 . . . . -29-
      2.4.16 McGuire 2 . . . . -31-
      2.4.17 Millstone 2 . . . . -32-
      2.4.18 North Anna 1 . . . . -33-
      2.4.19 North Anna 2 . . . . -33-
      2.4.20 Oconee 1, 2, and 3 . . . . -34-
      2.4.21 Palo Verde 2 . . . . -34-
      2.4.22 Point Beach 2 . . . . -35-
      2.4.23 Prairie Island 1 . . . . -36-

2.4.24 Sequoyah 1 . . . . -37-
2.4.25 South Texas Project 1 . . . . -38-
2.4.26 South Texas Project 2 . . . . -40-
2.4.27 St. Lucie 1 . . . . -42-
2.4.28 Summer . . . . -42-

3 THERMALLY TREATED ALLOY 690 STEAM GENERATOR TUBE
   OPERATING EXPERIENCE . . . . -92-
   3.1 Data-Gathering Methodology and Introduction . . . . -92-
   3.2 Unit Inspection Results . . . . -92-
      3.2.1 Arkansas Nuclear One Unit 2 . . . . -92-
      3.2.2 Braidwood 1 . . . . -94-
      3.2.3 Byron 1 . . . . -99-
      3.2.4 Calvert Cliffs 1 . . . . -101-
      3.2.5 Calvert Cliffs 2 . . . . -102-
      3.2.6 Catawba 1 . . . . -102-
      3.2.7 Cook 1 . . . . -104-
      3.2.8 Cook 2 . . . . -105-
      3.2.9 Farley 1 . . . . -109-
      3.2.10 Farley 2 . . . . -110-
      3.2.11 Ginna . . . . -111-
      3.2.12 Harris . . . . -113-
      3.2.13 Indian Point 3 . . . . -116-
      3.2.14 Kewaunee . . . . -119-
      3.2.15 McGuire 1 . . . . -121-
      3.2.16 McGuire 2 . . . . -122-
      3.2.17 Millstone 2 . . . . -123-
      3.2.18 North Anna 1 . . . . -126-
      3.2.19 North Anna 2 . . . . -128-
      3.2.20 Oconee 1 . . . . -129-
      3.2.21 Oconee 2 . . . . -131-
      3.2.22 Oconee 3 . . . . -131-
      3.2.23 Palo Verde 2 . . . . -132-
      3.2.24 Point Beach 2 . . . . -134-
      3.2.25 Prairie Island 1 . . . . -136-
      3.2.26 Sequoyah 1 . . . . -136-
      3.2.27 South Texas Project 1 . . . . -137-
      3.2.28 South Texas Project 2 . . . . -140-
      3.2.29 St. Lucie 1 . . . . -141-
      3.2.30 Summer . . . . -142-

4 SUMMARY . . . . -235-
   4.1 Design Summary . . . . -235-
   4.2 Operating Experience Summary . . . . -235-
      4.2.1 Forced Outages . . . . -236-
      4.2.2 Tube Pulls . . . . -237-
      4.2.3 Summary and Observations . . . . -237-

APPENDIX B: BIBLIOGRAPHY . . . . -259-

LIST OF FIGURES

Figure 2-1: Number of Units with Thermally Treated Alloy 690 Tubes per Year . . . . -46-
Figure 2-2: Number of Thermally Treated Alloy 690 Tubes in Service per Year . . . . -47-
Figure 2-3: Percentage of Units by Tube Manufacturer (for Units with Alloy 690 Tubes) . . -48-
Figure 2-4: Percentage of Alloy 690 Tubes by Manufacturer . . . . -49-
Figure 2-5: Pressurized-Water Reactor Recirculating Steam Generator . . . . -50-
Figure 2-6: Pressurized-Water Reactor Once-Through Steam Generator . . . . -51-
Figure 2-7: U-Bend Features . . . . -52-
Figure 2-8: Tube Support Naming Convention at ANO Unit 2 . . . . -53-
Figure 2-9: Tubesheet Map for Braidwood Unit 1 . . . . -54-
Figure 2-10: Tube Support Naming Convention at Braidwood Unit 1 . . . . -55-
Figure 2-11: Tubesheet Map for Byron Unit 1 . . . . -56-
Figure 2-12: Tube Support Naming Convention at Byron Unit 1 . . . . -57-
Figure 2-13: Tubesheet Map for Calvert Cliffs Units 1 and 2 . . . . -58-
Figure 2-14: Tube Support Naming Convention at Calvert Cliffs Units 1 and 2 . . . . -59-
Figure 2-15: Tubesheet Map for Catawba Unit 1 . . . . -60-
Figure 2-16: Tube Support Naming Convention at Catawba Unit 1 . . . . -61-
Figure 2-17: Tubesheet Map for Cook Unit 1 . . . . -62-
Figure 2-18: Tube Support Naming Convention at Cook Unit 1 . . . . -63-
Figure 2-19: Tubesheet Map for Cook Unit 2 . . . . -64-
Figure 2-20: Tube Support Naming Convention at Cook Unit 2 . . . . -65-
Figure 2-21: Tubesheet Map for Farley Units 1 and 2 . . . . -66-
Figure 2-22: Tube Support Naming Convention at Farley Units 1 and 2 . . . . -67-
Figure 2-23: Tubesheet Map for Ginna . . . . -68-
Figure 2-24: Tube Support Naming Convention at Ginna . . . . -69-
Figure 2-25: Tubesheet Map for Harris . . . . -70-
Figure 2-26: Tube Support Naming Convention at Harris . . . . -71-
Figure 2-27: Tubesheet Map for Indian Point Unit 3 . . . . -72-
Figure 2-28: Tube Support Naming Convention at Indian Point Unit 3 . . . . -73-
Figure 2-29: Tubesheet Map for Kewaunee . . . . -74-
Figure 2-30: Tube Support Naming Convention at Kewaunee . . . . -75-
Figure 2-31: Tubesheet Map for McGuire Units 1 and 2 . . . . -76-
Figure 2-32: Tube Support Naming Convention at McGuire Units 1 and 2 . . . . -77-
Figure 2-33: Tubesheet Map for Millstone Unit 2 . . . . -78-
Figure 2-34: Tube Support Naming Convention at Millstone Unit 2 . . . . -79-
Figure 2-35: Tubesheet Map for North Anna Units 1 and 2 . . . . -80-
Figure 2-36: Tube Support Naming Convention at North Anna Units 1 and 2 . . . . -81-
Figure 2-37: Tubesheet Map for Oconee Units 1, 2, and 3 . . . . -82-
Figure 2-38: Tube Support Naming Convention at Oconee Units 1, 2, and 3 . . . . -83-
Figure 2-39: Tubesheet Map for Palo Verde Unit 2 . . . . -84-
Figure 2-40: Tube Support Naming Convention at Palo Verde Unit 2 . . . . -85-
Figure 2-41: Tubesheet Map for Point Beach Unit 2 . . . . -86-
Figure 2-42: Tube Support Naming Convention at Point Beach Unit 2 . . . . -87-
Figure 2-43: Tubesheet Map for South Texas Project Units 1 and 2 . . . . -88-
Figure 2-44: Tube Support Naming Convention at South Texas Project Units 1 and 2 . . . -89-
Figure 2-45: Tubesheet Map for Summer . . . . -90-
Figure 2-46: Tube Support Naming Convention at Summer . . . . -91-
Figure 4-1: Number of Tubes Plugged per Year . . . . -253-
Figure 4-2: Percentage of Tubes Plugged per Year . . . . -254-
Figure 4-3: Causes of Tube Plugging . . . . -255-
Figure 4-4: Causes of Tube Plugging per Year . . . . -256-

LIST OF TABLES

Table 1-1: Units with Replacement Steam Generators (Part 1) . . . . -9-
Table 1-1: Units with Replacement Steam Generators (Part 2) . . . . -10-
Table 2-1: Units with Thermally Treated Alloy 690 Tubes (Part 1) . . . . -43-
Table 2-1: Units with Thermally Treated Alloy 690 Tubes (Part 2) . . . . -44-
Table 2-2: Steam Generator Design Information for Units with Thermally Treated Alloy 690
      Tubes . . . . -45-
Table 3-1: ANO 2 Full-Length Bobbin Exams . . . . -144-
Table 3-2: ANO 2 Causes of Tube Plugging . . . . -145-
Table 3-3: ANO 2: Tubes Plugged . . . . -146-
Table 3-4: Braidwood 1 Full-Length Bobbin Exams . . . . -147-
Table 3-5: Braidwood 1 Causes of Tube Plugging . . . . -148-
Table 3-6: Braidwood 1: Tubes Plugged . . . . -149-
Table 3-7: Byron 1 Full-Length Bobbin Exams . . . . -151-
Table 3-8: Byron 1 Causes of Tube Plugging . . . . -152-
Table 3-9: Byron 1: Tubes Plugged . . . . -153-
Table 3-10: Calvert Cliffs 1 Full-Length Bobbin Exams . . . . -154-
Table 3-11: Calvert Cliffs 1 Causes of Tube Plugging . . . . -155-
Table 3-12: Calvert Cliffs 1: Tubes Plugged . . . . -156-
Table 3-13: Calvert Cliffs 2 Full-Length Bobbin Exams . . . . -157-
Table 3-14: Calvert Cliffs 2 Causes of Tube Plugging . . . . -158-
Table 3-15: Calvert Cliffs 2: Tubes Plugged . . . . -159-
Table 3-16: Catawba 1 Full-Length Bobbin Exams . . . . -160-
Table 3-17: Catawba 1 Causes of Tube Plugging . . . . -161-
Table 3-18: Catawba 1: Tubes Plugged . . . . -162-
Table 3-19: Cook 1 Full-Length Bobbin Exams . . . . -163-
Table 3-20: Cook 1 Causes of Tube Plugging . . . . -164-
Table 3-21: Cook 1: Tubes Plugged . . . . -165-
Table 3-22: Cook 2 Full-Length Bobbin Exams . . . . -166-
Table 3-23: Cook 2 Causes of Tube Plugging . . . . -167-
Table 3-24: Cook 2: Tubes Plugged . . . . -168-
Table 3-25: Farley 1 Full-Length Bobbin Exams . . . . -169-
Table 3-26: Farley 1 Causes of Tube Plugging . . . . -170-
Table 3-27: Farley 1: Tubes Plugged . . . . -171-
Table 3-28: Farley 2 Full-Length Bobbin Exams . . . . -172-
Table 3-29: Farley 2 Causes of Tube Plugging . . . . -173-
Table 3-30: Farley 2: Tubes Plugged . . . . -174-
Table 3-31: Ginna Full-Length Bobbin Exams . . . . -175-
Table 3-32: Ginna Causes of Tube Plugging . . . . -176-
Table 3-33: Ginna: Tubes Plugged . . . . -177-
Table 3-34: Harris Full-Length Bobbin Exams . . . . -178-
Table 3-35: Harris Causes of Tube Plugging . . . . -179-
Table 3-36: Harris: Tubes Plugged . . . . -180-
Table 3-37: Indian Point 3 Full-Length Bobbin Exams . . . . -181-
Table 3-38: Indian Point 3 Causes of Tube Plugging . . . . -182-
Table 3-39: Indian Point 3: Tubes Plugged . . . . -183-
Table 3-40: Kewaunee Full-Length Bobbin Exams . . . . -184-
Table 3-41: Kewaunee Causes of Tube Plugging . . . . -185-
Table 3-42: Kewaunee: Tubes Plugged . . . . -186-
Table 3-43: McGuire 1 Full-Length Bobbin Exams . . . . -187-
Table 3-44: McGuire 1 Causes of Tube Plugging . . . . -188-
Table 3-45: McGuire 1: Tubes Plugged . . . . -189-
Table 3-46: McGuire 2 Full-Length Bobbin Exams . . . . -190-
Table 3-47: McGuire 2 Causes of Tube Plugging . . . . -191-
Table 3-48: McGuire 2: Tubes Plugged . . . . -192-

Table 3-49: Millstone 2 Full-Length Bobbin Exams . . . . -193-
Table 3-50: Millstone 2 Causes of Tube Plugging . . . . -194-
Table 3-51: Millstone 2: Tubes Plugged . . . . -195-
Table 3-52: North Anna 1 Full-Length Bobbin Exams . . . . -196-
Table 3-53: North Anna 1 Causes of Tube Plugging . . . . -197-
Table 3-54: North Anna 1: Tubes Plugged . . . . -198-
Table 3-55: North Anna 2 Full-Length Bobbin Exams . . . . -199-
Table 3-56: North Anna 2 Causes of Tube Plugging . . . . -200-
Table 3-57: North Anna 2: Tubes Plugged . . . . -201-
Table 3-58: Oconee 1 Full-Length Bobbin Exams . . . . -202-
Table 3-59: Oconee 1 Causes of Tube Plugging . . . . -203-
Table 3-60: Oconee 1: Tubes Plugged . . . . -204-
Table 3-61: Oconee 2 Full-Length Bobbin Exams . . . . -205-
Table 3-62: Oconee 2 Causes of Tube Plugging . . . . -206-
Table 3-63: Oconee 2: Tubes Plugged . . . . -207-
Table 3-64: Oconee 3 Full-Length Bobbin Exams . . . . -208-
Table 3-65: Oconee 3 Causes of Tube Plugging . . . . -209-
Table 3-66: Oconee 3: Tubes Plugged . . . . -210-
Table 3-67: Palo Verde 2 Full-Length Bobbin Exams . . . . -211-
Table 3-68: Palo Verde 2 Causes of Tube Plugging . . . . -212-
Table 3-69: Palo Verde 2: Tubes Plugged . . . . -213-
Table 3-70: Point Beach 2 Full-Length Bobbin Exams . . . . -214-
Table 3-71: Point Beach 2 Causes of Tube Plugging . . . . -215-
Table 3-72: Point Beach 2: Tubes Plugged . . . . -216-
Table 3-73: Prairie Island 1 Full-Length Bobbin Exams . . . . -217-
Table 3-74: Prairie Island 1 Causes of Tube Plugging . . . . -218-
Table 3-75: Prairie Island 1: Tubes Plugged . . . . -219-
Table 3-76: Sequoyah 1 Full-Length Bobbin Exams . . . . -220-
Table 3-77: Sequoyah 1 Causes of Tube Plugging . . . . -221-
Table 3-78: Sequoyah 1: Tubes Plugged . . . . -222-
Table 3-79: South Texas Project 1 Full-Length Bobbin Exams . . . . -223-
Table 3-80: South Texas Project 1 Causes of Tube Plugging . . . . -224-
Table 3-81: South Texas Project 1: Tubes Plugged . . . . -225-
Table 3-82: South Texas Project 2 Full-Length Bobbin Exams . . . . -226-
Table 3-83: South Texas Project 2 Causes of Tube Plugging . . . . -227-
Table 3-84: South Texas Project 2: Tubes Plugged . . . . -228-
Table 3-85: St. Lucie 1 Full Length Bobbin Exams . . . . -229-
Table 3-86: St. Lucie 1 Causes of Tube Plugging . . . . -230-
Table 3-87: St. Lucie 1: Tubes Plugged . . . . -231-
Table 3-88: Summer Full-Length Bobbin Exams . . . . -232-
Table 3-89: Summer Causes of Tube Plugging . . . . -233-
Table 3-90: Summer: Tubes Plugged . . . . -234-
Table 4-1: Total Number and Percentage of Tubes Plugged for All Models (12/2004)
        (Part 1) . . . . -239-
Table 4-1: Total Number and Percentage of Tubes Plugged for All Models (12/2004)
        (Part 2) . . . . -240-
Table 4-2: Plugging per Year (Part 1) . . . . -241-
Table 4-2: Plugging per Year (Part 2) . . . . -242-
Table 4-2: Plugging per Year (Part 3) . . . . -243-
Table 4-2: Plugging per Year (Part 4) . . . . -244-
Table 4-3: Cumulative Plugging per Year (Part 1) . . . . -245-
Table 4-3: Cumulative Plugging per Year (Part 2) . . . . -246-
Table 4-3: Cumulative Plugging per Year (Part 3) . . . . -247-
Table 4-3: Cumulative Plugging per Year (Part 4) . . . . -248-
Table 4-4: Number of Tubes Plugged as a Function of Mechanism, All Plants . . . . -249-
Table 4-5: Number of Tubes Plugged as a Function of Mechanism per Year (Detailed) . . -250-

Table 4-6: Number of Tubes Plugged as a Function of Mechanism per Year (Summary)
. . . . -251-
Table 4-7: Fraction of Tubes Plugged as a Function of Mechanism per Year (Summary)
. . . . -252-

EXECUTIVE SUMMARY

The susceptibility of steam generator tubes to degradation is affected by various factors, including the steam generator design, the operating environment (temperature and water chemistry), and operating and residual stresses. Two of the most important factors affecting the susceptibility of a tube to degradation are the tube material and the tube's heat treatment.

Tubes installed in U.S. nuclear steam generators placed in service in the 1960s and 1970s were usually only mill annealed (passed through a furnace at a high temperature). More than 25 years of operating experience have shown that mill-annealed Alloy 600 is susceptible to degradation in the steam generator operating environment. The degradation includes pitting, wear, thinning, wastage, and stress-corrosion cracking. Cracking has initiated from both the primary and secondary sides of the tubes.

The extensive tube degradation at pressurized-water reactors (PWRs) with mill-annealed Alloy 600 steam generator tubes resulted in numerous tube leaks, approximately nine tube ruptures, numerous midcycle steam generator tube inspections, and the replacement of steam generators at numerous plants. In addition, extensive tube degradation contributed to the permanent shutdown of other plants. Haddam Neck, Maine Yankee, Trojan, Zion 1, Zion 2, and San Onofre 1 ceased operation with significant amounts of tube degradation.

As mill-annealed Alloy 600 steam generator tubes began exhibiting degradation in the early 1970s, the industry pursued improvements in the design of future steam generators to reduce the likelihood of corrosion (including stress-corrosion cracking). In the late 1970s, some mill-annealed Alloy 600 tubes were subjected to high temperatures for 10 to 15 hours to relieve fabrication stresses and to improve the tubes' microstructure. This thermal treatment process was first used on tubes installed in replacement steam generators put into service in the early 1980s. The steam generators at 17 units use thermally treated Alloy 600 as of December 2004. At another unit, Callaway, the steam generators have thermally treated Alloy 600 tubes in the first 10 rows and mill-annealed Alloy 600 tubes in the remaining rows. Therefore, approximately 25 percent of the currently operating PWRs (18 of 69) use thermally treated Alloy 600 as of December 2004.

Another approach to improving the degradation resistance of steam generator tubes was to identify more corrosion-resistant alloys. Alloy 690 is a nickel alloy similar to Alloy 600; however, it has a higher chromium content, which reduces the likelihood of corrosion (including stress-corrosion cracking) in the steam generator operating environment. Thermally treated Alloy 690 was first used on tubes installed in replacement steam generators placed into service in 1989. As of December 31, 2004, 30 units use thermally treated Alloy 690 steam generator tubes. Therefore, approximately 43 percent of the currently operating PWRs (30 of 69) use thermally treated Alloy 690.

Various general communications, including U.S. Nuclear Regulatory Commission (NRC) Information Notices, Bulletins, Generic Letters, and NUREGs document the operating experience of plants with mill-annealed Alloy 600 steam generator tubes. NUREG-1771, "U.S. Operating Experience with Thermally Treated Alloy 600 Steam Generator Tubes," April 2003, describes the experience with thermally treated Alloy 600. The following sections document the design and operating experience of U.S. PWR steam generators with thermally treated Alloy 690 tubes as of December 2004.

A historical review of operating experience identified two unplanned outages as a result of steam generator issues in units with thermally treated Alloy 690 tubes as of December 31, 2004. These two outages occurred in 2004 when the units (Palo Verde 2 and Harris) were shut down after discovering primary-to-secondary leakage.

During the preparation of this report in the first half of 2005, several noteworthy events occurred in units with thermally treated Alloy 690 steam generator tubes. An additional two unplanned outages or inspections took place as a result of steam generator issues (bringing the total number of unplanned outages or inspections to four). One unit inadvertently introduced chemical contaminants into its steam generators, and a large number of tubes were found with wear scars in another unit. One of the unplanned outages/inspections occurred at Arkansas Nuclear One Unit 2, where the unit was shut down after the discovery of primary-to-secondary leakage. The other unplanned outage/inspection occurred at South Texas Project Unit 1, where tube inspections were performed during an outage in which none were planned because a large number of loose parts were found in a steam generator. At Kewaunee, chemical contaminants were inadvertently introduced into the steam generators and the unit remained shut down to restore proper water chemistry. At Oconee 1, several thousand indications of wear were detected in the replacement steam generators after one cycle of operation during its spring 2005 outage. None of the wear indications were safety significant; however, such a large number of wear flaws was not expected. The steam generators at Oconee 1 were the first once-through steam generators replaced in the United States (a very limited number of wear indications have been observed in recirculating steam generators with thermally treated Alloy 690 tubes). At the time the staff was preparing this report, a root cause investigation was ongoing.

Of the 577,070 thermally treated Alloy 690 tubes placed in service at 30 units between 1989 and 2004, only 333 tubes (0.06 percent) have been plugged. All together, these 30 units have operated for approximately 173 calendar years (as of December 2004) with steam generators with thermally treated Alloy 690 tubes. On average, each of these units has operated commercially for 6 calendar years (as of December 2004) with steam generators with thermally treated Alloy 690 tubes. Of the 333 tubes plugged, approximately 65 percent were plugged prior to placing the steam generators into service. Wear is the dominant inservice degradation mode for thermally treated Alloy 690 tubes. Tube wear occurs when the tube contacts a support structure (e.g., an antivibration bar) or a foreign object (e.g., a loose part). Wear accounts for approximately 24 percent of the plugged tubes.

Fewer tubes have been plugged in the steam generators with third-generation tube materials (i.e., thermally treated alloy 690) than in steam generators with mill-annealed or thermally treated Alloy 600 tubing. Improvements in the design and operation of the third-generation steam generators appear to have increased the corrosion resistance of the tubes, as evidenced by the general lack of corrosion degradation (including stress-corrosion cracking). The increased corrosion resistance results from the thermal treatment process and the improvements in the alloy.

Besides the heat treatment, several other factors also contribute to the relatively good operating experience of units with thermally treated Alloy 690 steam generator tubes, including hydraulic expansion of the tubes into the tubesheet, the design of the tube supports (e.g., lattice grids or quatrefoil-shaped holes), and the stainless steel material used to fabricate the tube supports. The residual stress levels at the expansion transition in tubes hydraulically expanded into the tubesheet are lower than those observed in units with tubes that were expanded mechanically or explosively. Since crack growth rate and time to crack initiation depend in part on the stress level, lower stresses may result in slower crack growth rates and/or longer times before crack initiation.

Although the operating experience with thermally treated Alloy 690 tubes has been favorable to date, licensees still need to monitor the tubes to detect the onset of tube degradation (including cracking) and assure the structural and leakage integrity of the tubes during the intervals between inspections. A better understanding of some of these issues would be useful in determining appropriate intervals for future monitoring of tube degradation.

1 INTRODUCTION

1.1 Safety Significance

In pressurized-water reactors (PWRs), the primary coolant removes the heat generated from the reactor core. Each primary coolant loop in U.S. PWR designs has one reactor coolant pump and one vertically mounted steam generator. Each plant contains two to four reactor coolant loops. The hot primary coolant enters and leaves the steam generator through nozzles in the hemispherical bottom head of the steam generator. The steam generator tubes provide the primary means for the transfer of heat from the primary system water to the water on the secondary side of the steam generator. This heat transfer boils the water on the secondary side of the steam generator. The primary coolant then returns to the reactor core via the reactor coolant pump, where it is reheated and the cycle is repeated.

Feedwater (secondary coolant) is pumped into the secondary or shell side of the steam generator, where it boils into steam. The steam exits the steam generator through an outlet nozzle and flows to the turbine generator, where it spins the turbine, generating electricity. After exiting the turbine, the steam is condensed into water and pumped back to the steam generator, where the cycle repeats.

Steam generator tubes constitute well over 50 percent of the surface area of the primary pressure boundary in a PWR. This pressure boundary is an important element in the defense in depth against a release of radioactive material from the reactor into the environment. Unlike other parts of the reactor coolant pressure boundary, the barrier to fission product release provided by the steam generator tubes is not reinforced by the reactor containment. That is, fission products released through leaking or ruptured steam generator tubes can escape directly into the environment through the secondary side of the steam generator. Consequently, the integrity of the steam generator tubes must be ensured with high confidence.

Because of the potential consequences of steam generator tube leakage, regulatory limits exist for the amount of primary-to-secondary leakage permitted during normal operation. In addition, the design of PWRs allows operators to rapidly and effectively respond to steam generator tube leakage during power operation. For postulated accidents, primary-to-secondary leakage is assumed to exist and is assessed in evaluating the radiological consequences of postulated accidents such as a feedwater or steam line break. In the event of leakage during normal operation or postulated accidents such as the rupture of the main steam line or feed line, leakage of reactor coolant through the tubes could contaminate the flow in these lines. In addition, leakage of primary coolant through openings in the steam generator tubes could deplete the inventory of water available for the long-term cooling of the core in the event of an accident.

For normal operation, a plant's technical specifications limit the amount of primary-to-secondary leakage. The limit is specific to each plant and ranges from approximately 150 to 720 gallons per day (gpd) through any one steam generator. Leakage through all steam generators is also limited typically to 1 gallon per minute (gpm). For postulated accidents such as the rupture of a main steam line or feed line, the design basis of the plant included an evaluation of the radiological dose consequences associated with primary-to-secondary leakage. Typically, plants were designed assuming that primary-to-secondary leakage during postulated accidents would be less than 1 gpm.

Although limits exist for the amount of primary-to-secondary leakage during normal operation (e.g., 150 gpd), it is possible for a tube to rupture during normal operation. Leakage from a ruptured tube can result in primary-to-secondary leak rates in the range of 100 to 700 gpm (depending on the severity of the tube rupture and the capacity of the safety injection/charging system pumps). The design of PWRs allows operators to rapidly and effectively respond to the accidental rupture of one steam generator tube during power operation. Although PWR

-1-

designs consider the rupture of a tube during normal power operation, they do not account for a tube rupture concurrent with a postulated accident.

1.2 Tube Integrity Program

1.2.1 Purpose of Inspections

Because of the importance of steam generator tube integrity, the NRC requires the performance of periodic inservice inspections of steam generator tubes. The requirements for the inspection of steam generator tubes are intended to ensure that this portion of the reactor coolant system maintains its structural and leakage integrity. Structural integrity refers to maintaining adequate margins against gross failure, rupture, and collapse of the steam generator tubes. Leakage integrity refers to limiting primary-to-secondary leakage during normal operation and postulated accidents to within acceptable limits.

Regulatory Guide 1.121, "Bases for Plugging Degraded PWR Steam Generator Tubes," issued August 1976, specifies the structural criteria that the tubes are intended to meet. Adequate leakage integrity during transients and postulated accidents is demonstrated by showing that the resulting leakage from the tubes will not exceed a rate that would violate offsite or control room dose criteria. Title 10, Part 100, "Reactor Site Criteria," of the *Code of Federal Regulations* (10 CFR Part 100) and General Design Criterion 19, "Control Room," of Appendix A, "General Design Criteria for Nuclear Power Plants," to 10 CFR Part 50, "Domestic Licensing of Production and Utilization Facilities," specify these criteria.

To provide assurance of adequate structural and leakage integrity, inspections are performed to detect mechanical or corrosive damage to the tubes from manufacturing and/or inservice conditions. In addition, the inservice inspections of the steam generator tubes provide a means of characterizing the nature and cause of any tube degradation so that corrective measures can be taken. Tubes that show an indication of degradation that exceeds the tube repair limits given in a plant's technical specifications are removed from service by plugging or are repaired by sleeving, as discussed in Section 1.2.3 of this report.

The frequency of the inservice inspections of the steam generator tubes depends primarily on the material used to fabricate them. For plants with tube materials such as mill-annealed Alloy 600, inservice inspections generally occur every 12 to 24 calendar months, as specified in a plant's technical specifications.[1] The specified maximum interval may be reduced to every 20 months in cases in which previous inspections have shown extensive degradation, and the interval may be increased to as much as every 40 months in cases in which previous inspections have revealed minor degradation. These intervals are reduced or extended based on the categorization of inspection results, as defined in the plant's technical specifications. In general, only plants with thermally treated steam generator tubes use the 40-month inspection interval. For units with thermally treated tubes, the time interval between inspections has been extended on a case-by-case basis.

Although many plants' technical specifications include a general provision to extend surveillances by 25 percent of the specified interval, this provision does not apply to steam generator tube inspections; the above criteria indicate the only conditions under which the surveillance interval for steam generator tube inspections may be changed. The NRC delineated this position in Generic Letter 91-04, "Changes in Technical Specification Surveillance Intervals to Accommodate a 24-Month Fuel Cycle," dated April 2, 1991. As a practical matter, however, utilities with extensive tube degradation (e.g., units with mill-annealed

---

[1]The technical specifications discussed in this report are based on those in place as of December 2004. Significant changes have been made beginning in late 2005 with the adoption of TSTF-449, Revision 4, "Steam Generator Tube Integrity," ADAMS Accession No. ML051090200.

Alloy 600 steam generator tubes) generally perform steam generator tube inspections at all refueling outages, which typically occur every 12 to 24 months.

The plant's technical specifications state the minimum number of steam generators inspected and the number of tubes inspected in these steam generators. The technical specifications typically permit a subset of steam generators to be examined provided all steam generators are performing in a similar manner. The steam generators inspected during a given outage are alternated so as to ensure that the material condition of each steam generator is monitored over time. Depending on the results of the inspections (i.e., the number and severity of the flaws identified), additional steam generators may require examination during an outage.

Since the purpose of the steam generator tube inspections is, in part, to ensure adequate structural and leakage integrity of the tube bundle, more frequent inservice inspections may be required, depending on the severity of the indications detected. To ensure that the frequency was adequate for the prior cycle, licensees for PWRs assess the inspection results following every outage to ensure that the tubes retained adequate structural and leakage integrity. This type of assessment is referred to as "condition monitoring." In addition, licensees project the condition of the tubes from the current inspection to the next inspection to ensure that the tubes will retain adequate integrity for the next operating interval. This type of assessment is referred to as an "operational assessment." Licensees perform these assessments in part because the inspection frequencies and tube repair criteria specified in the technical specifications were established based on specific assumptions concerning various parameters such as the forms of degradation (if any) to which the tubes may be susceptible, limitations of nondestructive examination techniques, and the rate of steam generator tube degradation. If any of these parameters exceeds those assumptions made during the development of the inspection intervals, the bases for the inspection frequency and the tube repair criteria are no longer considered valid.

In summary, the inservice inspection of steam generator tubes is to be conducted at appropriate intervals, to ensure that the structural and leakage integrity of the steam generator tubes is maintained with appropriate margins. These inspections should be adequate to detect degradation at a sufficiently early stage to preclude the progression of the degradation to the point that the regulatory criteria regarding steam generator tube structural and leakage integrity can no longer be met during the interval between inspections.

1.2.2 Eddy Current Testing

Eddy current testing (ECT) is the primary means for inspecting steam generator tubes. This method involves inserting a test coil inside the tube (i.e., the primary side of the tube) and pushing and pulling the coil so that it traverses the tube length. The test coil is then "excited" by alternating current, thereby creating a magnetic field that induces eddy currents in the tube wall. Disturbances of the eddy currents caused by flaws in the tube wall (such as cracks, holes, thinned regions, and other defects) produce corresponding changes in the electrical impedance as seen at the test coil terminals. Instruments translate these changes in test coil impedance into an output that the data analyst can monitor. The depth of certain types of flaws can be determined by the observed phase angle response of this output signal. The test equipment is calibrated using tube specimens containing artificially induced flaws of known depth. Geometric discontinuities (such as the expansion transition and dents) and support structures (such as the tubesheet and tube support plates) also produce eddy current signals, making it very difficult to discriminate defect signals at these locations. NUREG/CR-6365, "Steam Generator Tube Failures," issued April 1996, discusses some of the basic principles of ECT.

Bobbin coil eddy current probes are routinely used to inspect steam generator tubes. The bobbin coil probe permits a rapid screening of the tube for axially oriented and volumetric forms of degradation; however, it has several limitations:

- a general inability to permit characterization of identified degradation (e.g., axial, circumferential, or volumetric; single or multiple axial indications; etc.)

- relative insensitivity to detecting circumferentially oriented tube degradation

- limited capability to detect degradation in regions with geometric discontinuities (e.g., expansion transitions, U-bends, and dents) and deposits

As a result of the bobbin coil's limitations, the emergence of new forms of tube degradation (e.g., stress-corrosion cracking), and advancements in computer technology, additional inspection probes were used. Currently, inspections of steam generator tubes generally employ both a bobbin coil probe and an additional probe, such as a rotating probe. The bobbin coil probe permits rapid screening of the tube for degradation and can be pulled through a tube at speeds exceeding 40 inches per second, while the rotating probes are used to detect forms of degradation at specific locations since they do not suffer from many of the limitations of the bobbin coil (discussed above).

Rotating probes generally contain one to three specialized test coils. The coils used in the rotating probe head at a specific unit depend on many factors, including optimizing the coils for detecting the forms of degradation to which a tube may potentially be susceptible. The coils used on a rotating probe include (1) a pancake coil (which is sensitive to both axially and circumferentially oriented degradation), (2) an axially wound coil (which is sensitive to circumferentially oriented degradation), (3) a circumferentially wound coil (which is sensitive to axially oriented degradation), or (4) a plus-point coil (which reduces the effects of geometry variations in the tube and is sensitive to both axially and circumferentially oriented degradation).

Each of the above-mentioned test coils can be designed and driven at specific frequencies to ensure an optimal inspection of the tubing. In general, lower frequencies are better for detecting degradation initiating from the outside diameter of the tube, while higher frequencies are better for detecting degradation initiating from the inside diameter of the tube. The advantages of the rotating probes are that they are sensitive to circumferentially oriented degradation (which the bobbin coil probe is not), can better characterize the defect, and are less sensitive to geometric discontinuities. The major disadvantage of the rotating probes is their slow inspection speed (typically less than 1 inch per second). Because of this slow inspection speed, rotating probes are only used at specific locations (e.g., U-bends, sleeves, expansion transitions, dents, locations where there is a bobbin coil probe indication, locations where a more sensitive inspection is needed, and locations susceptible to circumferential cracking).

Tubes are generally selected for ECT on a random basis except where experience indicates critical areas requiring inspection and tubes previously found to contain detectable wall penetrations (greater than 20 percent) or imperfections. A preservice inspection of all steam generators is performed to establish a baseline condition of the tubes. The inservice inspection frequency is adjusted to account for the history of tube degradation encountered within the unit's steam generators.

1.2.3 Tube Repairs

The plant technical specifications set plugging and repair limits for the maximum allowable wall degradation beyond which the tubes must be removed from service by plugging or repaired by sleeving. Tube degradation is typically discovered during scheduled inservice examinations of steam generator tubes, and tube repair (plugging or sleeving) is required for all tubes with indications of tube degradation exceeding the tube repair limits. All plants have a depth-based repair limit that applies to all forms of steam generator tube degradation. The NRC has approved alternatives to this depth-based limit; however, the agency has not approved any alternatives for units with thermally treated Alloy 690 steam generator tubes. The depth-based repair limit varies from plant to plant, but it is typically 40 percent of the tube wall thickness. That is, if the depth of degradation is greater than or equal to 40 percent of the tube wall thickness, the licensee must plug or repair the tube. Some units have depth-based repair limits in their technical specifications with a limit other than 40 percent. For example, Robinson 2 has a depth-based repair limit of 47 percent.

The plugging and repair limits are established based on the minimum tube wall thickness necessary to provide adequate structural margins in accordance with Regulatory Guide 1.121 during normal operating and postulated accident conditions. These limits allow for eddy current error and incremental wall degradation that may occur before the next inservice inspection of the tube. These plugging and repair limits are conservatively established according to an assumed mode of degradation in which the walls are uniformly thinned over a significant axial length of tubing. These limits do not consider additional structural margins associated with defects such as small-volume thinning and pitting, and they do not consider the external structural constraints against gross tube failure provided by such support structures as the tubesheet and tube support plates.

Because of its conservative basis, the depth-based limit tends to be overly restrictive for highly localized flaws (such as stress-corrosion cracks) and flaws within the tubesheet. As a result, the industry has developed, and the NRC has approved, various alternative forms of repair criteria for specific forms of steam generator tube degradation.

The plugging technique involves installing plugs at the tube inlet and outlet. After plugging, the tube no longer functions as the boundary between the primary and secondary coolant systems. To prolong the life of severely degraded steam generator tubes, some utilities, with prior NRC approval, have repaired defective tubes by sleeving. After sleeving, the repaired tube may remain in service. Of the plants with thermally treated Alloy 690 steam generator tubes, only Ginna and Millstone 2 have NRC approval to sleeve tubes as of December 2004. However, the licensees have not installed any sleeves at these units.

1.2.4 Leakage Monitoring

Between tube inspections, plants monitor for a loss of tube integrity by monitoring for primary-to-secondary leakage. Licensees use various methods to monitor for tube leakage, including periodically sampling and analyzing the steam generator secondary water for radioactivity and continuously monitoring various streams (the steam generator blowdown, each main steam line, and the condenser air ejector exhaust) for the presence of or increases in radioactivity. The plant technical specifications limit the amount of primary-to-secondary leakage that can be present during plant operation. These limits vary from plant to plant, ranging from approximately 150 to 720 gpd. Additionally, technical specifications limit the specific activity of the secondary coolant (typically to 0.1 microcurie per gram of dose equivalent iodine-131). The licensee uses the specific activity to determine the radiological consequences of steam generator tube leakage.

1.3 Mill-Annealed Alloy 600 Steam Generator Operating Experience

A variety of steam generator designs exist in the United States. A number of factors affect the susceptibility of steam generator tubes to degradation, including the operating environment (temperature and water chemistry), the tube material and its heat treatment, and operating and residual stresses. Two of the most important factors affecting the susceptibility of a tube to degradation are the tube material and its heat treatment. Early steam generator designs used tubes fabricated from Alloy 600, which was typically mill annealed by passing the tubes through a furnace at a temperature high enough to recrystallize the material and dissolve the carbon. The carbon content and the mill-annealing temperature are important parameters for controlling the mechanical and corrosion properties of Alloy 600. As discussed in NUREG/CR-6365, the object of the mill annealing is to dissolve all the carbides, enlarge the grain size, and then cover the grain boundaries with carbides during slow cooling in air. Alloy 600 with insufficient carbides at the grain boundaries is more susceptible to primary water stress-corrosion cracking (PWSCC). Undissolved intragranular carbides are undesirable because they provide nucleation sites for the dissolved carbon and prevent precipitation of the carbides on the grain boundaries. Undissolved carbides also prevent the grains from growing. The smaller grains have a much larger grain boundary area per unit of volume, and the carbides do not properly cover the boundaries.

Tubes installed in U.S. nuclear steam generators placed in service in the 1960s and 1970s were usually only mill annealed. The annealing temperature depended on the manufacturer's practice at the time. More than 25 years of operating experience have shown that mill-annealed Alloy 600 is susceptible to various forms of degradation in the steam generator operating environment. The types of degradation affecting mill-annealed Alloy 600 steam generator tubes include pitting, wear, thinning, wastage, and stress-corrosion cracking. The stress-corrosion cracking can be either axially or circumferentially oriented. Degradation, of one form or another, has been observed on virtually every portion of the tube. Degradation has affected both recirculating steam generators with "U"-shaped, mill-annealed Alloy 600 tubes and once-through steam generators with straight, mill-annealed Alloy 600 tubes.

The extensive tube degradation at PWRs with mill-annealed Alloy 600 steam generator tubes resulted in numerous tube leaks, approximately nine domestic tube ruptures, numerous midcycle steam generator tube inspections, and the replacement of steam generators at numerous plants. In addition, extensive tube degradation has contributed to the shutdown of other plants. Haddam Neck, Maine Yankee, Trojan, Zion 1, Zion 2, and San Onofre 1 permanently ceased operation with significant amounts of tube degradation. As of December 2004, 39 plants in the United States had replaced their original mill-annealed Alloy 600 steam generators. With one exception (Palisades), the replacement steam generators typically had more advanced tube materials. The Palisades replacement SGs were obtained prior to the introduction of thermally treated Alloy 600. Table 1-1 lists the units that replaced their steam generators and provides the model and tube material of the replacement steam generator.

Operating experience for units with mill-annealed Alloy 600 steam generator tubes is well documented.

1.4 Thermally Treated Alloy 600 Tubes

As mill-annealed Alloy 600 steam generator tubes began exhibiting degradation in the early 1970s, engineers pursued improvements in the design of future steam generators to limit the likelihood of corrosion (including stress-corrosion cracking). Mill-annealed Alloy 600 tubes are generally resistant to chloride stress-corrosion cracking but are susceptible to caustic stress-corrosion cracking. The tube material and its heat treatment were of particular importance in these improved designs. The first major advance in limiting the corrosion susceptibility of the steam generator tubes was the use of a thermal treatment process to improve the tube's microstructure and thereby its corrosion resistance.

In the late 1970s, some mill-annealed Alloy 600 tubes were subjected to this thermal treatment process to relieve fabrication stresses and to further improve the tubes' microstructure. In this process, the tubes were subjected to high temperatures (approximately 705 °C) for 10 to 15 hours. This process promotes carbide precipitation at the grain boundaries and diffusion of chromium to the regions adjacent to the grain boundaries. Alloy 600 with insufficient carbides at the grain boundaries is more susceptible to PWSCC, and chromium depletion at the grain boundaries makes the material more susceptible to outside diameter stress-corrosion cracking (ODSCC).

This thermal treatment process was first used on tubes installed in replacement steam generators placed into service in the early 1980s. At present, 17 units use thermally treated

Alloy 600. Another unit, Callaway, has steam generators in which only the first 10 rows have thermally treated Alloy 600 tubes; the remaining rows have mill-annealed Alloy 600 tubes. Thermally treated Alloy 600 is considered to be highly resistant but not immune to PWSCC compared to mill-annealed Alloy 600 tubes. NUREG-1771, "U.S. Operating Experience with Thermally Treated Alloy 600 Steam Generator Tubes," April, 2003, documents the experience with thermally treated Alloy 600 tubes as of December 31, 2001. Since the issuance of NUREG-1771, inspections have identified additional crack-like indications at several units with thermally treated Alloy 600 tubes (i.e., in addition to those reported at Seabrook). These indications are primarily located in the portion of the tube within the tubesheet (although Braidwood 2 found several crack-like indications at the tube support plate elevations). NRC Information Notice 2005-09, "Indications in Thermally Treated Alloy 600 Steam Generator Tubes and Tube-to-Tubesheet Welds," dated April 7, 2005, contains additional information concerning the crack-like indications found in the portion of the tube within the tubesheet.

1.5 Thermally Treated Alloy 690 Tubes

Another approach to improving the degradation resistance of steam generator tubes was to identify more corrosion-resistant alloys. Alloy 690 is a nickel-chromium-iron alloy that is similar to Alloy 600 and therefore has similar high-temperature strength, thermal conductivity, and thermal expansion coefficient. These properties are important in steam generator functionality. The principal differences compared to Alloy 600 are the increase in the minimum chromium content from 14 percent to 27 percent, a decrease in the minimum nickel content from 72 percent to 58 percent, and a decrease in the maximum carbon content from 0.15 percent to 0.05 percent. The tensile strength, yield strength, elongation, thermal treatment (e.g., 700 °C for 15 hours), and metallurgical structure are nearly equivalent for the two alloys.

The higher chromium content in Alloy 690 when compared to Alloy 600 reduces the degree of sensitization (i.e., the amount of chromium depleted in areas adjacent to the metal grain boundaries), thus increasing resistance to corrosion attack at the metal grain boundaries. The heat treatment, which is intended to improve the stress-corrosion cracking resistance of the material, involves mill annealing at temperatures sufficient to put all the carbon into solution, followed by a thermal treatment to precipitate carbides on the metal grain boundaries into the tube metal microstructure. Resistance to stress-corrosion cracking is greatest when carbides fully populate the metal grain boundaries. Alloy 690 is more resistant to both primary- and secondary-side stress-corrosion cracking, pitting, and general corrosion. The improved resistance of Alloy 690 to intergranular attack, pitting corrosion, PWSCC, and ODSCC has been attributed mainly to the higher chromium content.

Various projects have been undertaken to assess the susceptibility of thermally treated Alloy 690 to degradation. Some investigators report that thermally treated Alloy 690 is immune to stress-corrosion cracking in acidic sulfate and chloride environments but could be susceptible to pitting or general corrosion (wastage). If copper or its oxides are present, thermally treated Alloy 690 becomes very susceptible to stress-corrosion cracking in acidic sulfate environments. Others have reported that testing of various Alloy 600 and Alloy 690 specimens under acidified conditions enriched with sulfates indicates, in part, that materials in the thermally treated condition are much less susceptible to stress-corrosion cracking and intergranular attack. In caustic environments, thermally treated Alloy 690 is generally considered superior in resistance to stress-corrosion cracking and intergranular attack than both mill-annealed and thermally treated Alloy 600. In a chloride-containing environment, Alloy 690 is considered as good or slightly better than Alloy 600 in resistance to pitting corrosion. It is expected that lead will cause stress-corrosion cracking in thermally treated Alloy 690 regardless of the pH of the solution. Thermally treated Alloy 690 is considered very resistant to PWSCC and is generally not expected to crack regardless of metallurgical condition or stress level unless the material and mechanical properties are outside the bounds of typical procurement specifications or the material was improperly heat treated. Some investigators report that thermally treated Alloy 690 appears to have the same resistance to wastage as Alloy 600.

One plant has reported that extensive testing has demonstrated that thermally treated Alloy 690 tubing is superior to mill-annealed Alloy 600 tubing in its resistance to both primary- and secondary-system stress-corrosion cracking, pitting, and general corrosion. One source of testing data this plant cited was the proceedings from the 1986 Electric Power Research Institute (EPRI) Workshop on Thermally Treated Alloy 690 Tubes for Nuclear Steam Generators (EPRI NP-4665S-SR). The plant indicated that primary-side corrosion testing at 680°F performed with statically loaded reverse U-bend specimens showed that cracking was observed within approximately 300 hours for mill-annealed Alloy 600 tubing and 800 hours for thermally treated Alloy 600 tubing. Cracking was not observed for the thermally treated Alloy 690 tubing even after 12,000 hours. Studies of statically loaded tensile specimens tested in 680°F primary water indicated that mill-annealed Alloy 600 tubing exhibited cracking within 2900 hours while thermally treated Alloy 690 did not exhibit cracking after 7000 hours of testing. Thermally treated Alloy 690 was also compared to mill-annealed Alloy 600 tubing in 760°F steam tests to produce accelerated PWSCC. These test results showed that mill-annealed Alloy 600 tubing exhibited cracking within 1000 hours, while thermally treated Alloy 690 did not exhibit any signs of cracking after 6000 hours. The environments considered were pure water, primary water, and uncontaminated all-volatile treatment secondary-system water. The thermally treated Alloy 690 improvement factor for stress-corrosion cracking in primary water and in uncontaminated all-volatile treatment environments exceeded 10. The improvement factors for other possible secondary-side environments were greater than 10 for near-neutral uncontaminated all-volatile treatment water, greater than 6 to about 20 for chlorides, approximately 5 for caustics (i.e., pH greater than 10), greater than 5 for other environments (e.g., resin liquor polluted or complex alumina silica), approximately 2 for sulfur-contaminated environments, and approximately 2 for lead-contaminated caustic environments. Service experience indicates that the more aggressive test environments that result in low improvement factors for thermally treated Alloy 690 rarely occur in actual plant service.

The performance of steam generators with Alloy 690 tubes cannot be attributed entirely to the material selection for the tubes. Steam generators that are new enough to have Alloy 690 tubing also have design and operational improvements, such as improved techniques for expanding the tube into the tubesheet, thermal stress relief of low-radius U-bends, the use of stainless steel in tube support structures, noncircular holes in tube support structures, and improved secondary-water chemistry. The tube performance in these steam generators therefore reflects the combined effect of these changes.

One plant has reported that no improvement is expected in the resistance to outside-diameter initiated degradation in the vicinity of the expansion transition for plants that have expanded their tubes with the hydraulic expansion method relative to the mechanical roll expansion method. As its basis, the plant indicated that the residual stresses at the expansion transition on the tube outside surface are primarily determined by the shape of the transitions, which are not very different for all of the various expansion methods. The method of expansion affect the residual stresses and cold work of the material on the tube inside surface.

Thermally treated Alloy 690 was first used for tubes installed in replacement steam generators placed into service in the late 1980s. As of December 31, 2004, 30 units use thermally treated Alloy 690 steam generator tubes.

It is important to evaluate the operating experience of thermally treated Alloy 690 because it will provide insights into the behavior of newer steam generator materials. Thermally treated Alloy 690 is currently the preferred material for tubes in new and replacement steam generators. Of the 69 operating PWRs in December 2004, approximately 32 percent have mill-annealed Alloy 600 steam generator tubes, approximately 25 percent have thermally treated Alloy 600 steam generator tubes, and approximately 43 percent have thermally treated Alloy 690 steam generator tubes.

Table 1-1: Units with Replacement Steam Generators (Part 1)

Unit Name	No. of Loops	SG Manufacturer/Model[1]		Completion Date	Tube Material[1]
		Original	Replacement		
Surry 2	3	W/51	W/51F	9/80	600 TT
Surry 1	3	W/51	W/51F	7/81	600 TT
Turkey Point 3	3	W/44	W/44F	4/82	600 TT
Turkey Point 4	3	W/44	W/44F	5/83	600 TT
Point Beach 1	2	W/44	W/44F	3/84	600 TT
Robinson 2	3	W/44	W/44F	10/84	600 TT
Cook 2	4	W/51	W/54F	3/89	690 TT
Indian Point 3	4	W/44	W/44F	6/89	690 TT
Palisades	2	CE	CE	3/91	600 MA
Millstone 2	2	CE/67	BWI	1/93	690 TT
North Anna 1	3	W/51	W/54F	4/93	690 TT
Summer	3	W/D3	W/D75	12/94	690 TT
North Anna 2	3	W/51	W/54F	5/95	690 TT
Ginna	2	W/44	BWI	6/96	690 TT
Catawba 1	4	W/D3	BWI	9/96	690 TT
Point Beach 2	2	W/44	W/D47	12/96	690 TT
McGuire 1	4	W/D2	BWI	5/97	690 TT
Salem 1	4	W/51	W/F	7/97	600 TT
McGuire 2	4	W/D3	BWI	12/97	690 TT
St. Lucie 1	2	CE/67	BWI	1/98	690 TT

[1]ABB = Asea Brown Boveri; B&W = Babcock and Wilcox; BWI = Babcock and Wilcox International; CE = Combustion Engineering; Fr = Framatome; MA = mill annealed; TT = thermally treated; W = Westinghouse

Table 1-1: Units with Replacement Steam Generators (Part 2)

| Unit Name | No. of Loops | SG Manufacturer/Model[1] | | Completion Date | Tube Material[1] |
		Original	Replacement		
Byron 1	4	W/D4	BWI	1/98	690 TT
Braidwood 1	4	W/D4	BWI	11/98	690 TT
South Texas Project 1	4	W/E	W/D94	5/00	690 TT
Farley 1	3	W/51	W/54F	5/00	690 TT
Cook 1	4	W/51	BWI	12/00	690 TT
Arkansas Nuclear One 2	2	CE/2815	W/D109	12/00	690 TT
Indian Point 2	4	W/44	W/44F	12/00	600 TT
Farley 2	3	W/51	W/54F	5/01	690 TT
Kewaunee	2	W/51	W/54F	12/01	690 TT
Harris	3	W/D4	W/D75	12/01	690 TT
Calvert Cliffs 1	2	CE	BWI	6/02	690 TT
South Texas Project 2	4	W/E	W/D94	12/02	690 TT
Calvert Cliffs 2	2	CE	BWI	5/03	690 TT
Sequoyah 1	4	W/51	ABB/Doosan	6/03	690 TT
Palo Verde 2	2	CE/80	ABB/Ansaldo	12/03	690 TT
Oconee 1	2	B&W	BWI	1/04	690 TT
Oconee 2	2	B&W	BWI	6/04	690 TT
Prairie Island 1	2	W/51	Fr 56/19	11/04	690 TT
Oconee 3	2	B&W	BWI	12/04	690 TT

[1]ABB = Asea Brown Bovari; B&W = Babcock and Wilcox; BWI = Babcock and Wilcox International; CE = Combustion Engineering; Fr = Framatome; MA = mill annealed; TT = thermally treated; W = Westinghouse

2 STEAM GENERATOR DESIGNS IN UNITS WITH THERMALLY TREATED ALLOY 690 TUBES

2.1 Introduction

Steam generators with thermally treated Alloy 690 tubes were first placed into service in 1989. Figure 2-1 graphs the deployment of steam generators with thermally treated Alloy 690 tubes. As of December 2004, 30 units have steam generators with thermally treated Alloy 690 tubes. Table 2-1 lists these units along with the approximate date the replacement steam generators were put into service, the steam generator model (or designer), the number of steam generators at the unit, the operating time of the original steam generators, and the operating time of the replacement steam generators. Figure 2-2 depicts the number of thermally treated Alloy 690 steam generator tubes placed into service per year.

As shown in Table 2-1, two units, Cook 2 and Indian Point 3, have now operated longer with their replacement steam generators than with their original steam generators. The average age of steam generators with thermally treated Alloy 690 tubes is approximately 6 calendar years as of December 31, 2004.

There are 577,070 thermally treated Alloy 690 tubes used in the 30 units with this tube material. Either Sandvik (in Sweden), Sumitomo (in Japan), or Valinox (in France) fabricated these tubes. As shown in Figure 2-3, of the 30 units with thermally treated Alloy 690 tubes, Sandvik has supplied the tubes for 15 units (50 percent), Sumitomo has supplied the tubes for 12 units (40 percent), and Valinox has supplied the tubes for 3 units (10 percent). As shown in Figure 2-4, of the 577,070 thermally treated Alloy 690 tubes, Sandvik fabricated 251,950 (44 percent), Sumitomo fabricated 291,360 (50 percent), and Valinox fabricated 33,760 (6 percent).

2.2 General Steam Generator Description

Steam generators in units with thermally treated Alloy 690 tubes are either recirculating, U-tube heat exchangers or once-through, straight-tube heat exchangers (refer to Figures 2-5 and 2-6, respectively).

For recirculating steam generators, heat is transferred from the hot primary coolant to the water on the secondary side of the steam generator as the primary coolant flows through the inverted U-tubes. The primary coolant enters and leaves the steam generators through nozzles in the hemispherical bottom head of the steam generator. Heat transfer from the primary system to the water on the secondary side of the steam generator is accomplished primarily through the steam generator U-tubes. After the primary coolant flows through the U-tubes, it exits the lower plenum of the steam generator through an outlet nozzle. A plate in the lower plenum below the tubesheet, called a "divider plate," separates the inlet and outlet primary coolant and directs the flow through the tubes.

Recirculating steam generators are designed with an evaporator section and a steam drum section. The steam drum section is the upper part of the steam generator containing the moisture separators. The evaporator section, sometimes called the "tube bundle," is an inverted U-tube heat exchanger containing the tubes. Figure 2-7 shows typical features of a U-tube. The evaporator section may have a preheater region, depending on the model. The preheater, which is a series of baffle plates around a portion of the cold-leg side of the steam generator, enhances heat transfer to the incoming feedwater. As of December 2004, Palo Verde 2 is the only unit with thermally treated Alloy 690 steam generator tubes that has a preheater region.

The number of tubes in each steam generator varies from unit to unit. For recirculating steam generators with thermally treated Alloy 690 tubes, the number of tubes varies from approximately 3,200 to 12,000 per steam generator. For recirculating steam generators, the tubes are welded to a thick plate, called a tubesheet, with a hole for each tube end. The weld is near the end of the tube. The tubesheet is approximately 2 feet thick. The tubes are hydraulically expanded against the tubesheet walls for the full depth of the tubesheet. The tube-to-tubesheet joint physically fastens the tube to the steam generator vessel and provides axial load resistance and a leaktight barrier between the primary and secondary sides of the steam generator. The tubes are supported with either plates or lattice grids at a number of fixed axial locations along the tube bundle and with bars/strips in the U-bend region of the tube bundle. These bars/strips are typically called antivibration bars (AVBs) or fan bars.

For the once-through steam generators, heat is transferred from the hot primary coolant to the water on the secondary side of the steam generator as the hot primary coolant flows downward through the straight tubes. The primary coolant enters the steam generator through a nozzle in the top of the steam generator. The transfer of heat from the primary system to the water on the secondary side of the steam generator primarily through the steam generator tubes. After the primary coolant flows through the tubes, it exits the steam generator through outlet nozzles at the bottom of the steam generator.

Each once-through steam generator with thermally treated Alloy 690 tubes has nearly 16,000 tubes. Similar to the recirculating steam generators, the tubes are welded to a thick plate or tubesheet, with a hole for each tube end. Unlike the recirculating steam generators, the once-through steam generators have two tubesheets—one at the top (or hot-leg) and the other at the bottom (or cold-leg) of the steam generator. The tubesheets are approximately 2 feet thick. The tubes are hydraulically expanded against the tubesheet walls for either the full depth or a portion of the depth of the tubesheet. Plates at a number of fixed axial locations along the tube bundle support the tubes. Oconee Units 1, 2, and 3 are the only units with once-through steam generators with Alloy 690 tubes as of December 31, 2004.

In addition to the advanced tubing material, steam generators with thermally treated Alloy 690 tubes have other features to increase the tubes' resistance to degradation. For some units, these features improve upon their original steam generator design. For other units, the original steam generator design already incorporated some of these features. These advanced design features include the use of hydraulic means to expand the tube within the tubesheet, expanding the tube for the full-depth of the tubesheet, use of stainless steel tube supports, and the use of non-circular tube support holes or lattice grid tube supports.

The tubes in early recirculating steam generator designs encountered severe corrosion problems within the tubesheet crevice when the tubes were only expanded for a portion of the tubesheet thickness (i.e., partial-depth expansion). In addition, tubes in early steam generator designs experienced stress-corrosion cracking at the transition zone between the roller or explosively expanded and unexpanded tube, since the residual stresses were high at this location. As a result, units with thermally treated Alloy 690 tubes expand their tubes into the tubesheet by hydraulic means rather than by roll expansion or explosive expansion methods. Hydraulic expansion reduces the residual stresses at the expansion transition region, reducing the potential for stress-corrosion cracking in this region. Hydraulic expansions typically produce 20–40 percent less stress than mechanical (hard) roll expansions. Except for the once-through steam generators at Oconee 1, 2, and 3, the tubes were also expanded for the full depth of the tubesheet. By performing a full-depth expansion, the crevice between the tube and the tubesheet hole was closed, essentially eliminating a region where dryout can concentrate chemicals that could lead to denting or corrosion. The tubes at Oconee were only expanded for a partial distance within the tubesheet because of the unique conditions associated with a once-through steam generator design.

Tubes with early steam generator tube bundle support systems encountered several corrosion problems, including (1) stress-corrosion cracking as a result of denting of the tubes from the corrosion of carbon steel tube supports and (2) stress-corrosion cracking or intergranular attack under deposits as a result of dryout in the crevices of drilled hole tube support plates. As a result, units with thermally treated Alloy 690 tubes use stainless steel tube supports rather than carbon steel tube supports. Stainless steel is less susceptible to corrosion than carbon steel in the operating environment of a steam generator. The carbon steel supports corroded and formed magnetite, which filled the crevice between the tubes and the tube supports, denting the tubes. Stainless steel also has acceptable characteristics to limit the potential for tube wear. In addition to using stainless steel tube supports, units with thermally treated Alloy 690 tubes use either tube support plates with noncircular holes (e.g., trifoil-, quatrefoil-, or other similarly shaped holes) or lattice grid tube supports. These advanced tube support designs promote high-velocity flow along the tube, sweeping impurities away from the support locations. The hole designs of various shapes and the lattice grid also limit the contact between the tube and the support, limiting local dryout and chemical concentration.

2.3 Grouping of Steam Generator Designs

Steam generators with thermally treated Alloy 690 tubes can be divided into three major categories—(1) Westinghouse Delta and "F" model steam generators, (2) Babcock and Wilcox International (BWI)-designed steam generators, and (3) other. Table 2-2 summarizes readily available design information for each unit with thermally treated Alloy 690 tubes.

2.3.1 Westinghouse Delta and "F" Model Steam Generators

As of December 31, 2004, 13 units have Westinghouse Delta and "F" model steam generators, including Arkansas Nuclear One (ANO) Unit 2, Cook 2, Farley 1 and 2, Harris, Indian Point 3, Kewaunee, North Anna 1 and 2, Point Beach 2, South Texas Project 1 and 2, and Summer. Each of these units has recirculating steam generators. There are two types of "F" model steam generators (44F and 54F) and four types of Delta model steam generators (D47, D75, D94, D109) in units with thermally treated Alloy 690 tubes. The models differ primarily in heat transfer surface area, number of tubes, and tube spacing (pitch). These 13 units represent 43 percent of the units with thermally treated Alloy 690 tubes and 35 percent of the thermally treated Alloy 690 steam generator tubes.

The Delta and "F" model steam generators exhibit several similarities and differences. The tubes in these steam generators were all hydraulically expanded into the tubesheet, are supported by horizontal support plates (rather than by lattice grid tube supports), and have support structures fabricated with Type 405 stainless steel. The Delta and "F" model steam generators differ, however, in that the holes in the tube support for the "F" model steam generators are quatrefoil shaped while the Delta model steam generators have trifoil-shaped holes. In addition, the steam generator tubes in the "F" model steam generators have an outside diameter of 0.875 inch and a wall thickness of 0.050 inch, while the Delta model steam generator tubes (with the exception of the Point Beach 2 tubes) have a diameter of 0.6875 inch and a wall thickness of 0.040 inch. The steam generator tubes at Point Beach 2 have an outside diameter of 0.875 inch and a wall thickness of 0.050 inch.

Either Westinghouse (in the United States), Equipos Nucleares (ENSA) (in Spain), or Ansaldo (in Italy) fabricated the Delta and "F" model steam generators. With the exception of the tubes used in the Kewaunee steam generators, Sandvik fabricated the tubes used in these steam generators. Valinox fabricated the Kewaunee steam generator tubes.

As of July 2002, 12 units with thermally treated Alloy 690 steam generator tubes had advanced design AVBs (i.e., close tolerance U-bend fitup, Type 405 stainless steel AVBs, and a rectangular section AVB). These units include ANO 2, Cook 2, Farley 1 and 2, Harris, Indian Point 3, Kewaunee, North Anna 1 and 2, Point Beach 2, South Texas Project 1, and Summer. In addition to these units, Indian Point 2, which has thermally treated Alloy 600 steam generator tubes, has chrome plated Alloy 600 AVBs, but tightness control was emphasized during assembly. Very little wear at the AVBs has been reported at any of the units with thermally treated Alloy 690 tubes with the advanced-design AVBs.

2.3.2 BWI Steam Generators

As of December 31, 2004, 14 units have BWI steam generators, including Braidwood 1, Byron 1, Calvert Cliffs 1 and 2, Catawba 1, Cook 1, Ginna, McGuire 1 and 2, Millstone 2, Oconee 1, 2, and 3, and St. Lucie 1. All of these units have recirculating steam generators except for Oconee 1, 2, and 3. The models differ primarily in heat transfer surface area, number of tubes, and tube spacing (pitch). These 14 units represent 47 percent of the units with thermally treated Alloy 690 tubes and 55 percent of the thermally treated Alloy 690 steam generator tubes.

The BWI recirculating steam generators exhibit several similarities and differences. The steam generator tubes were all hydraulically expanded into the tubesheet, are supported by horizontal lattice grid tube supports (rather than by support plates), and have support structures fabricated with Type 410 stainless steel. The steam generators differ, however, in heat transfer surface area, number of tubes, tube spacing (pitch), tube outside diameter, and wall thickness. Some of the steam generators are essentially identical in terms of number of tubes, tube diameter, wall thickness, and tube spacing (e.g., McGuire 1 and 2, Catawba 1, Braidwood 1, and Byron 1).

The BWI steam generators were fabricated in Canada. Either Sumimoto or Valinox fabricated the tubes used in these steam generators. The steam generators use Sumitomo tubes at Braidwood 1, Byron 1, Calvert Cliffs 1 and 2, Catawba 1, Cook 1, McGuire 1 and 2, Oconee 1, 2, and 3, and St. Lucie 1. The steam generators at Ginna and Millstone 2 use Valinox tubes.

Several different types of wear mechanisms occur, or are postulated to occur, in BWI steam generators. These include typical fan bar wear, atypical U-bend wear, localized U-bend wear, and tube-to-tube contact wear. Typical fan bar wear is a result of thermal hydraulic conditions and tube-to-support clearances that can vary because of manufacturing tolerances. Typical wear results in either uniform or tapered wear scars on the tube. Several units with BWI steam generators have exhibited typical fan bar wear. Localized U-bend wear is a phenomenon "localized" to specific columns of tubes and possibly the adjacent column. It is theorized to result from arch-bar distortion instead of a more random manufacturing tolerance issue (which causes typical fan bar wear). The local distortion in the U-bend supports result in an increased tube-to-support gap. Localized U-bend wear has been reported to have occurred at St. Lucie 1 and McGuire 1. Atypical U-bend wear refers to pit-like indications found at flat-bar supports. These indications are theorized to result from asperities on the flat bars and are attributed to fabrication deficiencies. This mechanism has been reported at McGuire 1 and 2 and at St. Lucie 1. Tube-to-tube contact wear could possibly occur when tubes are in close proximity. A number of units, such as Braidwood 1, have tubes that were identified as being in close proximity. This condition (i.e., tube-to-tube proximity) was determined to be caused by gravitational effects on the tubes and the floating fan bar assembly while the steam generators were in a horizontal position. No wear resulting from the close proximity of tubes has been reported.

2.3.3 Other Steam Generators

As of December 31, 2004, three units have Alloy 690 steam generator tubes that do not fit into the previous two categories (i.e., Westinghouse Delta and "F" model or BWI steam generators). These units are Palo Verde 2, Sequoyah 1, and Prairie Island 1. Each of these units has recirculating steam generators designed by either Asea Brown Boveri—Combustion Engineering, which is now a part of Westinghouse, or Framatome (in France). The Palo Verde 2 and Sequoyah 1 steam generators are similar in that they have horizontal lattice grid tube supports that are fabricated with Type 409 stainless steel. The Palo Verde 2 steam generators have traditional inverted "U"-shaped tubes in the first 17 rows; however, in the higher rows (i.e., greater than row 17), the tubes have two 90-degree bends with a horizontal run of tube between the bends. The 90-degree bends are frequently referred to as square bends. Of the units with recirculating steam generators, only Palo Verde 2 uses tubes other than traditional inverted "U"-shaped tubes. These three units represent 10 percent of the units with thermally treated Alloy 690 tubes. In addition, these three units contain 10 percent of the thermally treated Alloy 690 steam generator tubes.

Ansaldo (in Italy) fabricated the Palo Verde 2 steam generators. Doosan (in Korea) fabricated the Sequoyah 1 steam generators. Framatome (in France) fabricated the Prairie Island 1 steam generators. Sandvik fabricated the steam generator tubes at these three units.

2.4 Individual Unit Steam Generator Designs

The following sections provide steam generator design information for each unit with thermally treated Alloy 690 tubes, based primarily on reports provided by licensees to the U.S. Nuclear Regulatory Commission (NRC). In some cases, the staff took the information directly (verbatim) from the licensee's reports. The level of detail provided in these reports varies from unit to unit. In addition, the information provided in these reports represents the design, analysis, and evaluations at the time the report was submitted and may have changed over time. In spite of these potential limitations, this report provides useful insights into the design of steam generators with thermally treated Alloy 690 tubes.

The staff also obtained some design information through regional inspection reports, summaries of conference calls with units, and meeting summaries. However, the staff did not conduct a detailed review of regional inspection reports or compile those data.

2.4.1 Arkansas Nuclear One Unit 2

ANO 2 has two recirculating steam generators designed by Westinghouse and fabricated by ENSA in Spain. The model Delta 109 steam generators were put into service in 2000 during refueling outage (RFO) 14.

Each steam generator has 10,637 thermally treated Alloy 690 tubes that have an outside diameter of 0.6875 inch and a nominal wall thickness of 0.040 inch. The tubes, manufactured by Sandvik, are arranged in a triangular pattern with a spacing of approximately 0.95 inch. The heat transfer surface area in each steam generator is 108,700 ft².

The tubes were hydraulically expanded at each end for the full depth of the tubesheet. The tubes are supported by support plates and AVBs. All supports are constructed from Type 405 stainless steel. The tubes pass through flat contact, trifoil-shaped holes in the tube support plates. The AVBs are arranged in a "V" shape and are staggered. Figure 2-8 illustrates the tube support configuration and numbering.

The U-bend region of the tubes in rows 1 through 17 received a supplemental thermal treatment (stress relieving) after bending.

Strict ovality control was implemented during the manufacture of the tubes to limit dimensional variability in the U-bend region. The thickness of the AVBs was also tightly controlled. To limit the potential for U-bend vibration and wear, AVBs support the U-bends. The AVBs provide sufficient support to the U-bend so that all the tubes remain elastically stable even if it is assumed that some of the support points are inactive. The AVBs in adjacent columns are inserted to different depths (i.e., staggered) to limit the U-bend pressure drop and to discourage the formation of flow stagnation regions. The AVBs are nearly perpendicular to the centerline of the tubes at all locations in the U-bend region to provide support without unnecessary tube contact. These features provide margin against flow stagnation, corrosion, and tube vibration.

2.4.2 Braidwood 1

Braidwood 1 has four recirculating steam generators designed and fabricated by BWI. The model 7720 steam generators were put into service in 1998 during RFO 7.

Each steam generator has 6633 thermally treated Alloy 690 tubes that have an outside diameter of 0.6875 inch and a nominal wall thickness of 0.040 inch. As illustrated in Figure 2-9, the tubes, manufactured by Sumitomo, are arranged in a triangular pattern with a spacing of approximately 0.930 inch.

The tubes were hydraulically expanded at each end for the full depth of the tubesheet. The tubesheet is 26.625 inches thick. The tubes are supported by lattice grid tube supports and fan bars. (The lowest fan bar is also referred to as a collector bar since all other fan bars connect to it.) All tube supports are constructed from Type 410 stainless steel. Figure 2-10 illustrates the tube support configuration and numbering.

The tubes in rows 1 through 21 (i.e., those with a bend radius less than 12 inches) received a supplemental thermal treatment (stress relieving) after bending. Row 3 tubes have the smallest U-bend radius, 3.632 inches.

The lattice grid tube support structure in the steam generator provides (1) high circulation rates through lower flow resistance, thus leading to a lower tendency to accumulate deposits when compared to a broached plate, (2) vibration restraint and fretting resistance, and (3) reduced denting potential due to the selection of a stainless steel material. The lattice grid tube supports are positioned within the steam generator at elevations selected to prevent flow-induced vibration (which could cause tube fretting), while not creating excessive flow resistance resulting in tube deposits. The tubes pass through diamond-shaped openings formed by the intersecting lattice grid bars. The line contact between the tube and the lattice grid support bar is designed to limit the crevice area between the tubes and the tube support, reducing the potential for corrosion products to accumulate in this region and the potential for dryout caused by local superheat. The lattice grid consists of high (approximately 3 inches high) and low (approximately 1 inch high) bars that form a lattice pattern. The high bars are located every sixth pitch.

A flat bar U-bend restraint system (FURS) supports the U-bend region. This system provides close tolerance support of the U-bend region to prevent flow-induced vibration, similar to the lattice grid tube support system. The potential for tube fretting in the U-bend region is also reduced due to material compatibility and a relatively long contact length as compared to the original steam generator U-bend support designs. By distributing the contact force, the FURS limits the possibility of fretting. The FURS is designed with all spaces oriented with an upward slope. This promotes continuous sweeping of fluid past the support during operation and avoids the potential for creating steam pockets and corresponding dryout leading to deposition. The FURS does not cross the bundle centerline, thereby avoiding the creation of horizontal tube contacts where deposits have the potential to collect. The fan bars are 1.25 inches high. The nominal tube-to-fan bar clearance is 0.006 inch.

The steam generator circulation ratio is defined as the ratio of the mass flow rate in the evaporator section to the mass flow rate at the steam outlet. Increasing the circulation ratio of the steam generator may improve heat transfer performance, generator sludge management, corrosion product transfer, and tube dryout, among other factors. The replacement steam generator design has a circulation ratio that exceeds 5, which is more than double that experienced in the original Braidwood 1 steam generators. Maintaining a high circulation ratio encourages the secondary bulk water contaminants to remain in suspension, thus benefitting the effectiveness of blowdown cleanup and reducing sludge pile height on the tubesheet. A high circulation ratio also limits the potential for low-flow areas, where impurity hideout may occur and produce local corrosive environments. Reducing deposit/sludge loadings and limiting low-flow areas limits the potential to develop a chemical environment that can promote stress-corrosion cracking initiation and growth.

A power uprate at Braidwood 1 in the early 2000s is expected to lower the designed circulation ratio. As a result of the uprate, wear rates are expected to decrease due to a lower circulation ratio and increased secondary-side pressure caused by the corresponding average temperature increase.

The replacement steam generator design also incorporates features intended to limit the development of loose parts during operation and maintenance. Specific design measures were taken to limit the corrosion potential of small thickness metal parts and to incorporate mechanisms for capturing or eliminating fasteners.

During fabrication of the Braidwood 1 steam generators, a number of tubes were noticed to be in contact (or in close proximity) with other tubes while the steam generator vessel was in a horizontal position. A total of 508 tubes were identified as being in contact (or in close proximity) during the preservice inspection, which was performed with the steam generators in a horizontal position. The preservice inspections did not reveal any tube damage in this area. The tube-to-tube contact condition was determined to result from gravitational effects on the tubes and floating fan bar assembly while the steam generator was in a horizontal position. The condition is expected to naturally correct itself after one or two cycles of operation with the steam generator in a vertical position. Replacement steam generators of similar design, including Byron 1, exhibited similar tube-to-tube contact during preservice inspection. After one cycle of operations, these units reported far fewer tubes in contact (or in close proximity) and no tube damage.

BWI evaluated the tube-to-tube contact condition and assessed the potential for fretting/wear damage and corrosion-induced degradation due to long-term tube-to-tube contact. Based on estimates of wear coefficients and work rates at the tube-to-tube contact area, a maximum tube wall loss of 40 percent was calculated to occur after 60 years of continuous full-power operation. As a result of this work, tube contact fretting is not expected to result in exceeding the 60-percent through-wall structural criteria during the life of the steam generators. The conditions that exist at the top of the tubesheet (i.e., at the expansion transition region) are considered to bound the potential for corrosion-induced degradation as a result of excessive fouling or deposit bridging compounded by the tube contact condition. This region was previously qualified by extensive testing; therefore, it was concluded that there is not an additional tube degradation risk due to tube contact. The tube contact condition will be monitored over time through the normal steam generator inspection program.

2.4.3 Byron 1

Byron 1 has four recirculating steam generators designed and fabricated by BWI. The model 7720 steam generators were put into service in 1998 during RFO 8.

Each steam generator has 6633 thermally treated Alloy 690 tubes that have an outside diameter of 0.6875 inch and a nominal wall thickness of 0.040 inch. The tubes, manufactured by Sumitomo, are arranged in a triangular pattern as illustrated in Figure 2-11.

The tubes were hydraulically expanded at each end for the full depth of the tubesheet. The tubes are supported by lattice grid tube supports and fan bars. (The lowest fan bar is also referred to as a collector bar since all other fan bars connect to it.) All supports are constructed from Type 410 stainless steel. Figure 2-12 illustrates the tube support configuration and numbering.

The tubes (or U-bend region of the tubes) in rows 1 through 21 received a supplemental thermal treatment (stress relieving) after bending.

The Byron 1 steam generator design is essentially identical to the Braidwood 1 steam generators.

2.4.4 Calvert Cliffs 1

Calvert Cliffs 1 has two recirculating steam generators designed and fabricated by BWI. The model 7811 steam generators were put into service in 2002 during RFO 15. The replacement steam generator consists of a new tube bundle, new steam drum internals, and a new feedring; however, the steam drum from the original steam generator was reused.

Each steam generator has 8471 thermally treated Alloy 690 tubes that have an outside diameter of 0.75 inch and a nominal wall thickness of 0.042 inch. The tubes, manufactured by Sumitomo, are arranged in a triangular pattern as illustrated in Figure 2-13 with a spacing of approximately 1.0 inch. The heat transfer surface area in each steam generator is 92,000 ft².

The tubes were hydraulically expanded at each end for the full depth of the tubesheet. The tube-to-tubesheet seal welds are flush with the tubesheet. The tubesheet is 21.875 inches thick (with the austenitic stainless steel clad, the tubesheet is 22.25 inches thick). The tubes are supported by lattice grid tube supports and fan bars. (The lowest fan bar is also referred to as a collector bar since all other fan bars connect to it.) All supports are constructed from Type 410 stainless steel. Figure 2-14 illustrates the tube support configuration and numbering.

The U-bend region of the tubes in rows 1 through 18 received a supplemental thermal treatment (stress relieving) after bending. Row 1 tubes have the smallest U-bend radius, 3.5 inches.

The tubes were expanded to approximately 0.125 inch below the secondary face of the tubesheet. The expansion process was controlled to ensure that expansion of the tube is as close as possible to the secondary face of the tubesheet without going past the face. For peripheral tubes, where the curvature of the primary bowl limits access, expansion is performed in two overlapping zones with a shorter mandrel. The expansion zones overlap near the center of the tubesheet to ensure full-depth expansion. Following expansion, the profile of each tube was measured through the entire expanded area of the tubesheet, including the transition zone, using eddy current profilometry.

The lattice grid tube support is made up of two intersecting arrays of Type 410S stainless steel high bars (approximately 3 inches high) oriented at 30 and 150 degrees to the tube-free lane. The bars are located every fourth pitch to accommodate steam generator loading conditions. Type 410S stainless steel low bars (approximately 1 inch high) are located at every pitch location between the high bars. All low bars that are flush to the top of the high bars are oriented at 30 degrees to the tube-free lane, and all low bars that are flush to the bottom plane of the high bars are oriented at 150 degrees to the tube-free lane. All of the lattice bars are nominally 0.100 inch thick. The bar ends are fitted into precise slots of a specially designed peripheral support ring. The lattice grids are positioned within the steam generator shroud (wrapper) at elevations selected to prevent flow-induced vibration while not creating excessive flow resistance. The tubes are held in position within the diamond shaped bar opening, which provides line support contact. This limits the crevice area between the tubes and bars, which could trap corrosion products.

The FURS incorporates a series of flat bar fan assemblies on each side of the tube bundle. The fans are positioned so that all U-bends are supported at close intervals. The supports are made of Type 410 stainless steel that provides resistance to wear, acceptable strength, and resistance to corrosion and related tube denting.

2.4.5 Calvert Cliffs 2

Calvert Cliffs 2 has two recirculating steam generators designed and fabricated by BWI. The model 7811 steam generators were put into service in 2003 during RFO 14. The replacement steam generator consists of a new tube bundle, new steam drum internals, and a new feedring; however, the steam drum from the original steam generator was reused.

Each steam generator has 8471 thermally treated Alloy 690 tubes that have an outside diameter of 0.75 inch and a nominal wall thickness of 0.042 inch. The tubes, manufactured by Sumitomo, are arranged in a triangular pattern as illustrated in Figure 2-13 with a spacing of approximately 1.0 inch. The heat transfer surface area in each steam generator is 92,000 ft^2.

The tubes were hydraulically expanded at each end for the full depth of the tubesheet. The tube-to-tubesheet seal welds are flush with the tubesheet. The tubesheet is 21.875 inches thick (with the austenitic stainless steel clad, the tubesheet is 22.25 inches thick). The tubes are supported by lattice grid tube supports and fan bars. (The lowest fan bar is also referred to as a collector bar since all other fan bars connect to it.) All supports are constructed from Type 410 stainless steel. Figure 2-14 illustrates the tube support configuration and numbering.

The U-bend region of the tubes in rows 1 through 18 received a supplemental thermal treatment (stress relieving) after bending. Row 1 tubes have the smallest U-bend radius, 3.5 inches.

The tubes were expanded to approximately 0.125 inch below the secondary face of the tubesheet. The expansion process was controlled to ensure that expansion of the tube is as close as possible to the secondary face of the tubesheet without going past the face. For peripheral tubes, where the curvature of the primary bowl limits access, expansion is performed in two overlapping zones with a shorter mandrel. The expansion zones overlap near the center of the tubesheet to ensure full-depth expansion. Following expansion, the profile of each tube was measured through the entire expanded area of the tubesheet, including the transition zone, using eddy current profilometry.

The lattice grid tube support is made up of two intersecting arrays of Type 410S stainless steel high bars (approximately 3 inches high) oriented at 30 and 150 degrees to the tube-free lane. The bars are located every fourth pitch to accommodate steam generator loading conditions. Type 410S stainless steel low bars (approximately 1 inch high) are located at every pitch location between the high bars. All low bars that are flush to the top of the high bars are oriented at 30 degrees to the tube-free lane, and all low bars that are flush to the bottom plane of the high bars are oriented at 150 degrees to the tube-free lane. All of the lattice bars are nominally 0.100 inch thick. The bar ends are fitted into precise slots of a specially designed peripheral support ring. The lattice grids are positioned within the steam generator shroud (wrapper) at elevations selected to prevent flow-induced vibration while not creating excessive flow resistance. The tubes are held in position within the diamond shaped bar opening, which provides line support contact. This limits the crevice area between the tubes and bars, which could trap corrosion products.

The FURS incorporates a series of flat bar fan assemblies on each side of the tube bundle. The fans are positioned so that all U-bends are supported at close intervals. The supports are made of Type 410 stainless steel that provides resistance to wear, acceptable strength, and resistance to corrosion and related tube denting.

2.4.6 Catawba 1

Catawba 1 has four recirculating steam generators designed and fabricated by BWI. The model CFR 80 steam generators were put into service in 1996 during RFO 9.

Each steam generator has 6633 thermally treated Alloy 690 tubes that have an outside diameter of 0.6875 inch and a nominal wall thickness of 0.040 inch. The tubes were manufactured by Sumitomo and are arranged in a triangular pattern as illustrated in Figure 2-15 with a spacing of approximately 0.930 inch. The heat transfer surface area in each steam generator is 79,800 ft^2.

The tubes were hydraulically expanded at each end for the full depth of the tubesheet. The tubesheet is 27.1 inches thick with the clad. The tubes are supported by lattice grid tube supports and fan bars. (The lowest fan bar is also referred to as a collector bar since all other fan bars connect to it.) All supports are constructed from Type 410 stainless steel. Figure 2-16 illustrates the tube support configuration and numbering.

The tubes in rows 1 through 27 received a supplemental thermal treatment (stress relieving) after bending. The bend radii of the inner row tubes were increased by crossing the tubes. That is, the origin and termination points of the tubes in the first several rows differ between the hot-leg and cold-leg. As a result, the U-bend region of these tubes is in a plane skewed from the tube-free lane (rather than in a plane perpendicular to the tube-free lane). These tubes are referred to as crossover tubes. The radius of a row 1 tube is 3.973 inches.

The steam generator tube bundle wrapper (shroud) is attached to the main shell by robust lugs with full penetration welds at the lower end and by radial pins at various tube support elevations along the wrapper height. These components along with the tube supports are arranged (and analyzed) to accommodate thermal motions during operation as well as loads imposed during accident conditions. The post-weld heat treatment of the welds in the lower part of the steam generator vessel is performed before installation of the internals, and no full-vessel post-weld heat treatment is performed.

After completion of tubing and insertion of the steam drum internals, the steam drum to main shell closure weld (at the top of the transition cone section) was performed. A local post-weld heat treatment of the steam drum to transition cone weld was performed after welding. To isolate any post-weld heat treatment or related effects from the internals, the inside of the steam generator was insulated and evacuated and temperature and temperature differential limits were adhered to during the post-weld heat treatment process.

Each lattice grid tube support consists of interlocking high (approximately three inches high) and low (approximately one inch high) bars that form a lattice pattern. The high bars are located every sixth pitch to accommodate the steam generator loading conditions. The low bars are located at every pitch between the high bars. All of the lattice grid tube supports are the same, except the lowest, which incorporates a differential resistance lattice grid. This lattice grid tube support differs in that medium bars (approximately 2.5 inches high) replace the low bars on the periphery. Each tube has four contact points.

Fan bars and connector bars, which are flat, support the U-bends. The fan bars are 1.25 inches wide. The fan bars on either side of the tube are offset from one another such that the fan bar on one side of the tube touches a different axial location along the length of the tube than the fan bar on the other side of the tube. The fan bar assembly is free floating and therefore rests on the tubes for support.

During the fabrication of the BWI steam generators for other utilities, it was noticed that the positioning of the U-bend support components could have resulted in peripheral tubes coming in contact (or in close proximity). The U-bend support structure, which is free to move with the U-bend during operating transients, is supported off of the peripheral tubes by "L"- or "J"-shaped elements called J-tabs. The J-tabs are made from 316 stainless steel. It was determined that the positioning of some of the J-tabs during manufacture may cause contact between certain pairs of vertically adjacent peripheral tubes in the U-bend region. This contact may occur because the J-tabs may be pushed in too far, which can cause two tubes in the same column to be closer than the ideal design spacing. The potential for, and effect of, this condition was evaluated. The evaluation confirmed that while some fretting may occur at contact locations, it will be less than that predicted at the tube support locations and will not be sufficient to limit operation of the tubing. Inservice inspection of the steam generators has indicated that tube proximity (less than the desired clearance or possible contact) affects a relatively small number of tubes on a number of the replacement steam generators.

2.4.7 Cook 1

Cook Unit 1 has four recirculating steam generators designed and fabricated by BWI. The model 51R steam generators were put into service in 2000 during RFO 17. The replacement steam generator consists of a new tube bundle and a new moisture separation unit; however, the steam dome from the original steam generator was refurbished and reused.

Each generator has 3496 thermally treated Alloy 690 tubes that have an outside diameter of 0.875 inch and a nominal wall thickness of 0.049 inch. The tubes, manufactured by Sumitomo, are arranged in a triangular pattern as illustrated in Figure 2-17 with a spacing of approximately 1.1875 inches.

The tubes were hydraulically expanded at each end for the full depth of the tubesheet. The tube-to-tubesheet seal welds are flush with the tubesheet. The tubesheet is 21.25 inches thick. The tubes are supported by lattice grid tube supports and flat fan bars. (The lowest fan bar is also referred to as a collector bar since all other fan bars connect to it.) All supports are constructed from Type 410 stainless steel. Figure 2-18 illustrates the tube support configuration and numbering.

The U-bend region of the tubes in rows 1 through 13 (i.e., those with a bend radius less than 12 inches) received a supplemental thermal treatment (stress relieving) after bending. To reduce the stresses in the U-bend area, the U-bend radius for the tube with the smallest radius was increased from 2.19 inches in the original steam generators. The radius of a row 1 tube is 4.75 inches.

Each lattice grid tube support consists of interlocking high (3.15 inches high and 0.135 inch thick) and low (1.0 inch high and 0.135 inch thick) bars that form a lattice pattern. This lattice provides lateral support in the straight section of the tube.

2.4.8 Cook 2

Cook 2 has four recirculating steam generators designed and fabricated by Westinghouse. The model 54F steam generators were put into service in 1989 during RFO 6. The replacement steam generators consist of a new tube bundle and a refurbished upper assembly and internals.

Each steam generator has 3592 thermally treated Alloy 690 tubes that have an outside diameter of 0.875 inch and a nominal wall thickness of 0.050 inch. The tubes were manufactured by Sandvik. The tubes are arranged in a square pattern as illustrated in Figure 2-19 with a spacing of approximately 1.225 inches.

With the exception of seven tubes that lack a hydraulic expansion in either the hot- or cold-leg tubesheet, the tubes were hydraulically expanded at each end for the full depth of the tubesheet. The tubesheet is 21 inches thick. The tubes are supported by a flow distribution baffle, support plates, and AVBs. All supports are constructed from Type 405 stainless steel. The flow distribution baffle is 0.75 inch thick, and the tubes pass through octafoil-shaped holes in the baffle. The tube support plates are 1.12 inches thick, and the tubes pass through quatrefoil-shaped holes in the plates. The AVBs are arranged in a "V" shape. Figure 2-20 illustrates the tube support configuration and numbering.

The tubes in rows 1 through 8 received a supplemental thermal treatment (stress relieving) after bending. To reduce the stresses in the U-bend area, the bend radius for the smallest radius tube was increased from 2.19 inches (in the original steam generators) to 3.141 inches (in the replacement steam generators).

The feedrings are constructed of carbon steel, and the J-nozzles are made from Alloy 600, which is less susceptible to erosion damage than the typical carbon steel J-nozzles.

A Digital Metal Impact Monitoring System provides early detection of loose parts in the reactor coolant system which may occur as a result of material wear, component failure, or outage/maintenance work. The system alerts operators by an alarm to significant impacts from loose parts. The system provides operators with a digital display of the current impact status, printed reports summarizing the past day's events, and the amplified sound played through a speaker. Accelerometers are mounted above and below the tubesheet of each steam generator to supply a signal to the system.

2.4.9 Farley 1

Farley Unit 1 has three recirculating steam generators designed by Westinghouse and fabricated by ENSA. The model 54F steam generators were put into service in 2000 during RFO 16.

Each generator has 3592 thermally treated Alloy 690 tubes which have an outside diameter of 0.875 inch and a nominal wall thickness of 0.050 inch. The tubes were manufactured by Sandvik. The tubes are arranged in the tubesheet as illustrated in Figure 2-21.

The tubes were hydraulically expanded at each end for the full depth of the tubesheet. The tubes are supported by a flow distribution baffle, support plates, and AVBs. All supports are constructed from Type 405 stainless steel. The tubes pass through octafoil-shaped holes in the flow distribution baffle and through quatrefoil-shaped holes in the tube support plates. The AVBs are rectangular in cross-section and are arranged in a "V" shape. The AVBs were designed and assembled to limit tube vibration and to allow for thermal growth of the tubes. The AVBs stiffen the tubes in the U-bend region and maintain proper tube spacing and alignment to reduce tube vibration. Figure 2-22 illustrates the tube support configuration and numbering.

The U-bend region of the tubes in rows 1 through 8 (i.e., those with a bend radius less than 12 inches) received a supplemental thermal treatment (stress relieving) after bending. To reduce the stresses in the U-bend area, the bend radius for the tube with the smallest radius was increased from 2.188 inches (in the original steam generators) to 3.141 inches (in the replacement steam generators).

The flow distribution baffle is located between the top of the tubesheet and the lowest tube support plate and is largely open in the center. This increases the flow velocity across the tubesheet surface and places the low flow-velocity region in the center of the tube bundle near the blowdown intake. The purpose of this design is to reduce sludge accumulation and mitigate corrosion.

The steam generators were designed to improve secondary-side access for sludge and foreign object removal capabilities. Each steam generator contains six secondary-side handholes and two inspection ports. Each steam generator also has a sludge collection system. Because the cross-flow velocity in the sludge collector is less than the settling velocity, the sludge collector captures the suspended particles in the secondary-side fluid. The sludge collector is designed to operate passively during normal operating conditions.

An online acoustical monitoring system is used to detect loose parts.

2.4.10 Farley 2

Farley Unit 2 has three recirculating steam generators designed by Westinghouse and fabricated by ENSA. The model 54F steam generators were put into service in 2001 during RFO 14.

Each generator has 3592 thermally treated Alloy 690 tubes which have an outside diameter of 0.875 inch and a nominal wall thickness of 0.050 inch. The tubes were manufactured by Sandvik. The tubes are arranged in a square pattern as illustrated in Figure 2-21 with a spacing of approximately 1.225 inches.

The tubes were hydraulically expanded at each end for the full depth of the tubesheet. The tubesheet is 21.42 inches thick. The tubes are supported by a flow distribution baffle, support plates, and AVBs. All supports are constructed from Type 405 stainless steel. The flow distribution baffle is 0.75 inch thick, and the tubes pass through octafoil-shaped holes in the baffle. The tube support plates are 1.125 inches thick and have quatrefoil-shaped holes. The AVBs are rectangular in cross-section and are arranged in a "V" shape. The AVBs were designed and assembled to limit tube vibration and to allow for thermal growth of the tubes. The AVBs stiffen the tubes in the U-bend region and maintain proper tube spacing and alignment to reduce tube vibration. Figure 2-22 illustrates the tube support configuration and numbering.

The U-bend region of the tubes in rows 1 through 8 (i.e., those with a bend radius less than 12 inches) received a supplemental thermal treatment (stress relieving) after bending. To reduce the stresses in the U-bend area, the bend radius for the tube with the smallest radius was increased from 2.188 inches (in the original steam generators) to 3.141 inches (in the replacement steam generators).

The flow distribution baffle is located between the top of the tubesheet and the lowest tube support plate and is largely open in the center. This increases the flow velocity across the tubesheet surface and places the low flow-velocity region in the center of the tube bundle near the blowdown intake. The purpose of this design is to reduce sludge accumulation and mitigate corrosion.

The steam generators were designed to improve secondary-side access for sludge and foreign object removal capabilities. Each steam generator contains six secondary-side handholes and two inspection ports. Each steam generator also has a sludge collection system. Because the cross flow velocity in the sludge collector is less than the settling velocity, the sludge collector captures the suspended particles in the secondary-side fluid. The sludge collector is designed to operate passively during normal operating conditions.

An online acoustical monitoring system is used to detect loose parts.

2.4.11 Ginna

Ginna has two recirculating steam generators designed and fabricated by BWI. The steam generators were put into service in 1996 during RFO 25.

Each steam generator has 4765 thermally treated Alloy 690 tubes which have an outside diameter of 0.749 inch and a nominal wall thickness of 0.044 inch. The tubes were manufactured by Valinox. The tubes are arranged in a triangular pattern as illustrated in Figure 2-23. The heat transfer surface area in each steam generator is 54,000 ft^2.

The tubes were hydraulically expanded at each end for the full depth of the tubesheet. The tubesheet is 25.25 inches thick. The tubes are supported by lattice grid tube support and fan bars. (The lowest fan bar is also referred to as a collector bar since all other fan bars connect to it.) All supports are constructed from Type 410 stainless steel. Figure 2-24 illustrates the tube support configuration and numbering.

The U-bend region of the tubes in rows 1 through 18 (i.e., those with a bend radius less than 12 inches) received a supplemental thermal treatment (stress relieving) after bending. The bend radii of the inner row tubes (i.e., rows 1 and 2) were increased by crossing the tubes. That is, the origin and termination points of the tubes in the first several rows differ between the hot-leg and cold-leg. As a result, the U-bend region of these tubes is in a plane skewed from the tube-free lane (rather than in a plane perpendicular to the tube-free lane). These tubes are referred to as crossover tubes. The nominal tube-to-tube gap is 0.369 inch, although tube bending tolerances can reduce this to a design minimum of 0.269 inch.

As of 1998, the BWI replacement steam generators (Braidwood 1, Byron 1, Cook 1, Catawba 1, Ginna, McGuire 1 and 2, Millstone 2, and St. Lucie 1) were essentially identical to one another in concept and in almost all materials of construction, including the tubing and the tube support materials. However, they did differ in size since they replaced different models of original steam generators.

The wrapper (shroud) in the steam generator separates the downward-flowing recirculating liquid from the rising two-phase mixture within the tube bundle. The shroud is a steel cylinder (typically 0.5 inch thick) extending from just below the lowest lattice grid tube support up to a slip joint in the upper U-bend region. The wrapper design has two sections—a lower cylindrical section rigidly supported by shell/shroud lugs near the tubesheet, and an upper section semi-rigidly supported by the primary separator deck and deck lugs at the upper end. The primary separator deck lugs allow free radial differential thermal expansion of the primary separator deck and steam generator shell. The mating ends of the two wrapper sections have machined rings forming an overlapping slip joint which provides restraint in only the lateral direction and allows free axial differential thermal expansion between the upper and lower wrapper section.

The shroud is typically supported at its lower edge with 12 support blocks (2 inches by 4.5 inches) that are welded to the steam generator shell with full penetration welds. The shroud is welded to these lugs by full penetration welds. As a result of this arrangement, the shroud, tubes, and main shell all thermally expand in the same direction (upward from the tubesheet). The magnitude of the expansion is determined by the temperature and expansion properties of the parts involved. Above the shroud support blocks, the shroud is laterally supported at various tube support elevations by shroud pins. The top two lattice grid tube supports have shroud pins, and at least every other lattice grid tube support elevation includes shroud pins, down to the shroud support blocks. Typically, each of these support elevations may have 16 shroud pins. Shroud pins are robust (typically 2 inches in diameter), threaded pins that screw outward through threaded sockets which are welded to the shroud. The pins, which contact the main shell, position the shroud within the shell, and they also accommodate lateral loads from handling, shipping, and seismic-loading conditions. Axial differential expansion is accommodated by sliding of the shroud pin ends along the shell inner surface. The shroud pins are designed to accommodate the frictional loads involved. Local flexing of the shell accommodates radial differential expansions between the tube support peripheral rings, the shroud, and the shell. The shroud pins are offset from lattice grid tube support wedges so that shroud flexing can accommodate the necessary differential radial motion. The lattice grid tube supports are laterally positioned within the cylindrical wrapper by radial wedges between the lattice rings and wrapper and are vertically restrained by blocks above and below the lattice rings. Both the lattice wedges and support blocks are welded only to the wrapper. No welded connections exist between the lattice assemblies and the wrapper. The tube support rings move with the shroud in the longitudinal direction.

Differential thermal motion between the shroud and shell may occur on vessel heatup and cooldown. Since the shroud is thinner and wetted on both sides, it will follow the thermal fluid conditions much more closely than the thicker shell, which is wetted on one surface only.

During a cooldown, the shroud will shrink back from the shell, the shroud pin/shell load will be relaxed, and the differential thermal expansion will occur more easily. During heatup, the shroud pins will see increased radial load simultaneous with pin end sliding over the shell surface. Such radial motion is accommodated by local shroud flexure at the pin locations. The strength of the shroud pin/socket readily accommodates sliding drag forces.

The design requirements for the tube supports (1) preclude excessive flow-induced vibration, (2) limit the pressure loss in order to promote a high circulation ratio, (3) provide line support contact to reduce the potential for deposition of corrosion-causing impurities and localized dryout, (4) provide sufficient tube contact length to lower contact stress and hence limit fretting wear of tubes, (5) provide a strong tube support to withstand lateral seismic loads, loads caused by a loss-of-coolant accident (LOCA) and burst pipe events, and handling and shipping loads, (6) accommodate tube support motions during heatup/startup operation without risk of lockup or large thermally induced stresses, and (7) resist corrosion, denting, and stress-corrosion cracking due to normal operation and chemical cleaning.

The lattice grid tube support is made up of two intersecting arrays of Type 410S stainless steel high bars (approximately 3 inches high) oriented at 30 and 150 degrees to the tube-free lane. The bars are located every four to eight pitches, depending on the size of the bundle and the particular steam generator loading conditions. Type 410S stainless steel low bars (approximately 1 inch high) are located at every pitch location between the high bars. All low bars that are flush to the top of the high bars are oriented at 30 degrees to the tube-free lane, and all low bars that are flush to the bottom plane of the high bars are oriented at 150 degrees to the tube-free lane. The bar ends are fitted into precise slots of a specially designed peripheral support ring, which is then clamped by two outer retainer rings. Wedges and blocks welded only to the wrapper position these tube support assemblies within the tube bundle.

All of the lattice grid tube supports are the same except the lowest, which incorporates a differential resistance feature which is used to direct flow into the interior of the bundle above the tubesheet. The construction of the differential resistance lattice grid resembles that of a regular grid; however, medium bars replace the low bars located towards the bundle periphery. Because of the increased height of the medium bars (approximately 2.5 inches), all crossing bars in the outer regions intersect at each pitch location. As a result, these regions offer more resistance to flow and the fluid is preferentially directed to penetrate into the central region of the tube bundle. In summary, the lowest lattice grid tube support incorporates intersecting medium bars to increase the axial flow resistance around the periphery of the bundle, thus promoting flow penetration across the tubesheet. This tube support incorporates sealing strips to prevent bypass flow around the bundle in the event that flow passages within the bundle (i.e., the "crevices" between the tube and support) become plugged.

During normal operation, loads on the lattice tube support structures are relatively low (typically 1.0 to 1.5 pounds per square inch (psi) per support). The supports are exposed to heatup and cooldown effects and modest flow loads. The supports are designed to sustain accident loads due to seismic, LOCA, or steam/feed line break events. During heatup, the lattice grid tube supports and shroud heat more quickly than the shell, causing some radial load on the lattice rings at the lattice ring/shroud wedge supports. The tubes, which expand more than the vessel and shroud, expand vertically relative to the lattice grid tube support (typically about 0.1875 inch).

The primary objective of U-bend supports is to effectively restrain the tubing, thereby limiting flow-induced vibration and fretting wear. In addition, the U-bend support configuration must provide lateral support to U-tubes during fabrication, transportation, and service conditions such as seismic events.

The FURS consists of 410S stainless steel flat bar fan assemblies supported by 316L stainless steel J-tabs, carbon steel archbars, clamping bars, and tie tubes. Fan assemblies, which incorporate a number of flat bar "fingers," are positioned between each layer of tubes. The fan assemblies stagger in and out from tube layer to tube layer so that flat bars do not contact tubes on directly opposite sides. Depending on the radius of the tube bundle, the tube may be supported by up to five fan bars all connected at their lower ends to a connector bar by an autogenous full penetration weld that is post-weld heat treated.

The flat bars in a fan assembly are positioned so the U-bends are supported at close intervals (typically 19 to 22 inches). The actual span length and the number of support locations depend on the bundle size, tube size, and flow loadings. The U-bend region of all tubes is supported by at least two flat bars (one on the hot-leg and one the cold-leg). To limit the potential and amount of tube wall thinning due to fretting, there are wide, tangent contact regions and small, nominal gaps between the tubes and the flat bars.

The FURS allows free expansion of the U-bend tubes without the need for sliding to occur between the tubes and the bars. Free expansion avoids tube stress or damage. This is achieved by supporting the FURS assembly with the outermost layer of tubes and by avoiding other restraint points. The weight of the U-bend assembly is transferred to the outer U-bend tubes through J-tabs that are individually positioned to distribute the load onto the supporting tubes. If a supporting tube is taken out of service, its load is simply redistributed to the remaining tubes with no significant increase in deflection of the FURS assembly. The FURS assembly and U-tubes move up and down together on heatup and cooldown and also during operation at power when tube hot-legs and cold-legs have slightly unequal leg temperatures.

Testing was performed on the materials used to support the tubes to investigate their corrosion characteristics, especially the growth of the oxide film. It was shown that the 410S material oxide will not cause denting and that the material would not experience catastrophic oxidation even in the extreme test environment. The oxide layer was observed to be tightly adherent and nonvoluminous (i.e., thin). In addition, studies have shown that material containing more than 5 percent chromium (410S stainless steel has between 11.5 and 13.5 percent chromium) is not susceptible to flow-accelerated corrosion.

It was observed during manufacture that the positioning of the U-bend support components could result in contact between peripheral tubes of some BWI steam generators. That is, there was less than optimum radial spacing between the outermost tubes and the adjacent inner tubes in the U-bend region of some tubes. Proximity of peripheral tubes due to positioning of the U-bend supports during manufacture relates to the possibility that certain peripheral tubes (those with J-tabs) may be close to their vertically adjacent neighbors or even in contact. Tube proximity was deemed to be a condition of tube contact if the tube-to-tube clearance was less than 0.040 inch after the vessel was settled in its vertical orientation in the unit. Somewhat larger clearances were required in the horizontal position to allow for settling of the U-bend support assembly on uprighting.

As discussed above, the U-bend support structure, which is free to move with the U-bend tubes during operating transients, is supported off of the peripheral tubes by "L"- or "J"-shaped elements called J-tabs. These J-tabs support the FURS at a large number (several hundred) of locations. The J-tabs are inserted against the tube, then welded to the arch-bar/clamping-bar assembly. This process is performed with the steam generator in a horizontal position at the fabrication facility. Prior to the discovery of the tube-to-tube contact condition (or proximity issue), no special attention was paid to positioning the outermost tubes before setting and welding the J-tabs. If a tube was not properly positioned (i.e., not spaced consistently to maintain design clearances relative to the tube below it) prior to welding the J-tab, then the weight of the U-bend support structure may distort the tube shape when the steam generator is vertical. The proximity of one tube to the tube below it may differ between the cold conditions of the inspection and the hot normal operating conditions. This condition is limited to only those tubes at the periphery of the bundle that are in contact with J-tabs (i.e., vertically adjacent peripheral tubes).

The potential for, and effect of, this condition was assessed. The existence of tube-to-tube contact (or proximity) can potentially result in tube wear and can increase the potential for tube corrosion due to deposit buildup. Analysis confirmed that this situation does not result in a condition which adversely affects the integrity of the steam generator. It is bounded by a configuration normally present in the as-designed steam generators. The fretting wear damage assessment, which is based on estimates of wear coefficients and work rates at the tube-to-tube contact, shows that the maximum tube wall loss after 60 years of continuous full-power operation is 40 percent of the nominal wall thickness. Similar results can be expected for normal tube to flat bar contacts in the same area of the tube bundle. The potential for tube degradation as a consequence of excessive fouling is addressed by confirming that the U-bend condition is bounded by fouling at the tube-to-tubesheet joint region. This comparative assessment included consideration of heat flux, margin to critical heat flux, residual stresses, local environment, and material corrosion resistance. Potential tube touching, wall degradation, and deposition can be effectively monitored using normal eddy current inspection techniques carried out during routine inspections.

2.4.12 Harris

Harris has three recirculating steam generators designed and fabricated by Westinghouse. The model Delta 75 steam generators were put into service in 2001 during RFO 10.

Each steam generator has 6307 thermally treated Alloy 690 tubes which have an outside diameter of 0.688 inch and a nominal wall thickness of 0.040 inch. The tubes were manufactured by Sandvik. The tubes are arranged in a triangular pattern as illustrated in Figure 2-25 with a tube spacing of approximately 0.98 inch.

The tubes were hydraulically expanded at each end for the full depth of the tubesheet. The tubes are supported by a flow distribution baffle, support plates, and AVBs. All supports are constructed from Type 405 stainless steel. The flow distribution baffle is 0.75 inch thick, and the tubes pass through octafoil-shaped holes in the baffle. The tube support plates are 1.125 inches thick, and the tubes pass through trifoil-shaped holes in the plates. The AVBs are rectangular in cross-section, arranged in a "V" shape, staggered, and are 0.160 inch thick. Figure 2-26 illustrates the tube support configuration and numbering.

The tubes in rows 1 through 17 (i.e., those with a bend radius less than 12 inches) received a supplemental thermal treatment (stress relieving) after bending. The smallest U-bend radius is 3.25 inches.

The AVBs are staggered to limit the pressure drop in the U-bend region, thereby minimizing the impact on the secondary water circulation ratio and reducing the potential for steam blanketing. The rectangular AVB shape provides for a greater tube-to-AVB contact area, and the gap between the AVB and the tube is tightly controlled to reduce the potential for tube wear.

The steam generators have an internal sludge collector located above the tubes to reduce the amount of suspended solids in the steam generator bulk water and the amount of sludge deposited on the tubesheet. This is accomplished by circulating a portion of the secondary water over the sludge collector.

To limit the size of foreign material that the feedwater system might introduce into the steam generators, the feedwater must pass through the 0.38-inch holes of the vertical spray tubes. The spray tubes are made from Alloy 690 material and are spread uniformly around the feedwater distribution header. There is a metal impact monitoring system installed on the steam generators to detect loose parts in the event they are introduced into the steam generator.

2.4.13 Indian Point 3

Indian Point 3 has four recirculating steam generators designed and fabricated by Westinghouse. The model 44F steam generators were put into service in 1989 during RFO 6.

Each steam generator has 3214 thermally treated Alloy 690 tubes which have an outside diameter of 0.875 inch and a nominal wall thickness of 0.050 inch. The tubes were manufactured by Sandvik and are arranged in a square pattern as illustrated in Figure 2-27 with a spacing of approximately 1.2344 inches.

The tubes were hydraulically expanded at each end for the full depth of the tubesheet. The tube-to-tubesheet welds are flush with the tubesheet. The tubesheet is 21.81 inches thick (with the clad, the tubesheet is 21.96 inches thick). The tubes are supported by a flow distribution baffle, support plates, and AVBs. All supports are constructed from Type 405 stainless steel. The flow distribution baffle is 0.75 inch thick, and the tubes pass through octafoil-shaped holes in the baffle. The tube support plates are 1.125 inches thick and have quatrefoil shaped holes through which the tubes pass. The AVBs are arranged in a "V" shape and are 0.690 inch thick (on the side contacting the tubes). Since the AVBs are not perpendicular to the tubes, the contact length of the AVB with the tube varies from 0.697 inch to greater than 1.5 inches. The AVBs penetrate the tube bundle through row 9. Figure 2-28 illustrates the tube support configuration and numbering.

The tubes in rows 1 through 8 received a supplemental thermal treatment (stress relieving) after bending. The first three rows of tubes have bend radii of 2.187 inches, 3.421 inches, and 4.656 inches.

Thirty percent of the tubes in any steam generator may be plugged if the equivalent average plugging level in all steam generators is less than or equal to 24 percent.

2.4.14 Kewaunee

Kewaunee has two recirculating steam generators designed by Westinghouse and fabricated by Ansaldo Energia in Italy. The model 54F steam generators were put into service in 2001 during RFO 24.

Each steam generator has 3592 thermally treated Alloy 690 tubes which have an outside diameter of 0.875 inch and a nominal wall thickness of 0.050 inch. The tubes were manufactured by Valinox Nucleaire and are arranged in a square pattern as illustrated in Figure 2-29 with a spacing of approximately 1.225 inches.

The tubes were hydraulically expanded at each end for the full depth of the tubesheet. Hydraulic expansions typically produce 20 to 40 percent less stress than hard-rolled expansions. The tubesheet is 21.42 inches thick. The tubes are supported by a flow distribution baffle, support plates, and AVBs. All supports are constructed from Type 405 stainless steel. The tube support plates have quatrefoil shaped holes through which the tubes pass. The AVBs are rectangular in cross-section and are arranged in a "V" shape. Figure 2-30 illustrates the tube support configuration and numbering.

The U-bend region of the tubes in rows 1 through 8 (i.e., those with a bend radius less than 12 inches) received a supplemental thermal treatment (stress relieving) after bending. The smallest U-bend radius is 3.141 inches.

The flow distribution baffle is located between the top of the tubesheet and the lowest tube support plate and is largely open in the center. This increases the flow velocity across the tubesheet surface and places the low flow-velocity region in the center of the tube bundle near the blowdown intake. The purpose of this design is to reduce sludge accumulation and mitigate corrosion.

The AVBs are "V"-shaped, rectangular bars that stiffen the tubes in the U-bend region. They are designed to maintain proper tube spacing and alignment and to reduce tube vibration.

The circulation ratio for the replacement steam generators was increased from 2.71 to 4.28.

2.4.15 McGuire 1

McGuire 1 has four recirculating steam generators designed and fabricated by BWI. The model CFR 80 steam generators were put into service in 1997 during RFO 11.

Each steam generator has 6633 thermally treated Alloy 690 tubes which have an outside diameter of 0.688 inch and a nominal wall thickness of 0.040 inch. The tubes were manufactured by Sumitomo and are arranged in a triangular pattern as illustrated in Figure 2-31 with a spacing of approximately 0.930 inch. The heat transfer surface area in each steam generator is 79,800 ft^2.

The tubes were hydraulically expanded at each end for the full depth of the tubesheet. The tubesheet hole diameter is 0.6955 inch. The tubesheet is 26.63 inches thick (with the clad, the tubesheet is 27.1 inches thick). The tubes are supported by lattice grid tube supports and fan bars. (The lowest fan bar is also referred to as a collector bar since all other fan bars connect to it.) All supports are constructed from Type 410 stainless steel. Figure 2-32 illustrates the tube support configuration and numbering.

The tubes in rows 1 through 21 received a supplemental thermal treatment (stress relieving) after bending. The smallest U-bend radius, located in row 3, is 3.632 inches. (Row 1 has a radius of 3.973 inches.) The bend radii of the inner row tubes were increased by crossing the tubes. That is, the origin and termination points of the tubes in the first several rows differ between the hot-leg and cold-leg. As a result, the U-bend region of these tubes is in a plane skewed from the tube-free lane (rather than in a plane perpendicular to the tube-free lane). These tubes are referred to as crossover tubes.

The tube-to-tubesheet seal welds are flush with the tubesheet. The tubes are hydraulically expanded to increase the mechanical strength of the tube-to-tubesheet joint and to seal the tube-to-tubesheet crevice. The tubes are hydraulically expanded with a mandrel that has hydraulic seals positioned on the mandrel to control the length of the tube expanded. The elastomeric hydraulic seals are designed so that no metal parts are impressed upon the inside surface of the tube when the hydraulic pressure is applied. The position of the seal at the secondary face of the tubesheet is controlled to ensure that expansion of the tube is as close as possible to the secondary face of the tubesheet without going above the face. For peripheral tubes, curvature of the primary head limits access, and expansion is performed in two overlapping zones using a shorter expansion mandrel. The shorter mandrel can access the peripheral tubes without interfering with the primary heads. The expansion zones overlap near the center of the tubesheet. After hydraulic expansion, the inside profile of each tube was measured through the entire expanded area of the tubesheet (including the expansion transition area) using eddy current techniques.

The tube welding and hydraulic expansion process occurred after the steam generator lower shell and primary head assembly were welded and received their post-weld heat treatment. This sequence is intended to prevent tube sensitization and keep the tube-to-tubesheet joint from experiencing unnecessary exposure to thermal stresses, which could loosen the tubes or create a crevice as a result of relaxation of the expanded region.

The lattice grid tube supports provide lower flow resistance (resulting in higher circulation rates), greater strength (without requiring tie rods), and greater vibration control. The lattice grid tube supports also have less of a tendency to collect deposits when compared to tube support plates, because the lattice grid tube supports make contact with the tube in a line, rather than with an entire area as the plate would. The lattice grid tube support is made up of two intersecting arrays of a series of high bars (approximately 3 inches high) oriented at 30 and 150 degrees to the tube-free lane. The bars are located every sixth pitch to accommodate the steam generator loading conditions. Low bars (approximately 1 inch high) are located at every pitch location between the high bars. All low bars that are flush to the top of the high bars are oriented at 30 degrees to the tube-free lane, and all low bars that are flush with the bottom plane of the high bars are oriented at 150 degrees to the tube-free lane. The bar ends fit into precise slots in a peripheral support ring. The support ring is sandwiched by two retainer rings that are held together by welded studs and nuts. Tube-free lane support beams and span-breaker bars are secured on the upper and lower surfaces of the grid to enhance stability. All of the lattice grid tube supports are the same except the lowest, which incorporates a differential resistance lattice grid. This lattice grid tube support differs in that medium bars approximately 2 inches high replace the low bars on the periphery. The medium bars offer more resistance to periphery flow, which results in flow being directed to the central region of the tube bundle. The tubes pass through diamond-shaped openings formed by the intersecting lattice grid bars. These openings are designed to limit the crevice area between the tubes and the tube support, which reduces the potential for corrosion products to accumulate in this region and eliminates any stagnant spots responsible for dryout caused by local superheat. The lattice grid tube supports are positioned within the steam generator shroud at elevations selected to prevent flow-induced vibration.

The U-bends are supported by fan bars and connector bars, which are flat. The fan bars are connected to a nearly horizontal bar by full penetration, heat treated welds. This horizontal bar is referred to as a collector bar (some units also refer to this bar as a fan bar). The nearly horizontal bars provide support for the U-bends with the smallest radius. All U-bends are supported by either the collector bar or at least one fan bar. The U-bend support system provides open flowpaths and line contact support at all locations in the bundle, reducing the potential for sludge buildup. Each fan bar assembly is offset along the length of the tube such that the fan bar on one side of the tube touches a different axial location along the length of the tube than the fan bar on the other side of the tube.

The outermost layer of tubes supports the U-bend support system (fan bars and collector bars). This allows free expansion of the U-bend during operation of the steam generator. The arch bar assemblies transfer the weight of the U-bend support system to the outer layer of tubes. The U-bend support system and tube bundle move up and down together during heatup and cooldown. During power operation, the tube hot- and cold-legs have slightly different lengths (due to different leg temperatures), which results in the U-bend supports being at a slight angle. A clamping bar, which is welded to the arch bar, collects the upper ends of the U-bend support system. The weight of the fan bar assemblies is transferred from the arch bar to the outermost tubes by J tabs that are installed after the U-bend assembly is complete.

Fretting (tube wear) is a result of U-bend flow loading and support positioning, material, clearance, and contact length. The orientation and positioning of the U-bend support system are based on the flow-induced vibration analysis for fluidelastic instability and turbulent excitation.

During the fabrication of the BWI steam generators for other utilities, it was noticed that the positioning of the U-bend support components could have resulted in peripheral tubes coming in contact (or in close proximity). The U-bend support structure, which is free to move with the U-bend during operating transients, is supported off of the peripheral tubes by "L"- or "J"-shaped elements called J-tabs. The J-tabs are made from 316 stainless steel. The positioning of some of the J-tabs during manufacture may cause contact between certain pairs of vertically adjacent peripheral tubes in the U-bend region. This is a result of the J-tabs being pushed in too far, which can cause two tubes in the same column to be closer than the ideal design spacing (i.e., the tubes immediately under the outermost tube in the same column are free to move with only the friction of the fan bars and collector bars holding them in place). The potential for, and effect of, this condition was evaluated. The evaluation confirmed that while some fretting may occur at contact locations, it will be less than that predicted at the tube support locations and will not be sufficient to limit operation of the tubing. Inservice inspection of the steam generators has indicated that tube proximity (i.e., less than desired clearance or possible contact) affects a relatively small number of tubes on a number of the replacement steam generators.

The feedwater distribution system was designed, in part, to address water hammer, thermal stratification, erosion, and internal feedwater header collapse. The feedwater distribution system is a split ring design connected via a T-section to a goose neck assembly attached to the thermal sleeve in the feedwater piping. The header is supported by the thermal sleeve/feedwater piping weld interface, and by supporting lugs located on the header pipe at approximately 90 degrees to the feedwater nozzle, and at the header ends (near the split in the header ring) opposite the feedwater nozzle. The Alloy 690 J-tubes are positioned so as to reduce the possibility of feedwater impinging on internal surfaces, reducing the possibility of erosion.

The letter from M.S. Tuckman, Duke Power Company, to the NRC, dated September 30, 1994 (see Appendix B), provides additional design information (e.g., feedwater, blowdown, and moisture separating system).

The replacement steam generator design has a circulation ratio of 5.7. High circulation improves steam generator performance by promoting flow penetration across the tubesheet and reducing fluid quality and zones of low velocity, thereby reducing sludge accumulations.

2.4.16 McGuire 2

McGuire 2 has four recirculating steam generators designed and fabricated by BWI. The model CFR 80 steam generators were put into service in 1997 during RFO 11.

Each steam generator has 6633 thermally treated Alloy 690 tubes which have an outside diameter of 0.688 inch and a nominal wall thickness of 0.040 inch. The tubes were manufactured by Sumitomo and are arranged in a triangular pattern as illustrated in Figure 2-31 with a spacing of approximately 0.930 inch. The heat transfer surface area in each steam generator is 79,800 ft^2.

The tubes were hydraulically expanded at each end for the full depth of the tubesheet. The tubesheet is 26.63 inches thick (with the clad, the tubesheet is 27.1 inches thick). The tubes are supported by lattice grid tube supports and fan bars. (The lowest fan bar is also referred to as a collector bar since all other fan bars connect to it.) All supports are constructed from Type 410 stainless steel. Figure 2-32 illustrates the tube support configuration and numbering.

The tubes in rows 1 through 21 received a supplemental thermal treatment (stress relieving) after bending. The row 1 tubes have a U-bend radius of 3.973 inches. The bend radii of the inner row tubes were increased by crossing the tubes. That is, the origin and termination points of the tubes in the first several rows differ between the hot-leg and cold-leg. As a result, the U-bend region of these tubes is in a plane skewed from the tube-free lane (rather than in a plane perpendicular to the tube-free lane). These tubes are referred to as crossover tubes.

The tube-to-tubesheet seal welds are flush with the tubesheet. The tube welding and hydraulic expansion process occurred after the steam generator lower shell and primary head assembly were welded and received their post-weld heat treatment. This sequence is intended to keep the tube-to-tubesheet joint from experiencing unnecessary exposure to thermal stresses that could loosen the tubes or create a crevice if the expanded region relaxes. After hydraulic expansion, the inside profile of each tube was measured through the entire expanded area of the tubesheet (including the expansion transition area) using eddy current techniques.

The lattice grid tube supports provide lower flow resistance, greater strength, greater vibration control, and less tendency to collect deposits when compared to tube support plates. The lattice grid tube support is made up of two intersecting arrays of a series of high bars (approximately 3 inches high) oriented at 30 and 150 degrees to the tube-free lane. The bars are located every sixth pitch to accommodate the steam generator loading conditions. Low bars (approximately 1 inch high) are located at every pitch location between the high bars. All low bars that are flush to the top of the high bars are oriented at 30 degrees to the tube-free lane, and all low bars that are flush with the bottom plane of the high bars are oriented at 150 degrees to the tube-free lane. The bar ends fit into precise slots in a peripheral support ring. The support ring is sandwiched by two retainer rings that are held together by welded studs and nuts. Tube-free lane support beams and span-breaker bars are secured on the upper and lower surfaces of the grid to enhance stability. All of the lattice grid tube supports are the same except the lowest, which incorporates a differential resistance lattice grid. This lattice grid tube support differs in that medium bars approximately 2 inches high replace the low bars on the periphery. The medium bars offer more resistance to periphery flow, which results in flow being directed to the central region of the tube bundle. The tubes pass through diamond-shaped openings formed by the intersecting lattice grid bars. These openings are designed to limit the crevice area between the tubes and the tube support, which reduces the potential for corrosion products to accumulate in this region.

The U-bends are supported by fan bars and connector bars, which are flat. The fan bars are connected to a nearly horizontal bar by full penetration, heat treated welds. This horizontal bar is referred to as a collector bar (some units also refer to this bar as a fan bar). The nearly horizontal bars provide support for the U-bends with the smallest radius. Either the collector bar or at least one fan bar supports all U-bends.

The outermost layer of tubes supports the U-bend support system (fan bars and collector bars). This allows free expansion of the U-bend during operation of the steam generator. Arch bar assemblies transfer the weight of the U-bend support system to the outer layer of tubes. A clamping bar, which is welded to the arch bar, collects the upper ends of the U-bend support system. After the U-bend assembly is complete, J-tabs are installed to transfer the weight of the fan bar assemblies from the arch bar to the outermost tubes.

2.4.17 Millstone 2

Millstone 2 has two recirculating steam generators designed and fabricated by BWI. The steam generators were put into service in 1993 during RFO 11. The replacement steam generator consisted of a new tube bundle. The steam drum from the original steam generators was reused.

Each steam generator has 8523 thermally treated Alloy 690 tubes that have an outside diameter of 0.750 inch and a nominal wall thickness of 0.0445 inch. The tubes were manufactured by Valinox using a pilgering process. The tubes are arranged in a triangular pattern as illustrated in Figure 2-33 with a spacing of approximately 1.0 inch. The heat transfer surface area in each steam generator is 93,500 ft^2.

The tubes were hydraulically expanded at each end for the full depth of the tubesheet. The tubesheet is 21.75 inches thick (with the clad, the tubesheet is 22.06 inches thick). The tubes are supported by lattice grid tube supports and fan bars. (The lowest fan bar is also referred to as a collector bar since all other fan bars connect to it.) All supports are constructed from Type 410 stainless steel. Figure 2-34 illustrates the tube support configuration and numbering.

The U-bends in rows 1 through 8 received a supplemental thermal treatment (stress relieving) after bending. The smallest U-bend radius is 3.905 inches and corresponds to the tubes in row 3. Tubes in rows 1 and 2 have radii of 4.272 inches and 3.968 inches, respectively. The bend radii of the inner three rows of tubes were increased by crossing the tubes. That is, the origin and termination points of the tubes in the first several rows differ between the hot-leg and cold-leg. As a result, the U-bend region of these tubes is in a plane skewed from the tube-free lane (rather than in a plane perpendicular to the tube-free lane). These tubes are referred to as crossover tubes.

Each lattice grid tube support consists of interlocking high (approximately 3 inches high) and low (approximately 1 inch high) bars that form a lattice pattern. The low bars are located at every pitch between the high bars. All of the lattice grid tube supports are the same except the lowest, which incorporates medium bars (approximately 2.5 inches high).

The fan bars are of various widths (1.0 inch, 2.6 inches, and 3.2 inches). The fan bars on either side of the tube are offset from one another such that the fan bar on one side of the tube touches a different axial location along the length of the tube than the fan bar on the other side of the tube. This offset distance varies from one fan bar to another.

2.4.18 North Anna 1

North Anna 1 has three recirculating steam generators designed and fabricated by Westinghouse. The model 54F steam generators were put into service in 1993 during RFO 9.

Each steam generator has 3592 thermally treated Alloy 690 tubes that have an outside diameter of 0.875 inch and a nominal wall thickness of 0.050 inch. The tubes were manufactured by Sandvik and are arranged in a square pattern as illustrated in Figure 2-35 with a spacing of approximately 1.225 inches. The heat transfer surface area in each steam generator is 54,500 ft^2.

The tubes were hydraulically expanded at each end for the full depth of the tubesheet. The tubes are supported by a flow distribution baffle, support plates, and AVBs. All supports are constructed from Type 405 stainless steel. The flow distribution baffle is 0.75 inch thick, and has octafoil-shaped holes through which the tubes pass (with the exception of the row 1 tubes, most of which have pentafoil shaped holes). The tube support plates are 1.125 inches thick and have quatrefoil-shaped holes through which the tubes pass. The AVBs are arranged in a "V" shape and penetrate the tube bundle through row 8. Figure 2-36 illustrates the tube support configuration and numbering.

The U-bend region of the tubes in rows 1 through 8 received a supplemental thermal treatment (stress relieving) after bending.

The AVB-to-tube gaps were closely controlled and monitored during steam generator fabrication. The shop procedures covered insertion, alignment, and welding to provide for minimum clearance while avoiding compressive loading on the tubes. After installation, measurements were taken and reviewed to confirm that the conditions which affect flow-induced vibration were within expectations and that the conditions would continue to meet design assumptions.

2.4.19 North Anna 2

North Anna 2 has three recirculating steam generators designed and fabricated by Westinghouse. The model 54F steam generators were put into service in 1995.

Each steam generator has 3592 thermally treated Alloy 690 tubes that have an outside diameter of 0.875 inch and a nominal wall thickness of 0.050 inch. The tubes were manufactured by Sandvik and are arranged in a square pattern as illustrated in Figure 2-35 with a spacing of approximately 1.225 inches. The heat transfer surface area in each steam generator is 54,500 ft^2.

The tubes were hydraulically expanded at each end for the full depth of the tubesheet. The tubes are supported by a flow distribution baffle, support plates, and AVBs. All supports are constructed from Type 405 stainless steel. The flow distribution baffle is 0.75 inch thick, and has octafoil-shaped holes through which the tubes pass (with the exception of the row 1 tubes, most of which have pentafoil-shaped holes). The tube support plates are 1.125 inches thick and have quatrefoil-shaped holes through which the tubes pass. The AVBs are arranged in a "V" shape and penetrate the tube bundle through row 8. Figure 2-36 illustrates the tube support configuration and numbering.

The U-bend region of the tubes in rows 1 through 8 received a supplemental thermal treatment (stress relieving) after bending.

The AVB-to-tube gaps were closely controlled and monitored during steam generator fabrication. The shop procedures covered insertion, alignment, and welding to provide for minimum clearance while avoiding compressive loading on the tubes. After installation, measurements were taken and reviewed to confirm that the conditions which affect flow-induced vibration were within expectations and that the conditions would continue to meet design assumptions.

2.4.20 Oconee 1, 2, and 3

Each of the three Oconee units has two once-through steam generators designed and fabricated by BWI. These steam generators were put into service in the three Oconee units in the 2003–2004 timeframe during RFOs 21, 20, and 21, respectively.

Each steam generator has 15,631 thermally treated Alloy 690 tubes that have an outside diameter of 0.625 inch and a nominal wall thickness of 0.038 inch. The tubes were manufactured by Sumitomo and are arranged in a triangular pattern as illustrated in Figure 2-37 with a spacing of approximately 0.875 inch. The heat transfer surface area is 134,600 ft^2. The total length of a tube is 56.2 feet, with a heated length of 52.4 feet.

The tubes were hydraulically expanded into the tubesheet for 13 inches from the tube end. The tubesheet is 22 inches thick. The tubes are supported by tube support plates constructed from 410 stainless steel, with trifoil-shaped holes through which the tubes pass. The trifoils have an hour-glass profile to improve hydraulic resistance (i.e., reduce the pressure drop across the plate), provide a flat contact surface for the tube, facilitate tubing of the steam generator, and provides better accessibility for water lancing and chemical cleaning. Figure 2-38 illustrates the tube support configuration and numbering.

In the original Oconee once-through steam generators, a portion of the tube bundle was not tubed. This region was referred to as the open tube lane. There is no open tube lane in the replacement steam generator design (i.e., this portion of the tube bundle was tubed in the replacement steam generators).

2.4.21 Palo Verde 2

Palo Verde 2 has two recirculating steam generators designed by Combustion Engineering and fabricated by Ansaldo Energia. The steam generators were put into service in 2003 during RFO 11.

Each steam generator has 12,580 thermally treated Alloy 690 tubes that have an outside diameter of 0.75 inch and a nominal wall thickness of 0.042 inch. The tubes were manufactured by Sandvik and are arranged in a triangular pattern as illustrated in Figure 2-39 with a spacing of approximately 0.866 inch.

The tubes were hydraulically expanded at each end for the full depth of the tubesheet. The tubesheet is 25 inches thick (with the clad, the tubesheet is 25.25 inches thick). The Siemens method of tube expansion was used to expand the tube into the tubesheet, which includes a hard roll near the top of the tubesheet and the end of the tube.

The tubes are supported by a flow distribution baffle, horizontal lattice grid supports, batwing (diagonal) supports, and vertical straps. All tube supports are constructed from Type 409 stainless steel. The flow distribution baffle is on the cold-leg side of the steam generator. There is one batwing support on each side of the steam generator and 5 vertical straps. The batwing supports all tubes, whereas the vertical straps only support specific tubes (except for VS3 which is the central support and supports all tubes). Figure 2-40 illustrates the tube support configuration and numbering.

The U-bend region of the tubes in rows 1 through 17 received a supplemental thermal treatment (stress relieving) after bending.

The replacement steam generators have "U"-shaped tubes in rows 1 through row 17, and have tubes with two 90-degree bends (referred to as square bends) in all rows greater than row 17. The bend radius is 3 inches for row 1 and 11 inches for row 17. The square bends have a 10-inch bend radius.

The tube supports have three basic configurations—(1) horizontal grids (eggcrates/lattice) that provide support to the vertical run of the tubes, (2) vertical grids that provide vertical and horizontal support to the horizontal run of the tubes in the upper bend region, and (3) diagonal strips (batwings) that provide out-of-plane support to the 90-degree bends.

The upper tube bundle support system (1) supports the horizontal tube spans against high-velocity, two-phase cross flow, (2) permits an expanded vertical tube pitch (from 1.0 inch to 1.75 inches) so as to promote free flow through the bend region and prevent low-flow dryout regions, and (3) supports the upper tube bundle via structural beams against postulated accident condition loads, seismic loads, transportation loads, and dead weight. The U-bend support structure for the replacement steam generator differs from the original design in that it includes welded connections between the vertical grids and the diagonal (batwing) supports. Other features of the U-bend support system are that the batwings bisect the 90-degree bends, the bend region supports are perforated and narrower than the original design, and the bend region supports have ventilation holes. These changes in design improve the thermal/hydraulic conditions in the upper bundle region, preventing crevice dryout and reducing secondary-side fouling, as well as addressing tube-wear phenomena observed in the original steam generator. The diagonal strips (batwings) are located at every row and are designed to prevent out-of-plane deflection and thus preclude the deflection amplitude required for fatigue.

The replacement steam generator design has an increased circulation ratio when compared to the original steam generator.

The average heated length of the tubes is 63.9 feet per tube.

2.4.22 Point Beach 2

Point Beach 2 has two recirculating steam generators designed and fabricated by Westinghouse. The model Delta 47F steam generators were put into service in 1997 during RFO 22.

Each steam generator has 3499 thermally treated Alloy 690 tubes that have an outside diameter of 0.875 inch and a nominal wall thickness of 0.050 inch. The tubes were manufactured by Sandvik and are arranged in a triangular pattern as illustrated in Figure 2-41 with a spacing of approximately 1.234 inches. The heat transfer surface area in each steam generator is 47,500 ft^2.

The tubes were hydraulically expanded at each end for the full depth of the tubesheet. The tubesheet is 22.18 inches thick (with the clad, the tubesheet is 22.42 inches thick). The tubes are supported by a flow distribution baffle, support plates, and AVBs. All supports are constructed from Type 405 stainless steel. The flow distribution baffle is 0.74 inch thick. The tube support plates are 1.125 inches thick and have trifoil shaped holes through which the tubes pass. The AVBs are arranged in a "V" shape and are 0.565 inch thick. Figure 2-42 illustrates the tube support configuration and numbering.

The U-bend region of the tubes in rows 1 through 14 received a supplemental thermal treatment (stress relieving) after bending. The smallest U-bend radius is 3.25 inches.

Each steam generator has a digital metal impact monitoring system. This system provides an alarm to alert operating personnel of potential loose parts in the steam generators.

2.4.23 Prairie Island 1

Prairie Island 1 has two recirculating steam generators designed and fabricated by Framatome in France. The model 56/19 steam generators were put into service in 2004 during RFO 23.

Each steam generator has 4868 thermally treated Alloy 690 tubes that have an outside diameter of 0.750 inch and a nominal wall thickness of 0.043 inch. The tubes were manufactured by Sandvik and are arranged in a square pattern with a spacing of approximately 1.0425 inches. The heat transfer surface area in each steam generator is 61,281 ft^2.

The tubes were hydraulically expanded at each end for the full depth of the tubesheet. The tubesheet is 21.46 inches thick (with the clad, the tubesheet is 21.835 inches thick). The tubes are supported by support plates and AVBs. All tube support plates are constructed from Type 410 stainless steel. The tube support plates are 1.18 inches thick and have quatrefoil shaped holes through which the tubes pass. Except for the no-tube lane in the top tube support plate, the tube support plates are identical in configuration. The top tube support plate (i.e., number 8) does not have flow slots in the no-tube lane. Instead, the no-tube lane of the eighth tube support plate has small openings for capturing the straight anti-vibration bars and it has flow holes. The AVBs are constructed from Type 405 stainless steel and are rectangular in cross-section (0.5 inch by 0.3 inch). There are five sets of AVBs (four "V" shaped and one straight). Each of the four sets of "V"-shaped AVBs has its ends linked together by clamps that maintain the spacing of the AVBs and restrict U-bend out-of-plane motion. The clamped sets of AVB ends are fastened to hoops that maintain in-plane configuration and prevent lift off of the AVB system under transient and accident conditions. The "V"-shaped AVBs penetrate the tube bundle through rows 27 and 15. The straight AVBs have their upper ends linked together and fit into slots machined in the uppermost tube support plate, providing a sliding connection for their lower ends. The straight AVBs provide support for all U-bends in the center of the U-bend (i.e., they penetrate the tube bundle through row 1).

The tubes (i.e., the entire tube) in rows 1 through 9 received a supplemental thermal treatment (stress relieving) after bending. The smallest U-bend radius is 2.700 inches.

The feedwater ring and J-tube outlet nozzles are constructed from stainless steel and have a welded thermal sleeve.

2.4.24 Sequoyah 1

Sequoyah 1 has four recirculating steam generators designed by Combustion Engineering and fabricated by Doosan in Korea. The model 57AG steam generators were put into service in 2003 during RFO 12.

Each steam generator has 4983 thermally treated Alloy 690 tubes that have an outside diameter of 0.750 inch and a nominal wall thickness of 0.043 inch. The tubes were manufactured by Sandvik and are arranged in a triangular pattern with a spacing of approximately 1.0625 inches.

The tubes were hydraulically expanded at each end for the full depth of the tubesheet. The tubes are supported by lattice grid tube supports and AVBs. All supports are constructed from Type 409 stainless steel.

The tubes in rows 1 through 16 received a supplemental thermal treatment (stress relieving) after bending. The smallest U-bend radius is 3.1875 inches. The thermal stress relief and the larger minimum U-bend radius (compared to the original steam generators) provide added assurance that this region will not develop stress-corrosion cracking.

The tube-to-tubesheet seal welds are flush with the tubesheet. The tubes are hydraulically expanded for the full depth of the tubesheet to improve the mechanical strength of the joint and to limit the tube-to-tubesheet crevice. The tubes were installed into the tubesheet after the lower shell and tubesheet were welded and received a post-weld heat treatment. This precluded any possibility of tube sensitization and avoided subjecting the tube-to-tubesheet joint to thermal stresses from these operations. This also eliminated concerns over loosening of the tubes or the creation of crevices as a result of relaxation of the expanded region. A temperature limit imposed on the tubes during the post-weld heat treatment of the final primary head assembly to tubesheet weld further ensured the tubes were adequately protected during fabrication.

The hydraulic expansion process used hydraulic seals made from elastomeric material and designed such that no metal parts were impressed upon the inside surface of the tube when the hydraulic pressure was applied. The position of the seal at the secondary face of the tubesheet was controlled to ensure that expansion of the tube was as close as possible to the secondary face of the tubesheet without going past the face. After expansion, the inside profile of each tube was measured through the entire expanded area of the tubesheet (including the transition) using an eddy current method and recorded. The measurement indicated both the position and condition of the tube expansion, and it serves as a baseline for subsequent inservice inspections.

The advanced tube support grids provide a higher circulation ratio (through lower flow resistance) than the original steam generator. In addition, the support grids limit the tube-to-tube support contact, provide vibration restraint and fretting resistance, lower the tendency for deposits to accumulate when compared to a support plate or standard eggcrate design, and eliminate the potential for denting since the grids are constructed from stainless steel.

The U-bend supports were designed to support the U-bend region of the tube to limit wear and vibration of the tube, limit the potential for sludge deposition, and increase circulation. The upper bundle support system features diagonal and vertical strip assemblies that provide support to the U-bends. The U-bend supports were fabricated from perforated strips. As the tube bundle was assembled, these supports were positioned and interlocked by the center vertical strip. The upper ends of the vertical strips were captured by crescent plates above the tube bundle. The outer ends of the diagonal strips are also linked together. The lower ends of the assembly, where the vertical and diagonal strips intersect, are supported and spaced by slotted tees attached to the uppermost advanced tube support grid. A combination of increased vertical pitch and the perforation of the diagonal/vertical strips creates a low resistance flow through the upper bundle. The perforated strips promote local washing of tube surfaces and thereby limit the potential for local sludge deposition. The upper bundle support system is integral with the U-bends of the tube bundle and generally moves with the tube bundle during heatup and cooldown.

The upper bundle support system is arranged to meet the design limits established for fluid elastic instability and response to turbulence. The width of the diagonal and vertical strips optimizes the amount of live contact between tube and strip while taking into account the presence of the ventilating perforation. The potential for fretting was assessed through a flow-induced vibration sensitivity analysis. The tube bundle design achieves limited resistance to riser flow with an open flow configuration and support bar orientation that is compatible with the flow direction.

The feedring includes spray pipes with small holes that will preclude the introduction of significant loose parts from the feedwater.

The circulation ratio of the replacement steam generator secondary-side fluid (ratio of riser mass flow rate to steam outlet mass flow rate) is nearly double that of the original steam generator. A higher circulation ratio limits concerns regarding heat transfer performance, generator sludge management, corrosion product transfer, and tube dryout.

2.4.25 South Texas Project 1

South Texas Project 1 has four recirculating steam generators designed and fabricated by Westinghouse. The model Delta 94 steam generators were put into service in 2000 during RFO 9. The South Texas Project 1 steam generators were the last (along with the Harris steam generators) to be fabricated at the Westinghouse facility in Pensacola, Florida.

Each steam generator has 7585 thermally treated Alloy 690 tubes that have an outside diameter of 0.688 inch and a nominal wall thickness of 0.040 inch. The tubes were manufactured by Sandvik, are arranged in a triangular pattern as illustrated in Figure 2-43 with a spacing of approximately 0.980 inch. The gap between the tubes is 0.293 inch.

The tubes were hydraulically expanded at each end for the full depth of the tubesheet. The tubesheet is 25.43 inches thick. The tubes are supported by a flow distribution baffle, support plates, and AVBs. All supports are constructed from Type 405 stainless steel. The flow distribution baffle has nonafoil shaped holes through which the tubes pass. The tube support plates have trifoil-shaped holes through which the tubes pass. The AVBs are arranged in a "V" shape and are 0.480 inch thick. Figure 2-44 illustrates the tube support configuration and numbering.

The U-bend region of the tubes in rows 1 through 17 received a supplemental thermal treatment (stress relieving) after bending. The smallest U-bend radius is 3.250 inches (corresponding to a row 1 tube).

The steam generators were designed with sludge collector systems that reduce the amount of suspended solids and the amount of sludge deposited on the tubesheet. The sludge collector is an integral part of the primary moisture separator assembly and is designed to limit the amount of suspended particles in the secondary-side circulation flow. The sludge collector consists of a cylindrical drum divided into two levels by an internal horizontal plate. During normal operation, a controlled amount of circulation flow mixture enters the sludge collector through central entrance holes in the top, flows slowly and radially outward, and exits at the outlet holes near the periphery. This path provides a laminar flow settling zone for the suspended solids. The sludge collector design is based on the principle that suspended particles will settle if the flow velocity is less than the threshold settling velocity. The sludge collector has built-in cleaning jets and a suction line which are used during periodic maintenance to remove the sludge from the collector.

The blowdown pipe is located on top of the tubesheet in the tube lane. It extends essentially for the full length of the tube lane and has two end connections 180 degrees apart on the tubesheet. It is designed to accommodate a 1.0-percent feedwater flow continuous blowdown rate from a single pipe connection (two connections are provided) at full-power conditions. The primary function of the blowdown line is to remove bulk fluid from the flow entering the tube bundle. The tube lane is wide to enhance maintenance access to the tube bundle and the tube support plates have large flow slots that permit upper bundle cleaning and inspection tools to enter through the lower handholes.

The tube support plates have trifoil-shaped holes to reduce tube dryout and chemical concentration in the region where the tubes pass through. The tube support plates result in line contact of the tube at only three locations or "lands." The flat land tube support is designed to reduce the tube-to-tube support plate crevice area, while providing for maximum steam/water flow in the open areas adjacent to the tube. The flat land contact geometry provides increased dryout resistance over drilled hole configurations. The tube support lobes prevent sludge from widening the tube hole dryout zone, and the broached design directs flow adjacent to the tube and adds margin against dryout. The stainless steel oxide volume ratio is 1.0, whereas the carbon steel oxide volume ratio is 4. This led to the blocking of crevices and chemical concentration in steam generators with drilled hole carbon steel tube supports.

The design of the flow distribution baffle limits the number of tubes exposed to low-velocity flow in the vicinity of the tubesheet. The flow distribution baffle plate makes line contact with each tube at nine locations. The baffle is designed and located to produce a sweeping flow across the tubesheet in order to limit the area where sludge deposits. The center portion of the flow distribution baffle is cut out in order to control the velocity so that the low-velocity region (and sludge deposition zone) is located at the center of the tube bundle near the blowdown intake.

The steam generators have an advanced minimum gap U-bend support structure and wider bars. The AVB assemblies within each set are installed at staggered depths to limit the pressure drop in the U-bend region so as to increase the circulation ratio and reduce the potential for steam blanketing. This arrangement provides at least single-sided support above the top tube support plate for every tube in the tube bundle. The AVBs are inserted deeper at several peripheral locations of the U-bend in order to provide additional support.

Feedwater is introduced to the secondary side of the steam generator through the feedwater nozzle. The feedwater flows through a welded thermal sleeve made from thermally treated Alloy 690, into the elevated feedwater distribution ring pipe and pipe fittings, and out through 34 spray nozzle assemblies located on the top side of the ring. The thermally treated Alloy 690 spray nozzle assemblies are arranged to uniformly distribute the feedwater into the upper downcomer plenum. Water removed from the wet steam in the first- and second-stage moisture separators joins the feedwater in the upper downcomer plenum. The water mixture enters the downcomer annulus, travels down to the bottom of the annulus, and enters the tube bundle at the tubesheet.

All components of the feedwater distribution equipment are of all-welded construction and are made of materials that are resistant to erosion-corrosion, thermal fatigue, and corrosion cracking. The system is configured to avoid trapping steam, particularly at nonvented high points, which could result in water hammer. Feedwater discharges into the steam generator at the top of the feedring, which is entirely submerged during normal operation.

The steam generators have features to limit the development of loose parts during operation and maintenance. For example, the feedwater ring spray nozzle assemblies have a series of 0.29-inch outlet holes, which function to trap potential foreign objects that might otherwise be introduced from the feedwater system. Each of the 34 spray nozzles has 130 holes of this diameter. Likewise, the auxiliary feedwater is introduced through a single cylindrical nozzle with a diameter of 6.625 inches and a height of approximately 13.5 inches. The outer surface of the cylindrical nozzle has 560 holes with a diameter of 0.29 inch. Like the feedwater spray nozzles, the auxiliary feedwater spray nozzle will trap foreign objects. If parts are small enough to pass through the spray nozzle perforations, they will also pass between the tubes. The flow will transport such objects into the low-velocity region where they have the least potential to produce tube wear.

The circulation ratio is defined as the ratio of riser mass flow rate to steam outlet flow rate. Increasing the circulation ratio of the steam generator secondary-side fluid limits concerns regarding heat transfer performance, sludge management, corrosion product transfer, and tube dryout. The benefits of higher circulation ratio are that the void fraction in the upper bundle is slightly less, margin to dryout in the U-bend is slightly larger, fluid damping in the U-bend is slightly larger, and sweeping forces at the top of the tubesheet are slightly higher.

During the post-weld heat treatment of the channel head to tubesheet weld, the shell barrel elongated. Since the barrel holds the tube support plates in place and the tube support plate stayrods and the tubes are anchored into the tubesheet, the tube support plates deflected slightly at the outer edges with the growth of the shell barrel. The center of the plate did not deflect as much since the stayrods did not experience as much growth from the thermal effect. This led to denting of the tubes at the upper tube support plate in the outer periphery with some denting of the tubes occurring at lower tube support plates.

2.4.26 South Texas Project 2

South Texas Project 2 has four recirculating steam generators designed by Westinghouse and fabricated by ENSA. The model Delta 94 steam generators were put into service in 2002 during RFO 9.

Each steam generator has 7585 thermally treated Alloy 690 tubes that have an outside diameter of 0.688 inch and a nominal wall thickness of 0.040 inch. The tubes were manufactured by Sandvik and are arranged in a triangular pattern as illustrated in Figure 2-43 with a spacing of approximately 0.980 inch. The gap between the tubes is 0.293 inch.

The tubes were hydraulically expanded at each end for the full depth of the tubesheet. The tubesheet is 25.43 inches thick. The tubes are supported by a flow distribution baffle, support plates, and AVBs. All supports are constructed from Type 405 stainless steel. The flow distribution baffle has nonafoil shaped holes through which the tubes pass. The tube support plates have trifoil shaped holes through which the tubes pass. The AVBs are arranged in a "V" shape and are 0.480 inch thick. Figure 2-44 illustrates the tube support configuration and numbering.

The U-bend region of the tubes in rows 1 through 17 received a supplemental thermal treatment (stress relieving) after bending. The smallest U-bend radius is 3.250 inches (corresponding to a row 1 tube).

The steam generators were designed with sludge collector systems that reduce the amount of suspended solids and the amount of sludge deposited on the tubesheet. The sludge collector is an integral part of the primary moisture separator assembly and is designed to limit the amount of suspended particles in the secondary-side circulation flow. The sludge collector consists of a cylindrical drum divided into two levels by an internal horizontal plate. During normal operation, a controlled amount of circulation flow mixture enters the sludge collector through central entrance holes in the top, flows slowly and radially outward, and exits at the outlet holes near the periphery. This path provides a laminar flow settling zone for the suspended solids. The sludge collector design is based on the principle that suspended particles will settle if the flow velocity is less than the threshold settling velocity. The sludge collector has built-in cleaning jets and a suction line that are used during periodic maintenance to remove the sludge from the collector.

The blowdown pipe is located on top of the tubesheet in the tube lane. It extends essentially along the full length of the tube lane and has two end connections 180 degrees apart on the tubesheet. It is designed to accommodate a 1.0-percent feedwater flow continuous blowdown rate from a single pipe connection (two connections are provided) at full-power conditions. The primary function of the blowdown line is to remove bulk fluid from the flow entering the tube bundle. The tubelane is wide to enhance maintenance access to the tube bundle and the tube support plates have large flow slots that permit upper bundle cleaning and inspection tools to enter through the lower handholes.

The steam generator has features to limit the development of loose parts during operation and maintenance. For example, the feedwater ring spray nozzle assemblies have a series of 0.29-inch outlet holes, which function to trap potential foreign objects that might otherwise be introduced from the feedwater system. If parts are small enough to pass through the spray nozzle perforations, they will also pass between the tubes. The flow will transport such objects into the low-velocity region where they have the least potential to produce tube wear.

The tube support plates have trifoil-shaped holes to reduce tube dryout and chemical concentration in the region where the tubes pass through. The tube support plates result in line contact of the tube at only three points or "lands." The flat land tube support is designed to reduce the tube-to-tube support plate crevice area, while providing for maximum steam/water flow in the open areas adjacent to the tube. The flat land contact geometry provides increased dryout resistance over drilled hole configurations. The tube support lobes prevent sludge from widening the tube hole dryout zone, and the broached design directs flow adjacent to the tube and adds margin against dryout.

The design of the flow distribution baffle limits the number of tubes exposed to low-velocity flow in the vicinity of the tubesheet. The flow distribution baffle plate makes line contact with each tube at nine locations. The baffle is designed and located to produce a sweeping flow across the tubesheet in order to limit the area where sludge deposits. The center portion of the flow distribution baffle is cut out in order to control the velocity so that the low-velocity region (and sludge deposition zone) is located at the center of the tube bundle near the blowdown intake.

The steam generators have an advanced minimum gap U-bend support structure and wider bars. The AVB assemblies within each set are installed at staggered depths to limit the pressure drop in the U-bend region so as to increase the circulation ratio and reduce the potential for steam blanketing. This arrangement provides at least single-sided support above the top tube support plate for every tube in the tube bundle. The AVBs are inserted deeper at several peripheral locations of the U-bend in order to provide additional support.

The feedwater nozzle introduces feedwater to the secondary side of the steam generator. The feedwater flows through a welded thermal sleeve made from thermally treated Alloy 690, into the elevated feedwater distribution ring pipe and pipe fittings, and out through 34 spray nozzle assemblies located on the top side of the ring. The thermally treated Alloy 690 spray nozzle assemblies are arranged to uniformly distribute the feedwater into the upper downcomer plenum. Water removed from the wet steam in the first- and second-stage moisture separators joins the feedwater in the upper downcomer plenum. The water mixture enters the downcomer annulus, travels down to the bottom of the annulus, and enters the tube bundle at the tubesheet.

All components of the feedwater distribution equipment are of all-welded construction and are made of materials that are resistant to erosion-corrosion, thermal fatigue, and corrosion cracking. The system is configured to avoid trapping steam, particularly at nonvented high points, which could result in water hammer. Feedwater discharges into the steam generator at the top of the feedring, which is entirely submerged during normal operation.

The circulation ratio is defined as the ratio of riser mass flow rate to steam outlet flow rate. Increasing the circulation ratio of the steam generator secondary-side fluid limits concerns regarding heat transfer performance, sludge management, corrosion product transfer, and tube dryout. The benefits of higher circulation ratio are that the void fraction in the upper bundle is slightly less, margin to dryout in the U-bend is slightly larger, fluid damping in the U-bend is slightly larger, and sweeping forces at the top of the tubesheet are slightly higher.

2.4.27 St. Lucie 1

St. Lucie 1 has two recirculating steam generators designed and fabricated by BWI. The steam generators were put into service in 1998 during RFO 15.

Each steam generator has 8523 thermally treated Alloy 690 tubes manufactured by Sumitomo.

The tubes were hydraulically expanded at each end for the full depth of the tubesheet and are supported by lattice grid tube supports and fan bars. (The lowest fan bar is also referred to as a collector bar since all other fan bars connect to it.)

The bend radii of the inner row tubes were increased by crossing the tubes. That is, the origin and termination points of the tubes in the first several rows differ between the hot-leg and cold-leg. As a result, the U-bend region of these tubes is in a plane skewed from the tube-free lane (rather than in a plane perpendicular to the tube-free lane). These tubes are referred to as crossover tubes.

2.4.28 Summer

Summer has three recirculating steam generators designed and fabricated by Westinghouse. The model Delta 75 steam generators were put into service in 1994 during RFO 8.

Each steam generator has 6307 thermally treated Alloy 690 tubes that have an outside diameter of 0.6875 inch and a 0.040-inch nominal wall thickness. The tubes were manufactured by Sandvik and are arranged in a triangular pattern as illustrated in Figure 2-45 with a spacing of approximately 0.980 inch.

The tubes were hydraulically expanded at each end for the full depth of the tubesheet and are supported by a flow distribution baffle, support plates, and AVBs. All supports are constructed from Type 405 stainless steel. The tube support plates have flat contact, trifoil shaped holes through which the tubes pass. The AVBs are arranged in a "V" shape. Figure 4-26 illustrates the tube support configuration and numbering.

The tubes in rows 1 through 17 received a supplemental thermal treatment (stress relieving) after bending. The triangular tube pitch arrangement enhances heat transfer area and tube stability and provides additional margin against vibration. Tightly controlling AVB insertion depth and increasing the number of sets of AVBs reduce the number of tubes that are potentially affected by the vibration mechanism to which tube degradation has been attributed in some of the early steam generator designs.

A sludge collector tray assists in capturing impurities from the feedwater system. The design of the feedwater ring spray nozzle assemblies limits the potential for loose parts entering the steam generators. Each assembly consists of a series of 0.25-inch-diameter outlet holes. The 0.25-inch dimension is smaller than the spacing between the tubes; therefore, if parts are small enough to pass through these holes, they will also pass between the tubes and into the low-velocity regions of the tube bundle.

-42-

Table 2-1: Units with Thermally Treated Alloy 690 Tubes (Part 1)

Unit	Date	Model	Number of SGs	Operating Time[1] Original SG	Operating Time[1] Replacement SG
Arkansas Nuclear One 2	2000	W/D109	2	21	4
Braidwood 1	1998	BWI	4	10	6
Byron 1	1998	BWI	4	12	7
Calvert Cliffs 1	2002	BWI	2	27	3
Calvert Cliffs 2	2003	BWI	2	26	2
Catawba 1	1996	BWI	4	11	8
Cook 1	2000	BWI	4	25	4
Cook 2	1989	W/54F	4	11	16
Farley 1	2000	W/54F	3	22	5
Farley 2	2001	W/54F	3	20	4
Ginna	1996	BWI	2	26	9
Harris	2001	W/D75	3	15	3
Indian Point 3	1989	W/44F	4	13	16
Kewaunee	2001	W/54F	2	28	3
McGuire 1	1997	BWI	4	15	8
McGuire 2	1997	BWI	4	14	7
Millstone 2	1993	BWI	2	17	12
North Anna 1	1993	W/54F	3	15	12
North Anna 2	1995	W/54F	3	14	10

[1]Operating Time = calendar years of operation as of 12/31/04

BWI = Babcock and Wilcox International
CE = Combustion Engineering
OTSG = once-through steam generator
W = Westinghouse

Table 2-1: Units with Thermally Treated Alloy 690 Tubes (Part 2)

Unit	Date	Model	Number of SGs	Operating Time[1] Original SG	Operating Time[1] Replacement SG
Oconee 1	2004	BWI - OTSG	2	30	1
Oconee 2	2004	BWI - OTSG	2	30	1
Oconee 3	2004	BWI - OTSG	2	30	<1
Palo Verde 2	2003	Other (Ansaldo)	2	17	1
Point Beach 2	1996	W/D47F	2	24	8
Prairie Island 1	2004	Other (Framatome)	2	31	<1
Sequoyah 1	2003	Other (Doosan)	4	22	2
South Texas Project 1	2000	W/D94	4	12	5
South Texas Project 2	2002	W/D94	4	14	2
St. Lucie 1	1998	BWI	2	21	7
Summer	1994	W/D75	3	11	10

[1]Operating Time = calendar years of operation as of 12/31/04

BWI = Babcock and Wilcox International
CE = Combustion Engineering
OTSG = once-through steam generator
W = Westinghouse

Table 2-2: Steam Generator Design Information for Units with Thermally Treated Alloy 690 Tubes

Unit	Commercial Operation Date	SG Replacement	Designer	Fabricator	SG Model	Tube Manufacturer	Tube Material	Tube OD	Tube Wall	Number of Tubes	Tube Pitch	Tubesheet Thickness	Expansion Method	Expansion Extent	Support Material	FDB Design	TSP Design	AVB Design	Stress Relief After Bending	Notes
ANO 2	03/26/80	12/01/00	Westinghouse	ENSA	D109	Sandvik	690 TT	0.688	0.040	10687	0.85 Tr		Hydraulic	Full	405SS	N/A	Trifoil	AVB	R1-17	AVBs numbered A01-A20.
Braidwood 1	07/29/88	11/01/98	BWI	BWI	7720	Sumitomo	690 TT	0.688	0.040	6933	0.93 Tr	26.625"	Hydraulic	Full	410SS	N/A	Lattice	Fan Bar	R1-21	Smallest U-bend radius exist in row 3 tubes and is 3.832 inches versus 2.25 inches in original design. Stress relieved up to 12-inch centerline radius. Tube to tube gap is less then 0.25 inch.
Byron 1	09/16/85	01/01/98	BWI	BWI	7720	Sumitomo	690 TT	0.688	0.040	6933	0.93 Tr		Hydraulic	Full	410SS	N/A	Lattice	Fan Bar	R1-21	
Calvert Cliffs 1	05/08/75	04/01/02	BWI	BWI	7811	Sumitomo	690TT	0.750	0.042	8471	1.0 Tr	22.25" w/clad	Hydraulic	Full	410SS	N/A	Lattice	Fan Bar	R1-18	Smallest U-bend exists in row 1 tubes and is 3.5 inches.
Calvert Cliffs 2	04/01/77	05/01/03	BWI	BWI	7811	Sumitomo	690TT	0.750	0.042	8471	1.0 Tr	22.25" w/clad	Hydraulic	Full	410SS	N/A	Lattice	Fan Bar	R1-18	Smallest U-bend radius exists in row 1 tubes and is 3.5 inches.
Catawba 1	06/29/85	09/01/96	BWI	BWI	CFR80	Sumitomo	690 TT	0.688	0.040	6933	0.93 Tr	27.1" w/clad	Hydraulic	Full	410SS	N/A	Lattice	Fan Bar	R1-18	Crossover Tubing
Cook 1	08/28/75	12/01/00	BWI	BWI	51R	Sumitomo	690 TT	0.875	0.049	3490	1.1875 Tr	21.25"	Hydraulic	Full	410SS	N/A	Lattice	Fan Bar	R1-37	Row 1 radius is 4.750 inches versus 2.19 inches in original design. Westinghouse manufactured the steam domes.
Cook 2	07/01/78	03/01/89	Westinghouse	Westinghouse	54F	Sandvik	690 TT	0.875	0.050	3592	1.525 Sq	21"	Hydraulic	Full	405SS	0.75" Octofoil	1.125" Quatrefoil	AVB	R1-8	Row 1 radius is 3.141 inches versus 2.19 inches in original design.
Farley 1	12/01/77	05/01/00	Westinghouse	ENSA	54F	Sandvik	690 TT	0.875	0.050	3592	1.525 Sq	21"	Hydraulic	Full	405SS	Octofoil	Quatrefoil	AVB	R1-8	Stress relieved up to 12-inch centerline radius. Smallest U-bend radius is 3.141 inches versus 2.1875 inches in original design.
Farley 2	12/01/81	05/01/01	Westinghouse	ENSA	54F	Sandvik	690 TT	0.875	0.050	3592	1.525 Sq	21.42"	Hydraulic	Full	405SS	0.75" Octofoil	1.125" Quatrefoil	AVB	R1-8	Stress relieved up to 12-inch centerline radius. Smallest U-bend radius is 3.141 inches versus 2.1875 inches in original design. Row 8 radius is 11.7/16 inches.
Ginna	07/01/70	06/01/96	BWI	Valinox	690 TT		0.749	0.044	4765	Tr	25.25"	Hydraulic	Full	410SS	N/A	Lattice	Fan Bar	R1-18	Stress relieved up to 12-inch centerline radius. Row 1 and 2 tubes crossover. Nominal tube-to-tube gap is 0.369 inch.
Haris	05/02/87	12/01/01	Westinghouse	Westinghouse	D75	Sandvik	690 TT	0.688	0.040	6307	0.98 Tr		Hydraulic	Full	405SS	0.78"	1.125" Trifoil	AVB	R1-17	Smallest U-bend radius is 3.25 inches versus 2.25 inches in original design. Row 17 radius is 12 inches.
Indian Point 3	08/30/76	06/01/89	Westinghouse	Areseda	44F	Sandvik	690 TT	0.875	0.050	3214	1.2344 Sq	21.69" w/clad	Hydraulic	Full	405SS	0.75" Octofoil	1.125" Quatrefoil	AVB	R1-8	Flush tube-to-tubesheet weld. Row 1 radius is 2.167 inches.
Kewaunee	06/16/74	12/01/01	Westinghouse	Areseda	54F	Valinox	690 TT	0.875	0.050	3592	1.525 Sq	21.42"	Hydraulic	Full	405SS	Yes	Quatrefoil	AVB	R1-8	Stress relieved up to 12-inch centerline radius. Smallest U-band radius is 3.141 inches.
McGuire 1	12/01/81	05/01/97	BWI	BWI	CFR80	Sumitomo	690 TT	0.688	0.040	6933	0.93 Tr	27.1" w/clad	Hydraulic	Full	410SS	N/A	Lattice	Fan Bar	R1-21	Tubesheet is 28.63 inches thick without the clad. Smallest U-band radius exists in Row 3 tubes and is 3.832 inches.
McGuire 2	03/01/84	12/01/97	BWI	BWI	CFR80	Sumitomo	690 TT	0.688	0.040	6933	0.93 Tr	27.1" w/clad	Hydraulic	Full	410SS	N/A	Lattice	Fan Bar	R1-21	Tubesheet is 28.63 inches thick without the clad. Smallest U-band radius exists in Row 3 tubes and is 3.832 inches.
Millstone 2	12/26/75	01/01/93	BWI	BWI		Valinox	690 TT	0.750	0.045	8523	1.0 Tr	21.75"	Hydraulic	Full	410SS	N/A	Lattice	Fan Bar	R1-8	Clad thickness is 0.31 inch. Row 3 radius is 3.905 inches. Row 1 radius is 4.272 inches.
North Anna 1	06/06/78	04/01/93	Westinghouse	Westinghouse	54F	Sandvik	690 TT	0.875	0.050	3592	1.525 Sq	21"	Hydraulic	Full	405SS	0.75" Octofoil	1.125" Quatrefoil	AVB	R1-8	Controlled tolerance AVB to tube gaps.
North Anna 2	12/14/80	05/01/95	Westinghouse	Westinghouse	54F	Sandvik	690 TT	0.875	0.050	3592	1.525 Sq	21"	Hydraulic	Full	405SS	0.75" Octofoil	1.125" Quatrefoil	AVB	R1-8	Controlled tolerance AVB to tube gaps.
Oconee 1	07/15/73	01/01/04	BWI	BWI	OTSG	Sumitomo	690 TT	0.625	0.034	15531	0.875 Tr	22"	Hydraulic	13"	410SS	N/A	Trifoil	NA		
Oconee 2	09/09/74	06/01/04	BWI	BWI	OTSG	Sumitomo	690 TT	0.625	0.038	15531	0.875 Tr	22"	Hydraulic	13"	410SS	N/A	Trifoil	NA		
Oconee 3	12/16/74	12/31/04	BWI	BWI	OTSG	Sumitomo	690 TT	0.625	0.036	15531	0.875 Tr	22"	Hydraulic	13"	410SS	N/A	Trifoil	NA		
Palo Verde 2	09/19/86	12/01/03	ABB-CE	Areseda	D47	Sandvik	690 TT	0.750	0.042	12560	0.966 Tr	25"	Hydraulic w/ hard rolls	Full	410SS	N/A	Lattice	2 Batwings, 5 Vertical Straps	R1-17	Rows 1 through 17 contain U-band tubes. Tubes in Row 18 and higher are square band tubes.
Point Beach 2	10/01/72	12/01/96	Westinghouse	Framatome	56F/2	Sandvik	690 TT	0.875	0.050	3499	1.234 Tr	22.18" w/o clad	Hydraulic	Full	405SS	0.74"	1.125" Trifoil	AVB	R1-14	Row 1 radius is 3.25 inches.
Prairie Island 1	12/16/73	11/22/04	Framatome	Framatome	56/19	Sandvik	690 TT	0.750	0.043	4684	1.0423 Sq	21.839" w/o clad	Hydraulic	Full	405SS 410SS	N/A	Quatrefoil	AVB	R1-9	Row 1 radius is 2.7 inches. Tube support plates made from 410SS. AVBs made from 405SS.
Sequoyah 1	07/01/81	05/01/03	ABB-CE	Doosan	57A0	Sandvik	690 TT	0.750	0.043	4983	1.0662 Tr	21.839" w/clad	Hydraulic	Full	409SS	N/A	Lattice	Diagonal and Vertical Straps	R1-18	Smallest U-band radius is 3.1875 inches versus 2.1875 inches in original design.
South Texas 1	08/25/88	05/01/00	Westinghouse	Westinghouse	D94	Sandvik	690 TT	0.688	0.040	7585	0.98 Tr	25.43"	Hydraulic	Full	405SS	Nonofoil	Trifoil	AVB	R1-17	
South Texas 2	06/19/89	12/01/02	Westinghouse	ENSA	D94	Sandvik	690 TT	0.688	0.040	7585	0.98 Tr	25.43"	Hydraulic	Full	405SS	Nonofoil	Trifoil	AVB	R1-17	
St. Lucie 1	12/21/76	01/01/98	BWI	BWI		Sumitomo	690 TT	0.688	0.040	8523	0.98 Tr		Hydraulic	Full	SS	N/A	Lattice	Fan Bar		
Summer	01/01/84	12/01/04	Westinghouse	Westinghouse	D75	Sandvik	690 TT	0.688	0.040	6307	0.98 Tr		Hydraulic	Full	405SS	N/A	Trifoil	AVB	R1-17	

AVB = anti-vibration bar; BWI = Babcock and Wilcox International; CE = Combustion Engineering; FDB = flow distribution baffle; R = row; SG = steam generator; Sq = square; SS = stainless steel; Tr = triangular; TSP = tube support plate; TT = thermally treated

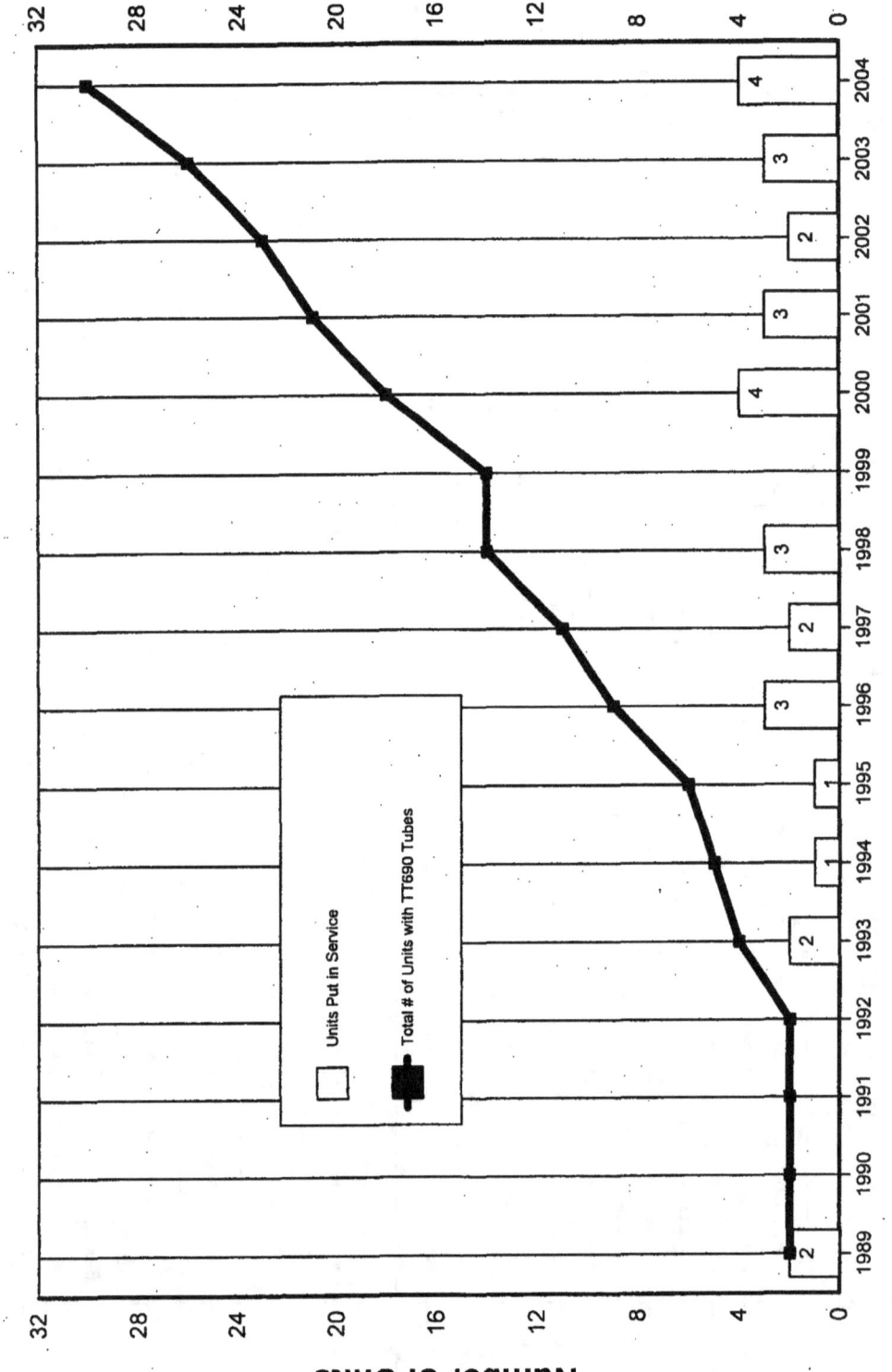

Figure 2-1: Number of Units with Thermally Treated Alloy 690 Tubes per Year

Figure 2-2: Number of Thermally Treated Alloy 690 Tubes in Service per Year

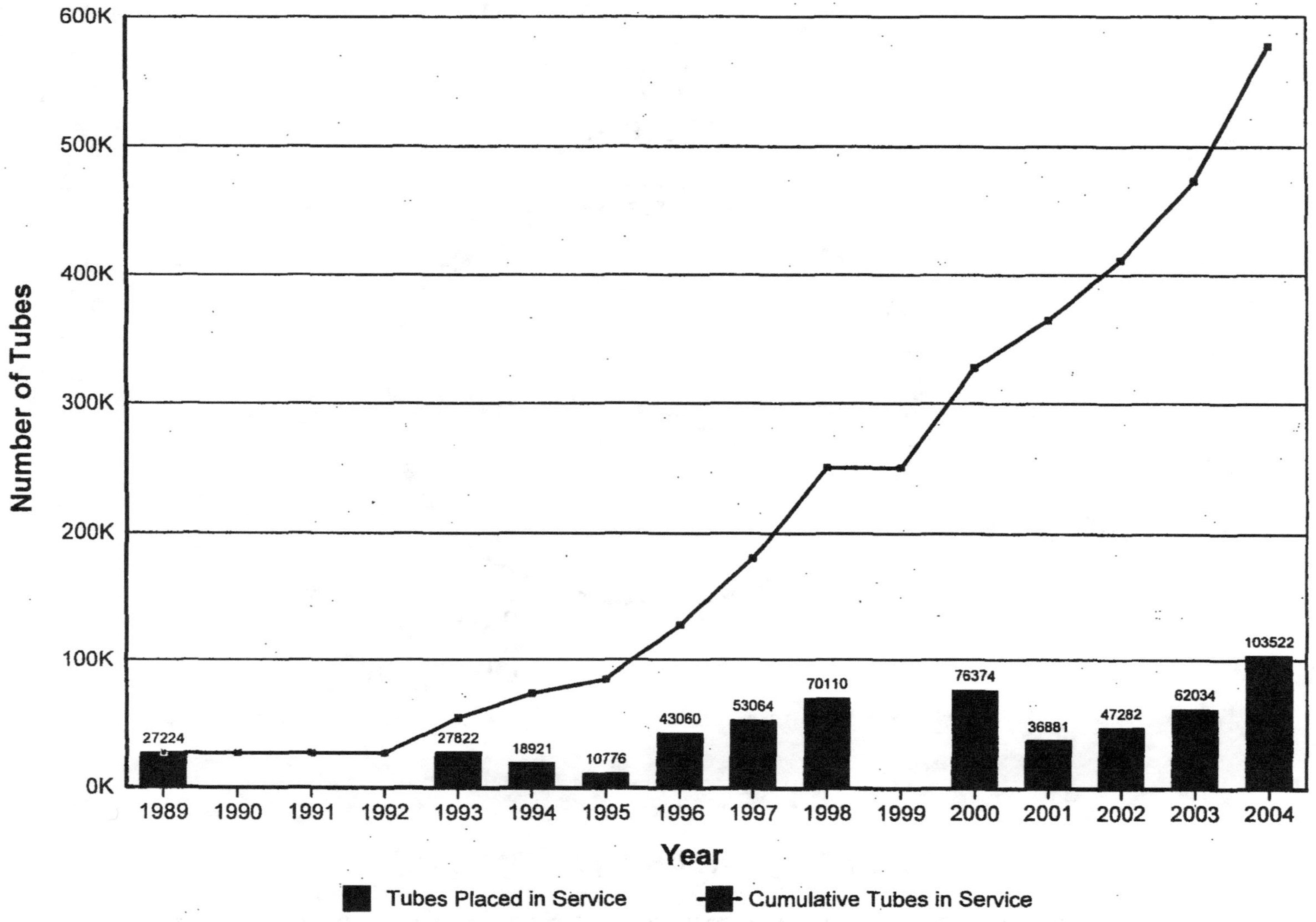

Legend:
- Tubes Placed in Service
- Cumulative Tubes in Service

Data values (Tubes Placed in Service):
- 1989: 27224
- 1993: 27822
- 1994: 18921
- 1995: 10776
- 1996: 43060
- 1997: 53064
- 1998: 70110
- 2000: 76374
- 2001: 36881
- 2002: 47282
- 2003: 62034
- 2004: 103522

Axis labels:
- Number of Tubes (0K, 100K, 200K, 300K, 400K, 500K, 600K)
- Year (1989–2004)

Figure 2-3: Percentage of Units by Tube Manufacturer
(for Units with Alloy 690 Tubes)

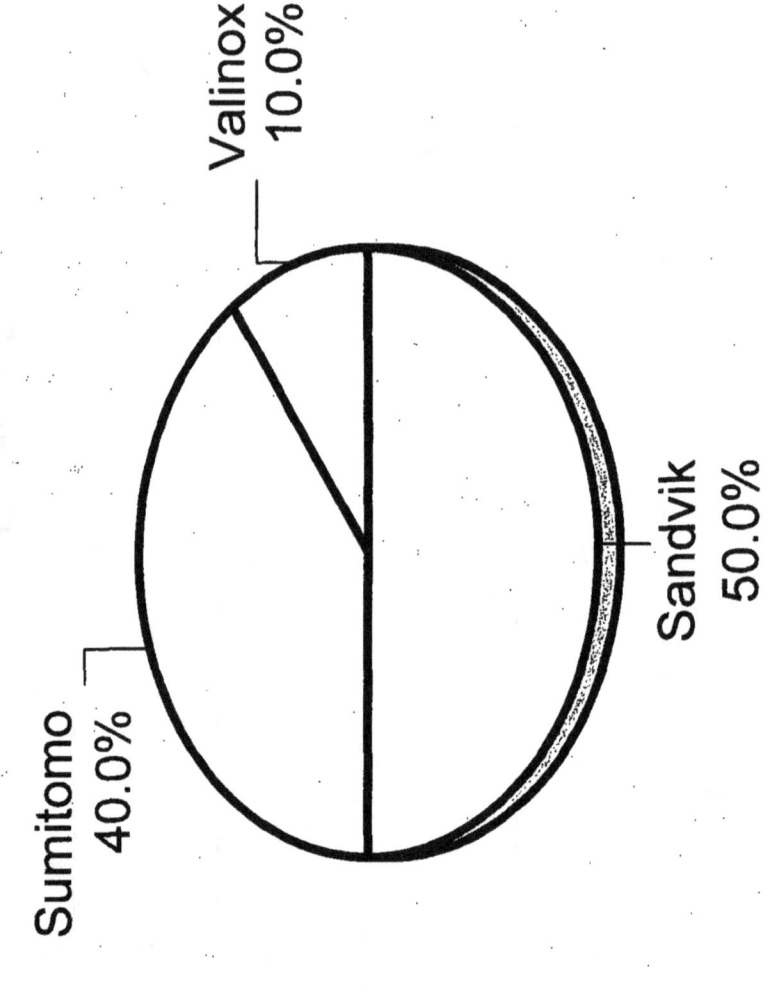

Valinox
10.0%

Sandvik
50.0%

Sumitomo
40.0%

Figure 2-4: Percentage of Alloy 690 Tubes by Manufacturer

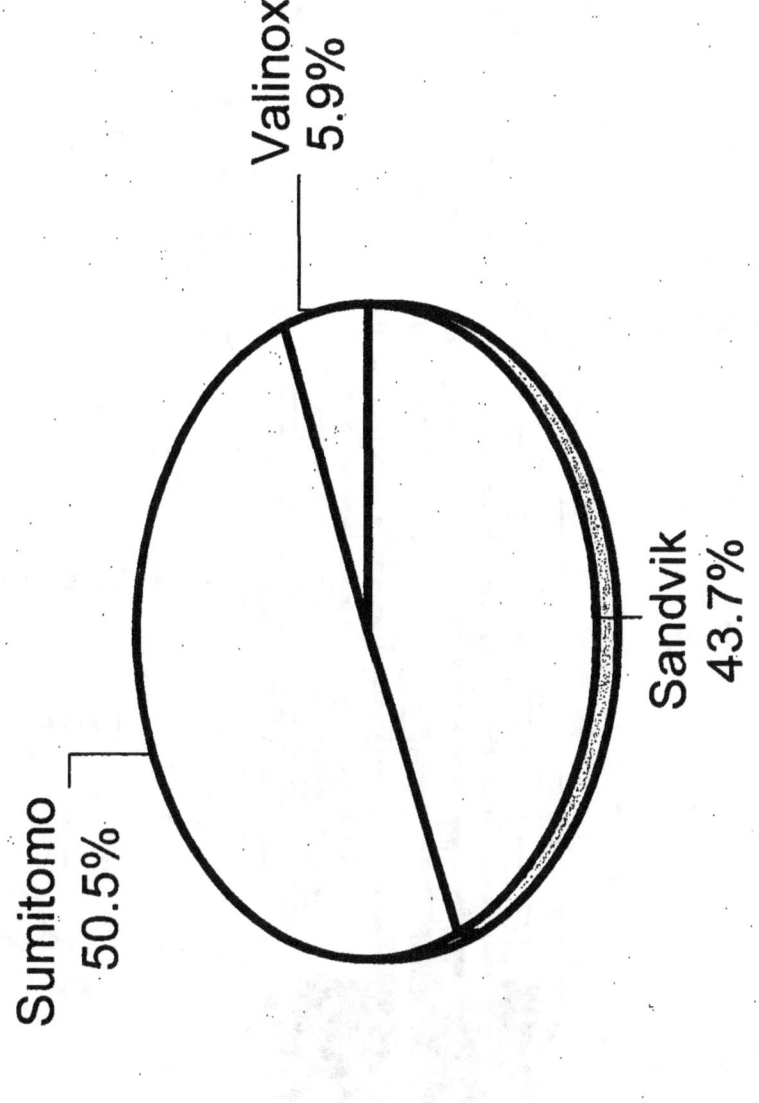

Valinox
5.9%

Sandvik
43.7%

Sumitomo
50.5%

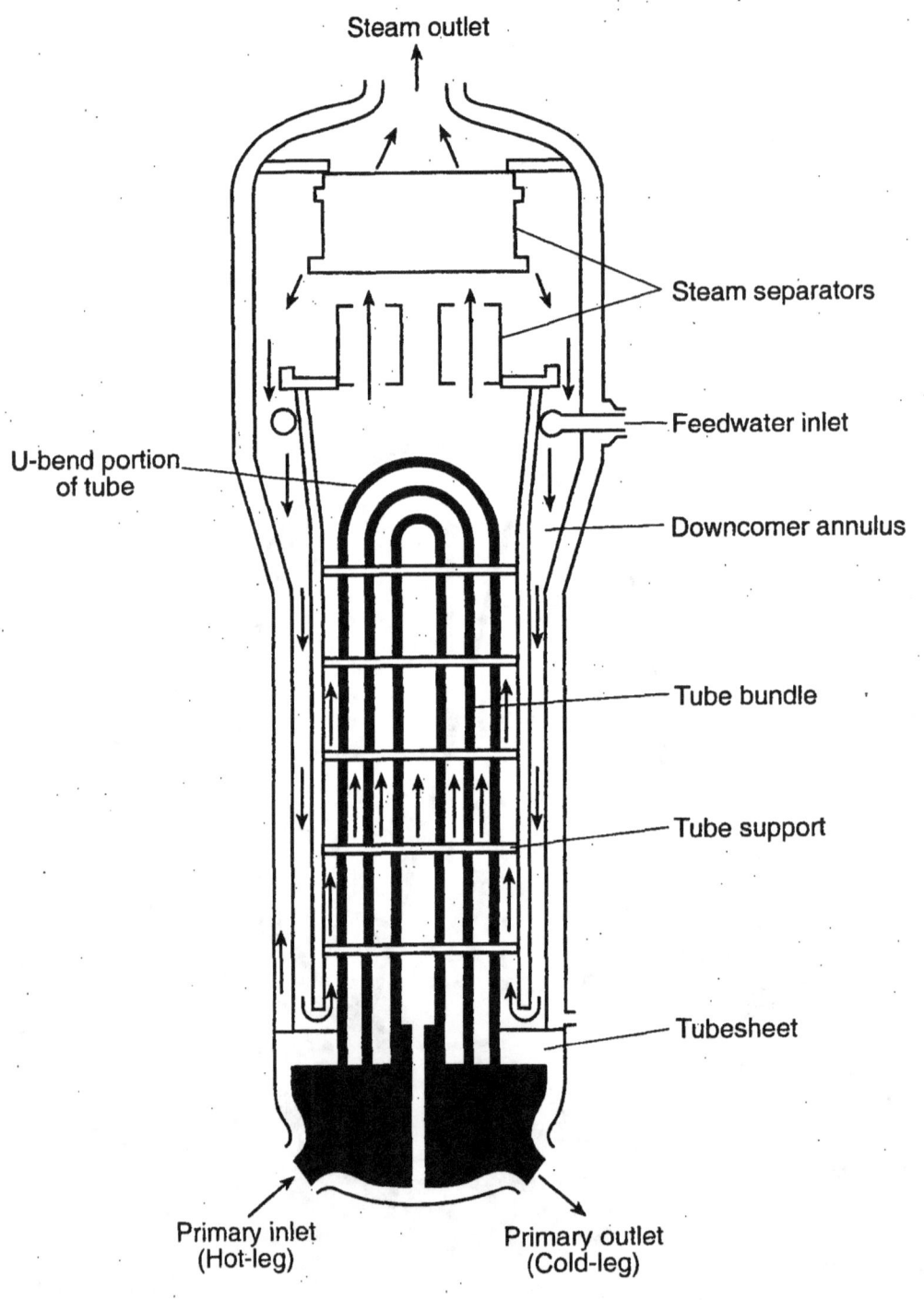

Figure 2-5: Pressurized-Water Reactor Recirculating Steam Generator

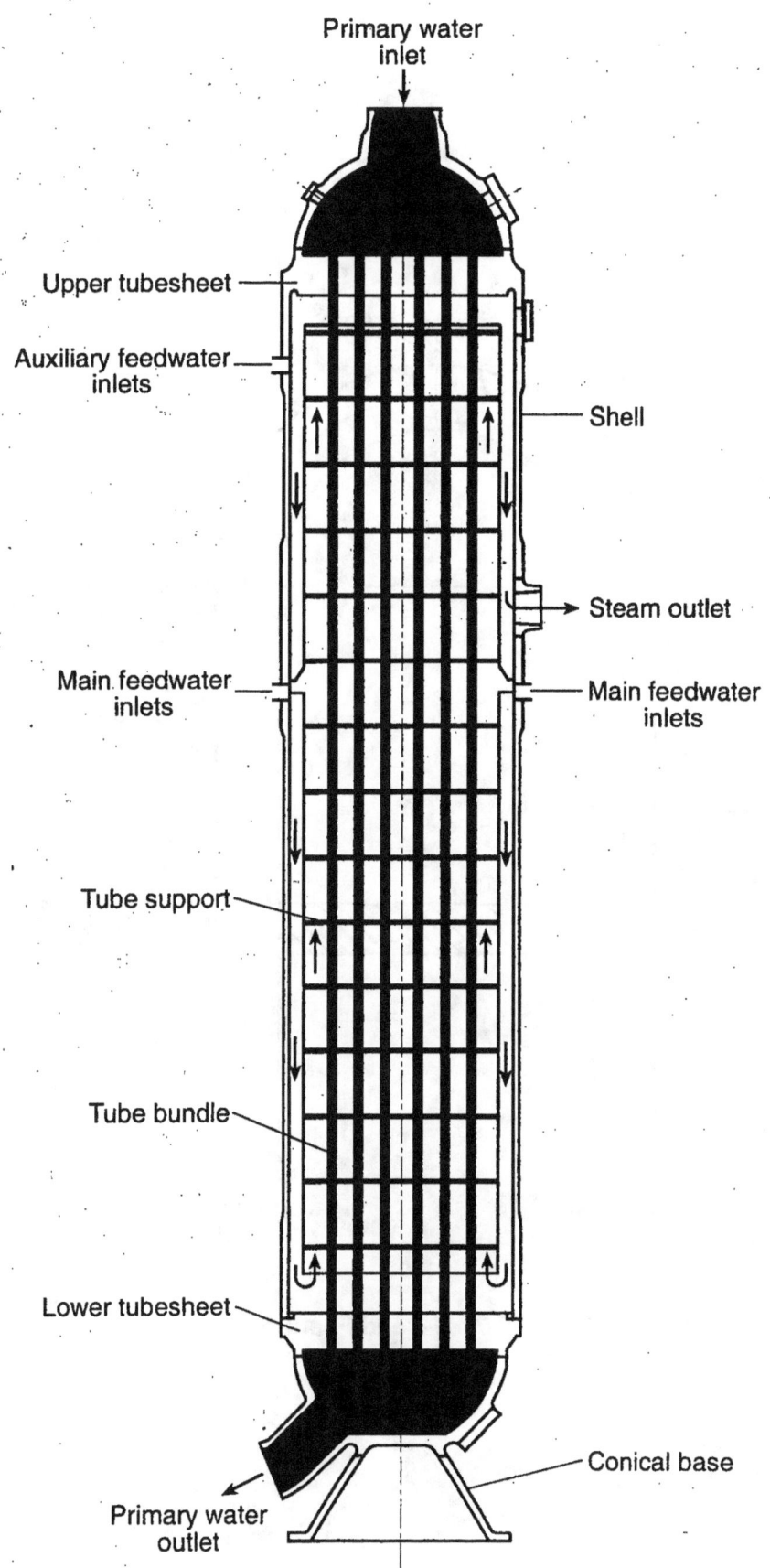

Figure 2-6: Pressurized-Water Reactor Once-Through Steam Generator

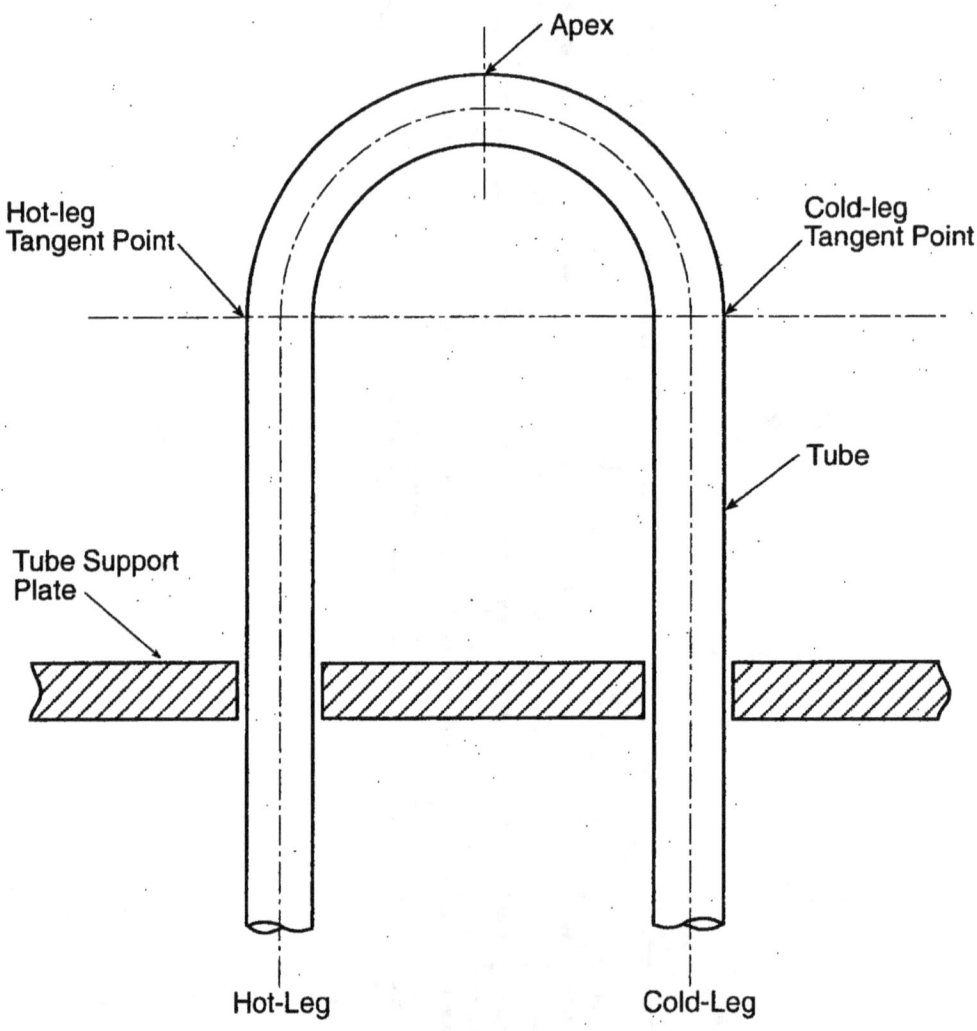

Figure 2-7: U-Bend Features

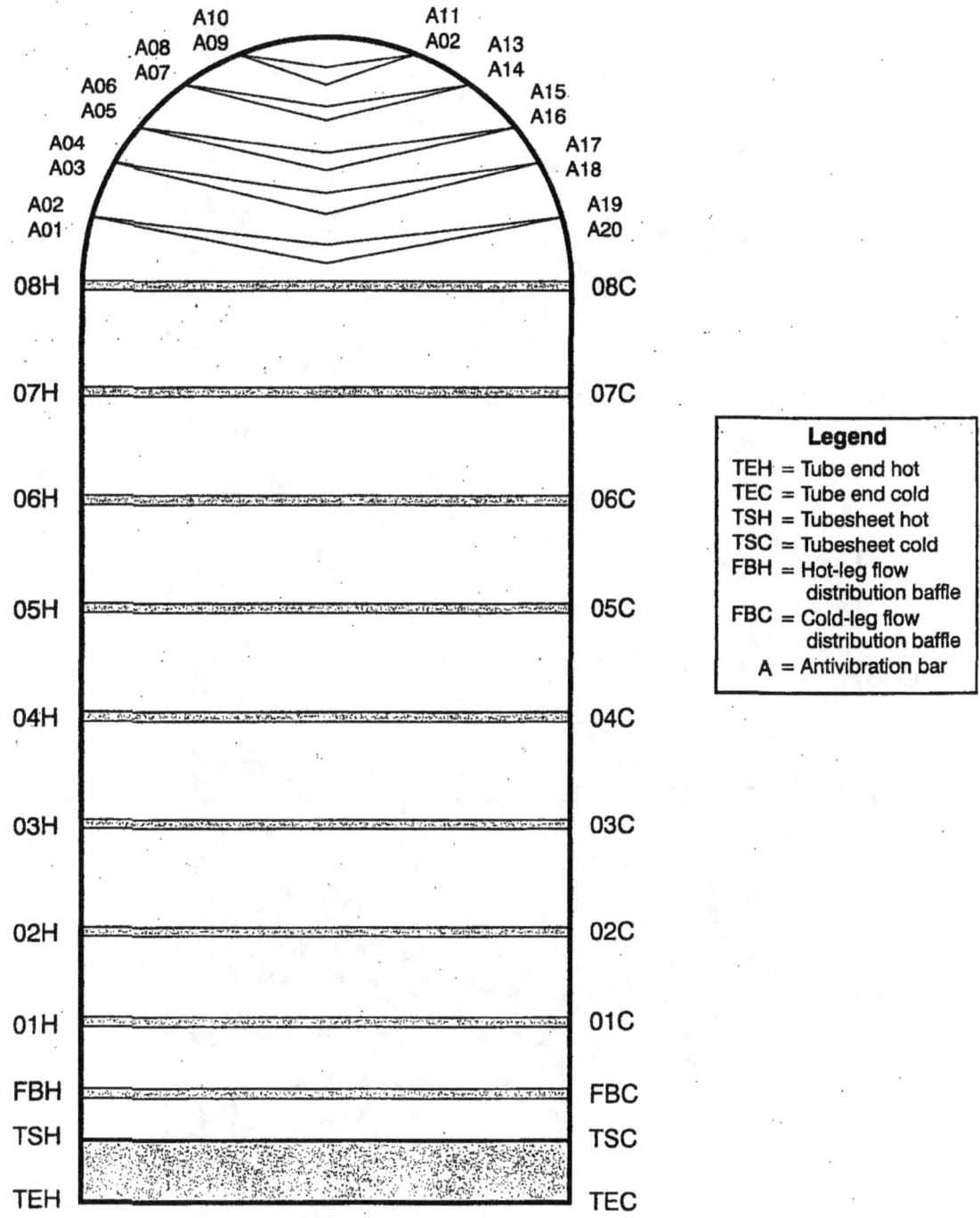

Figure 2-8: Tube Support Naming Convention at ANO Units 2

Legend
TEH = Tube end hot
TEC = Tube end cold
TSH = Tubesheet hot
TSC = Tubesheet cold
FBH = Hot-leg flow distribution baffle
FBC = Cold-leg flow distribution baffle
A = Antivibration bar

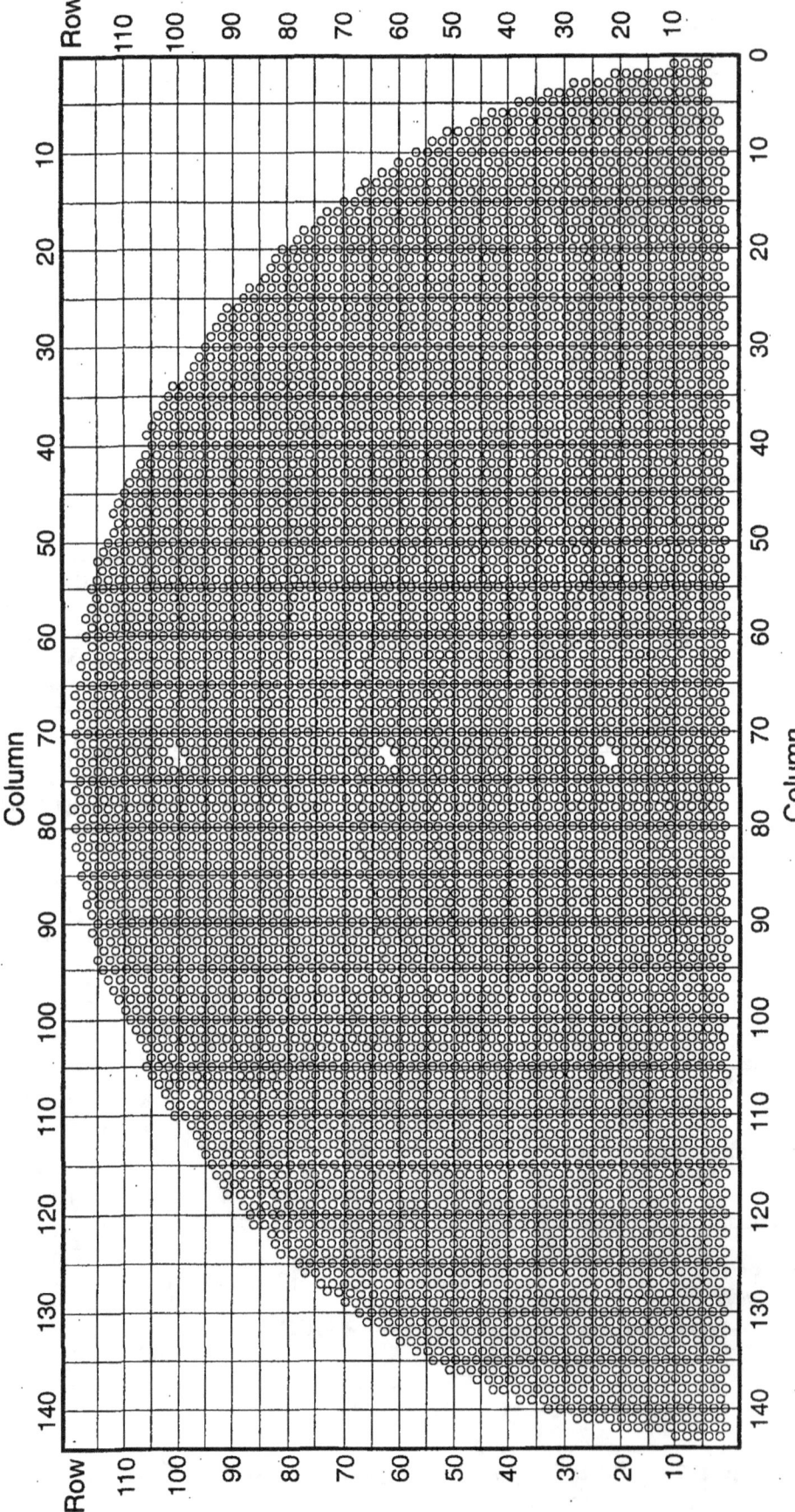

Figure 2-9: Tubesheet Map for Braidwood Unit 1

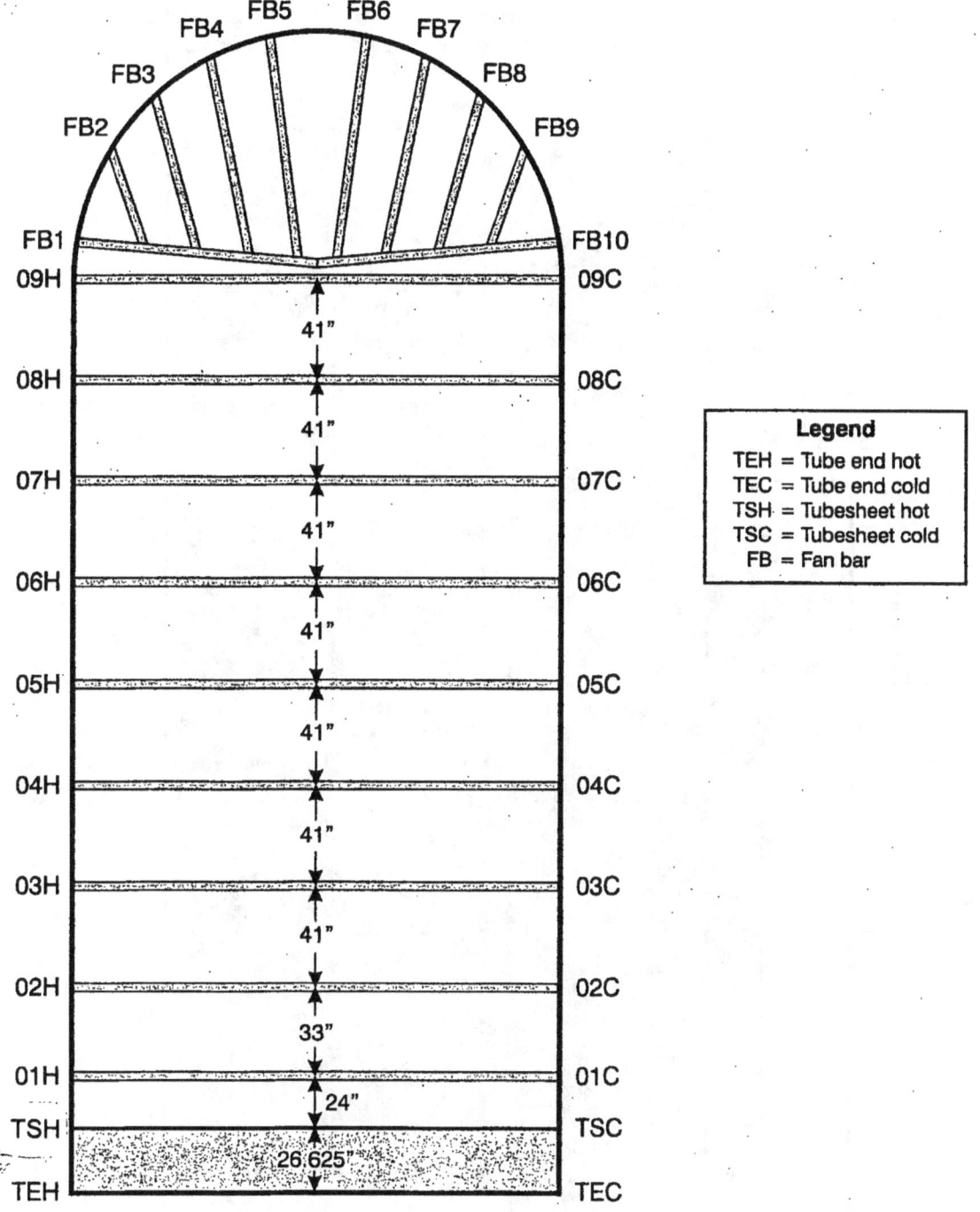

Figure 2-10: Tube Support Naming Convention at Braidwood Unit 1

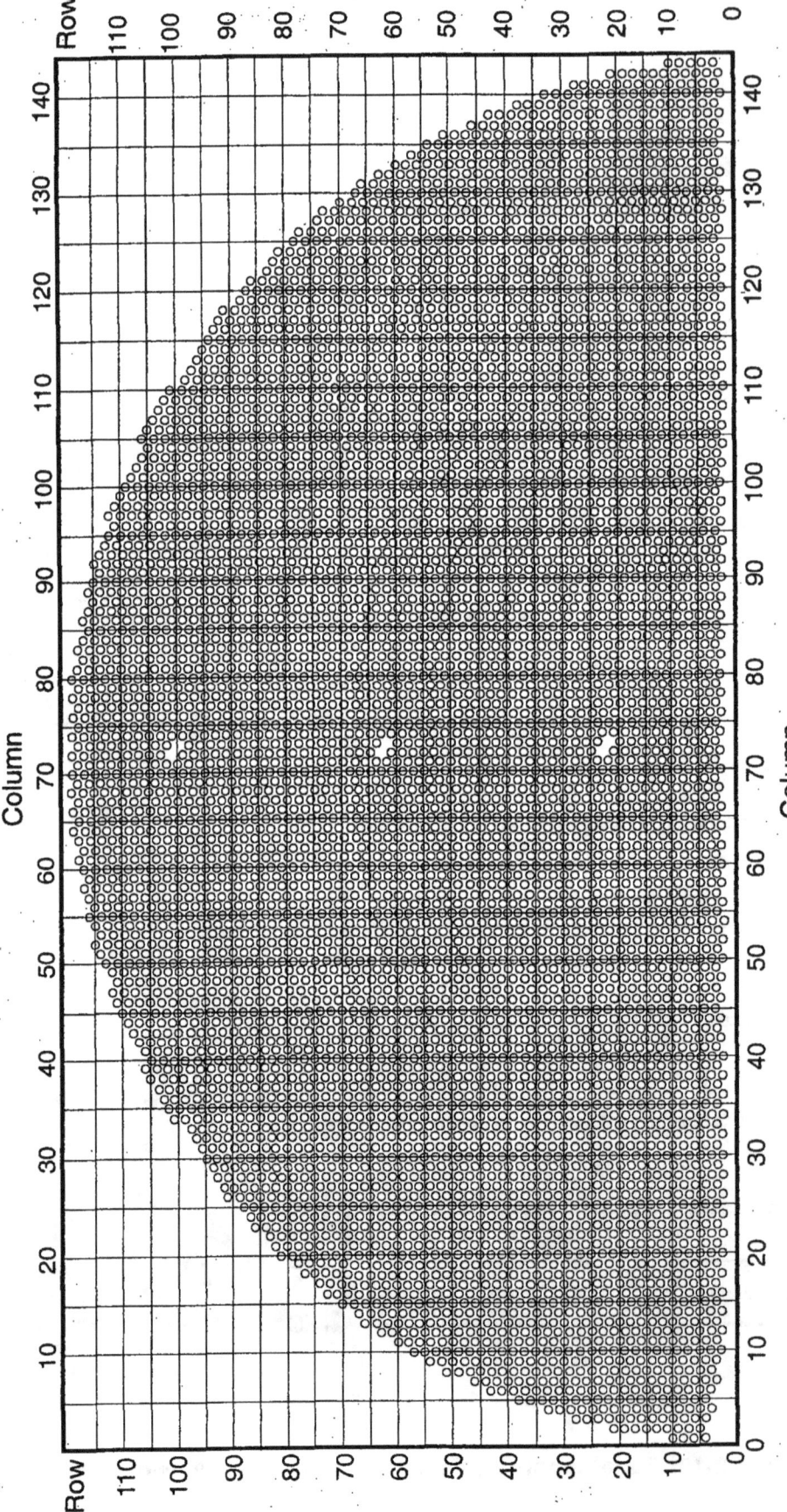

Figure 2-11: Tubesheet Map for Byron Unit 1

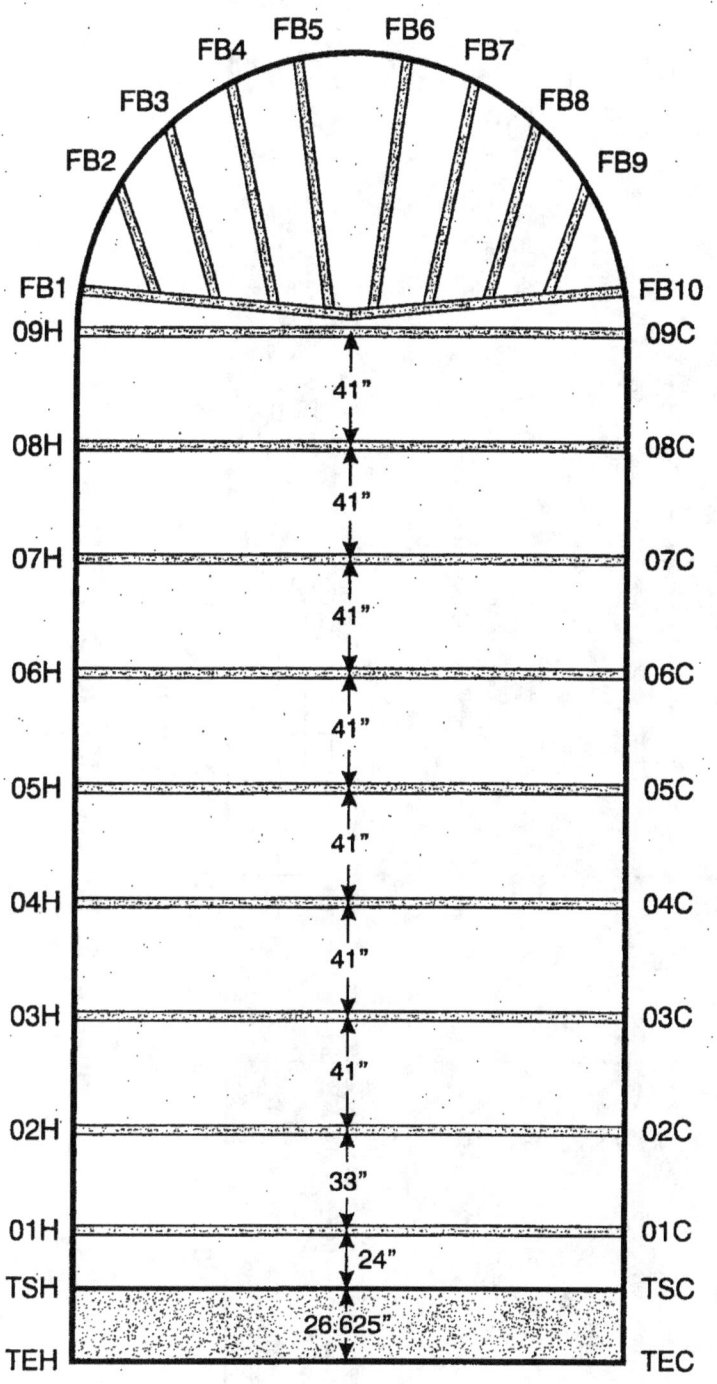

Figure 2-12: Tube Support Naming Convention at Byron Unit 1

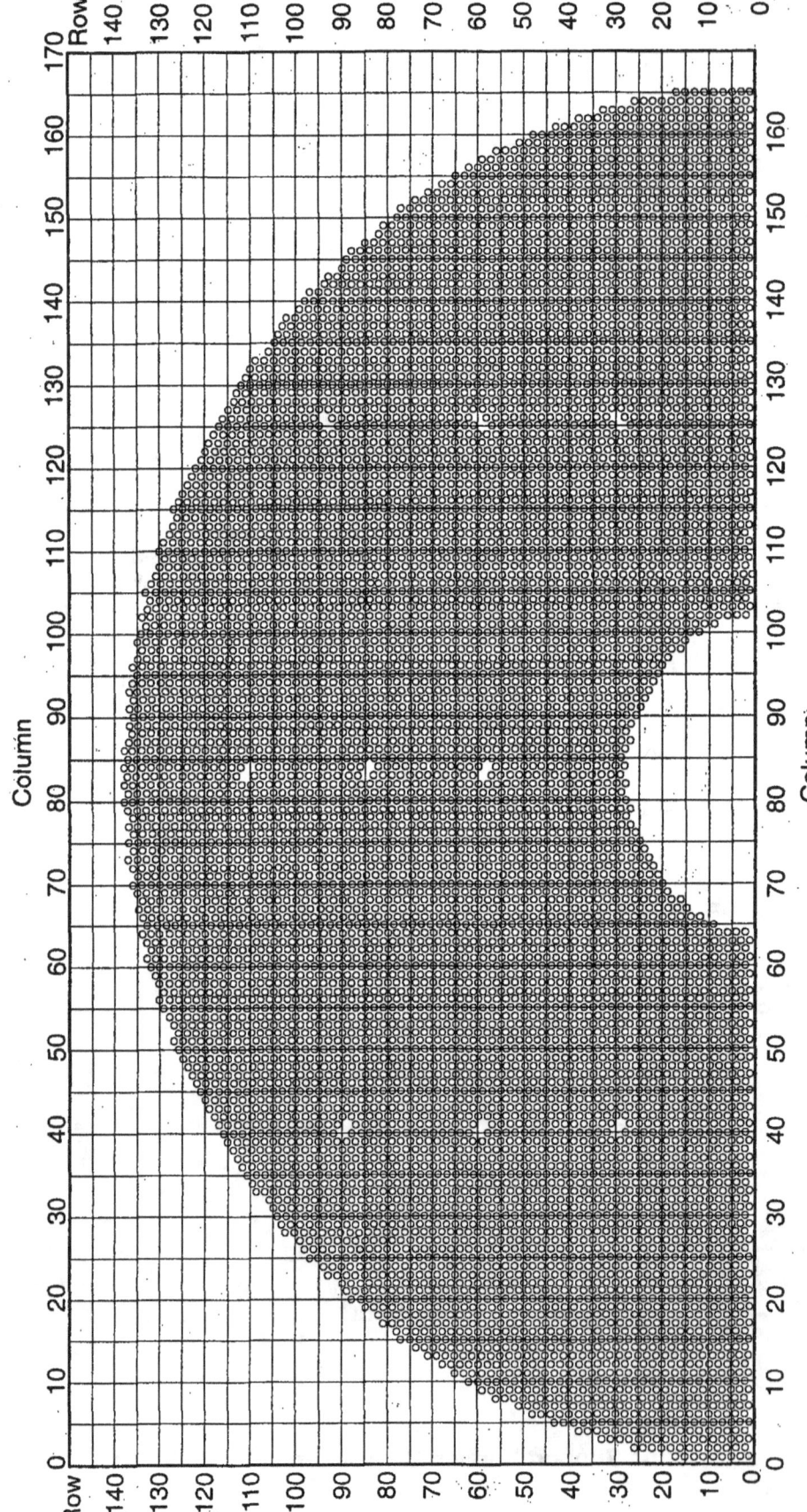

Figure 2-13: Tubesheet Map for Calvert Cliffs Units 1 and 2

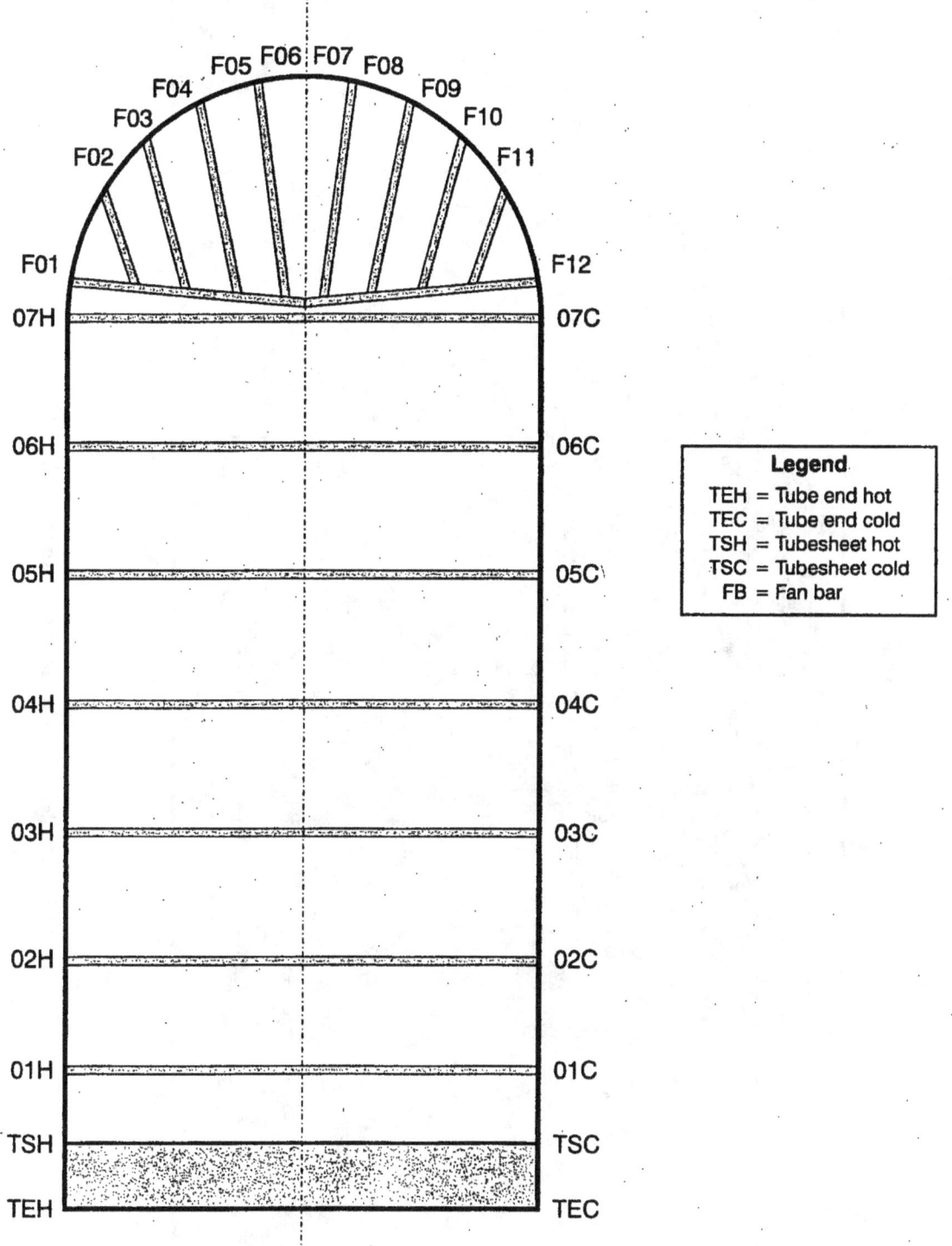

Figure 2-14: Tube Support Naming Convention at Calvert Cliffs 1 and 2

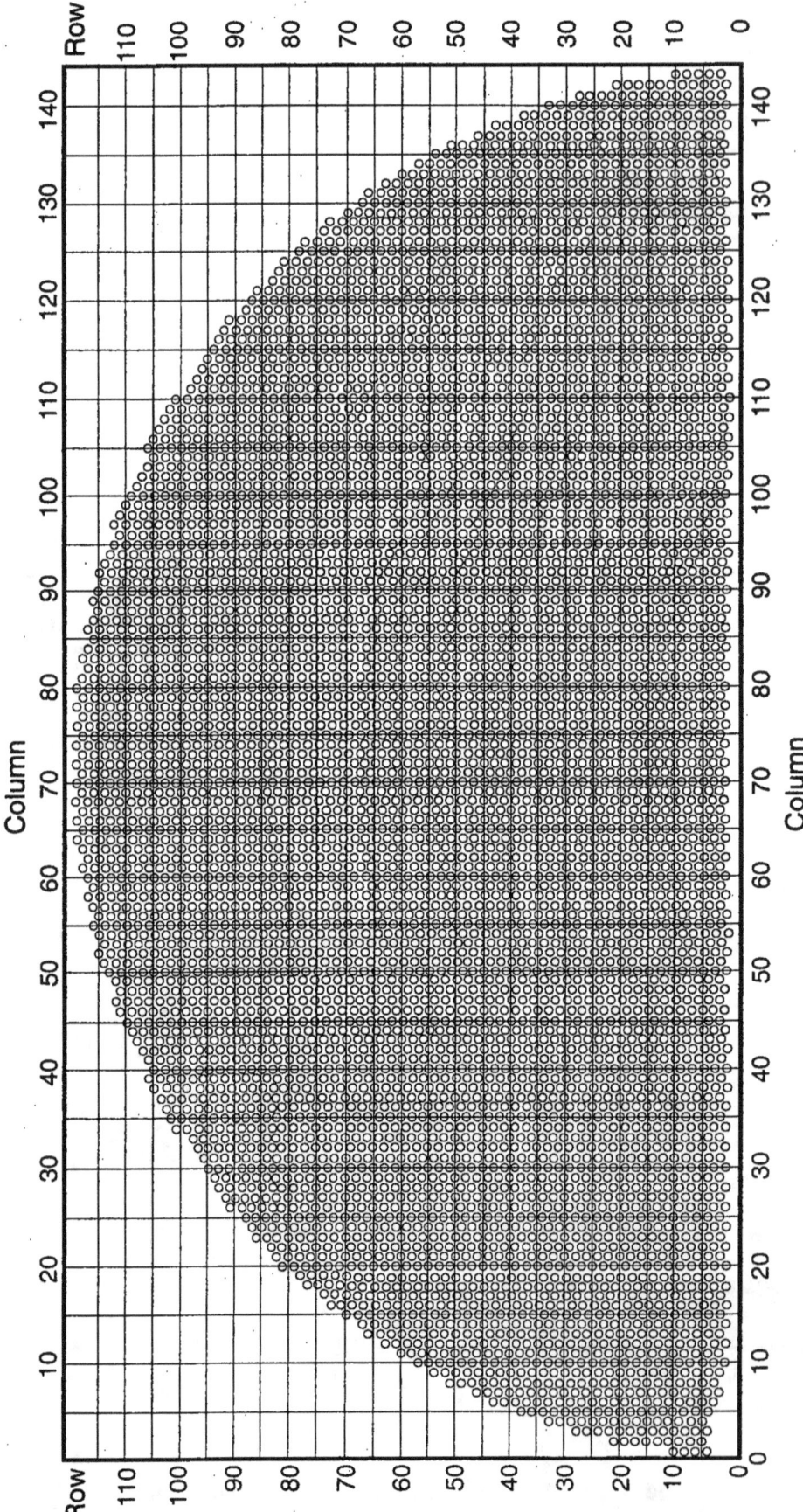

Figure 2-15: Tubesheet Map for Catawba Unit 1

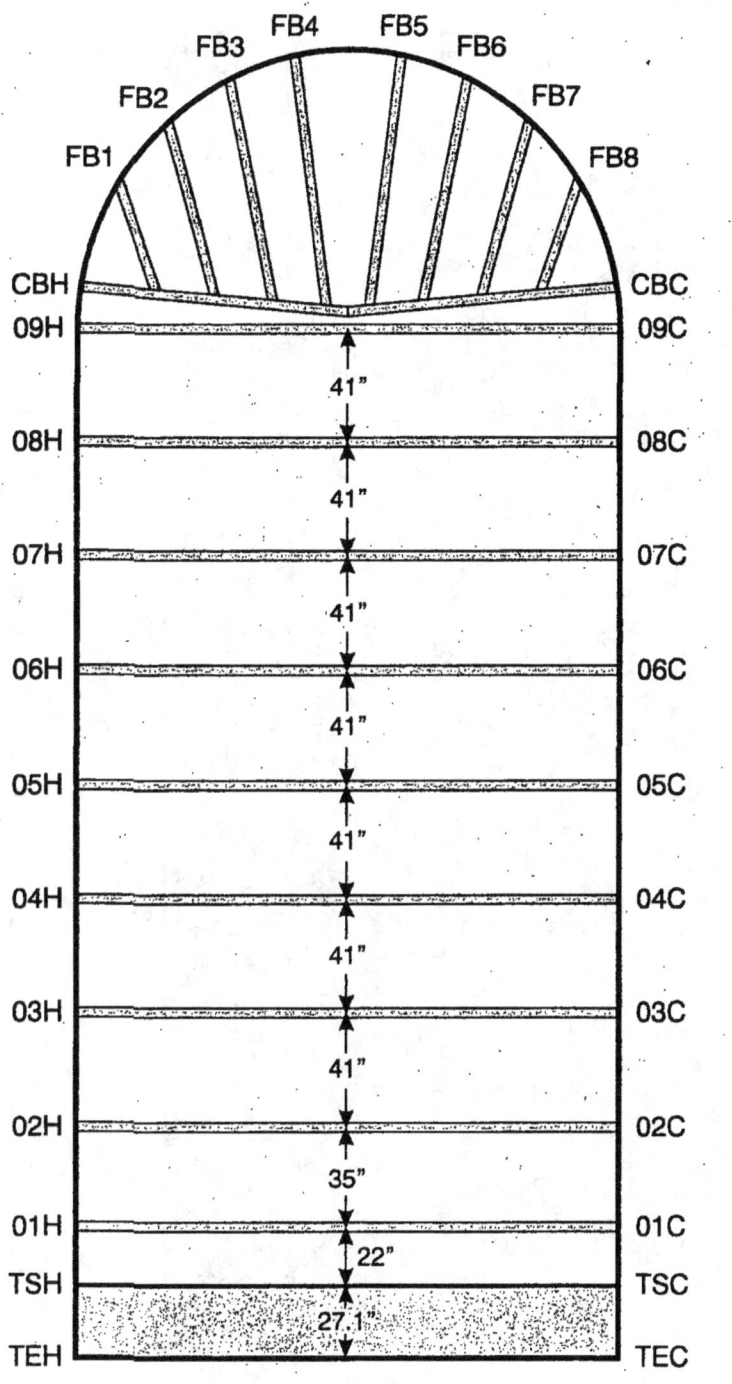

Legend

TEH = Tube end hot
TEC = Tube end cold
TSH = Tubesheet hot
TSC = Tubesheet cold
CBH = Connector bar hot
CBC = Connector bar cold
FB = Fan bar

Figure 2-16: Tube Support Naming Convention at Catawba Unit 1

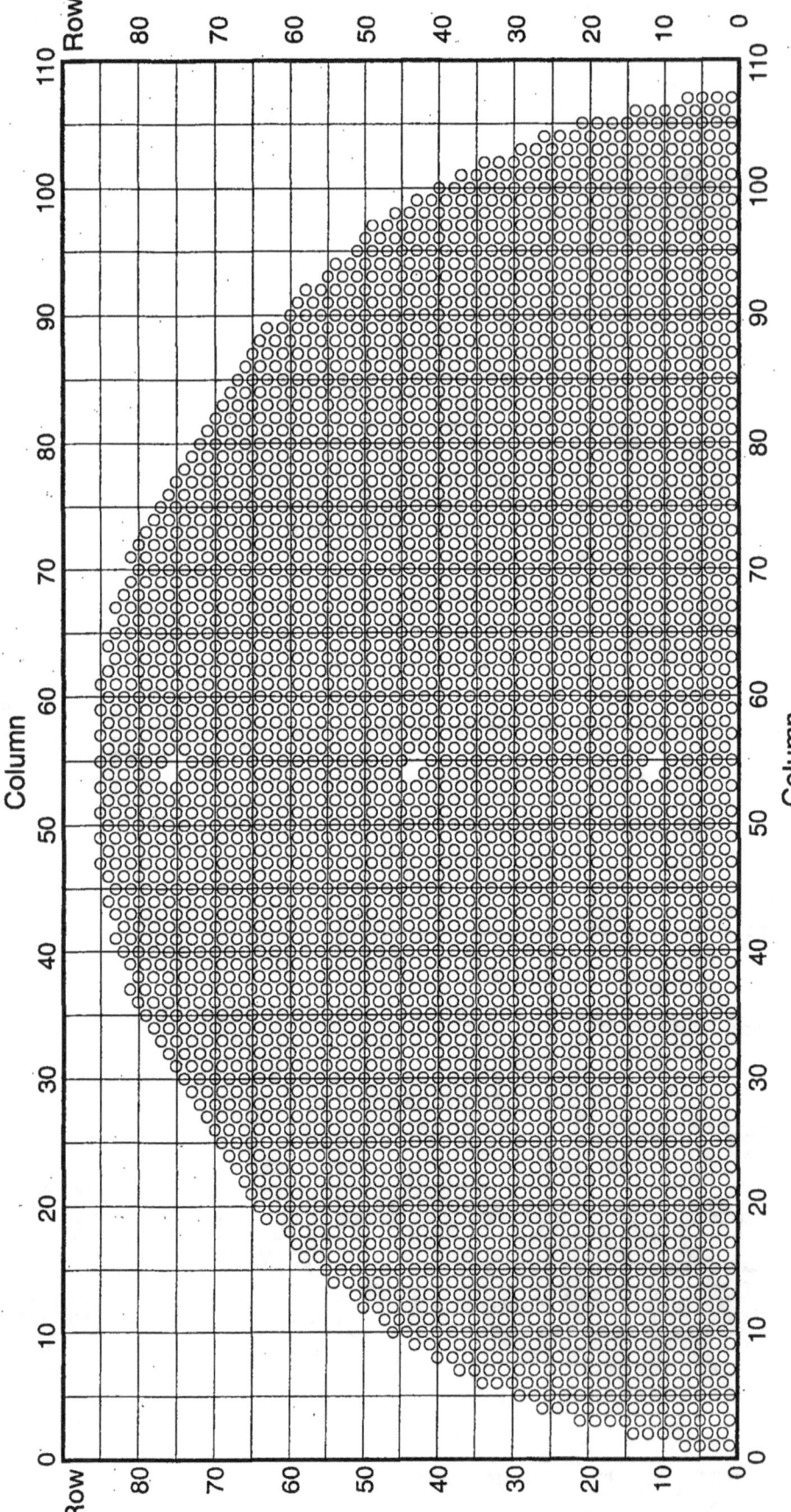

Figure 2-17: Tubesheet Map for Cook Unit 1

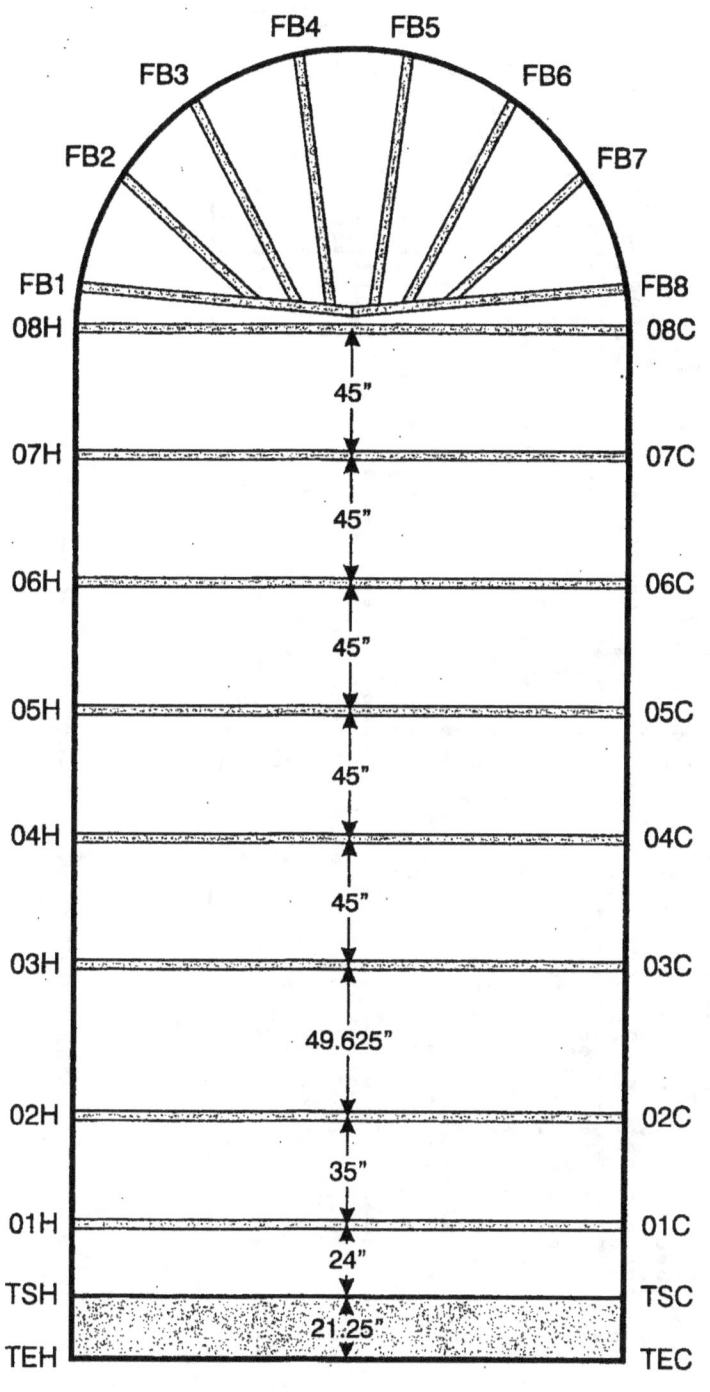

Legend
TEH = Tube end hot
TEC = Tube end cold
TSH = Tubesheet hot
TSC = Tubesheet cold
FB = Fan bar

Figure 2-18: Tube Support Naming Convention at Cook Unit 1

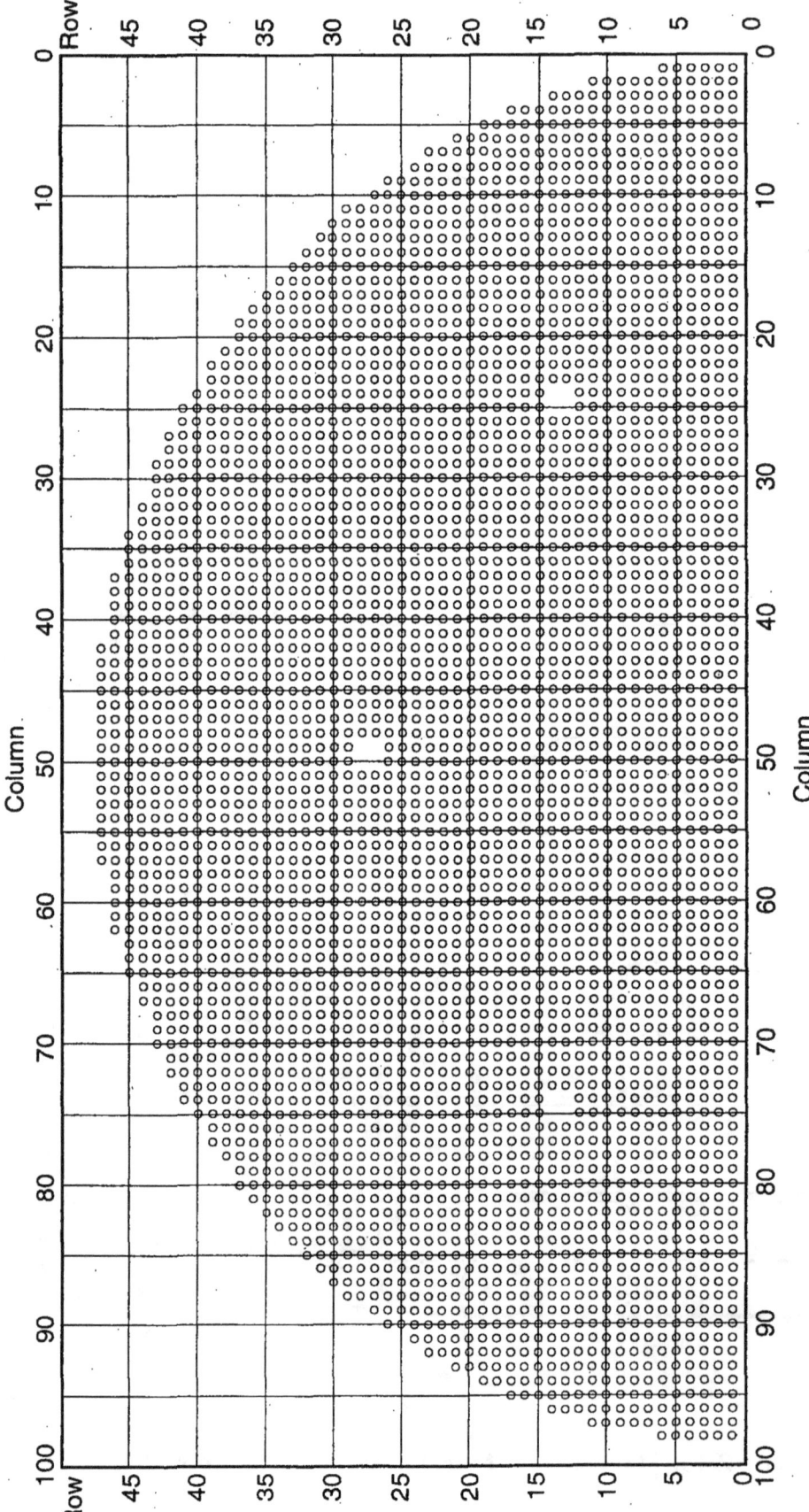

Figure 2-19: Tubesheet Map for Cook Unit 2

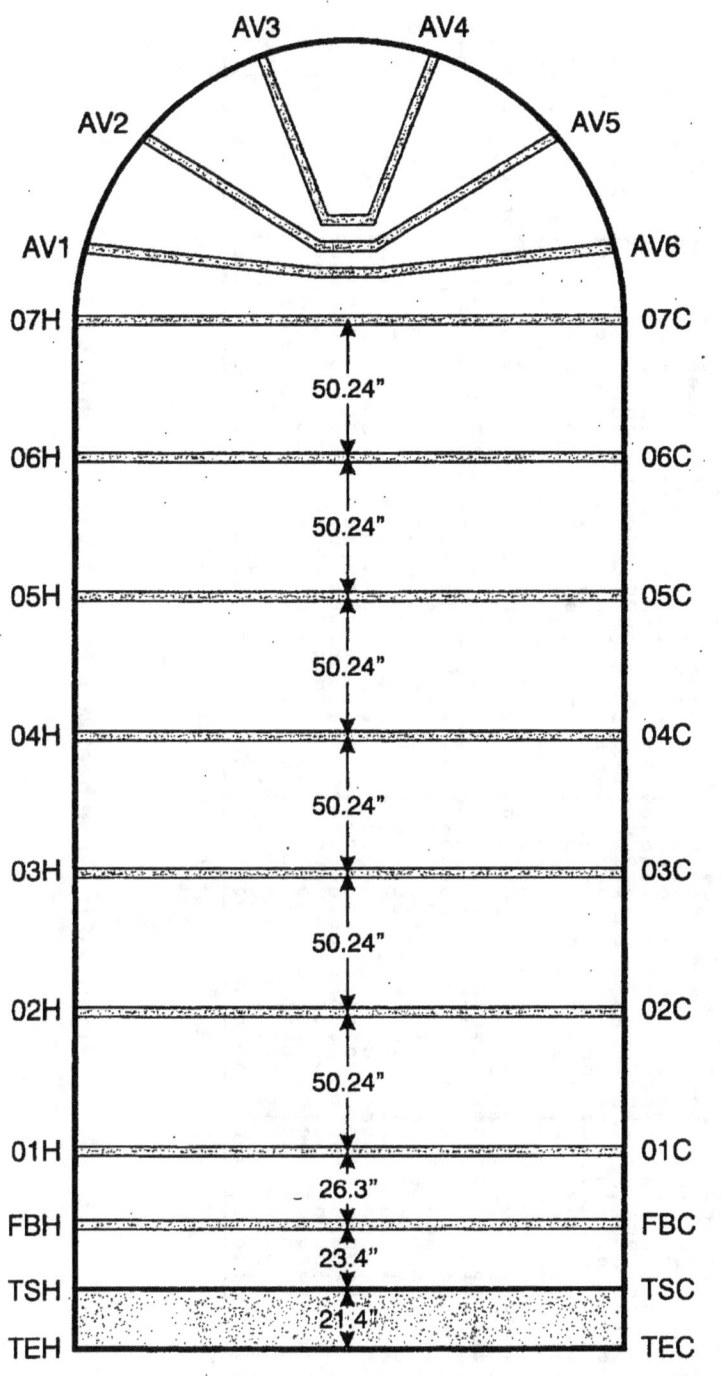

Legend

TEH = Tube end hot
TEC = Tube end cold
TSH = Tubesheet hot
TSC = Tubesheet cold
FBH = Hot-leg flow
      distribution baffle
FBC = Cold-leg flow
      distribution baffle
AV = Antivibration bar

Figure 2-20: Tube Support Naming Convention at Cook Unit 2

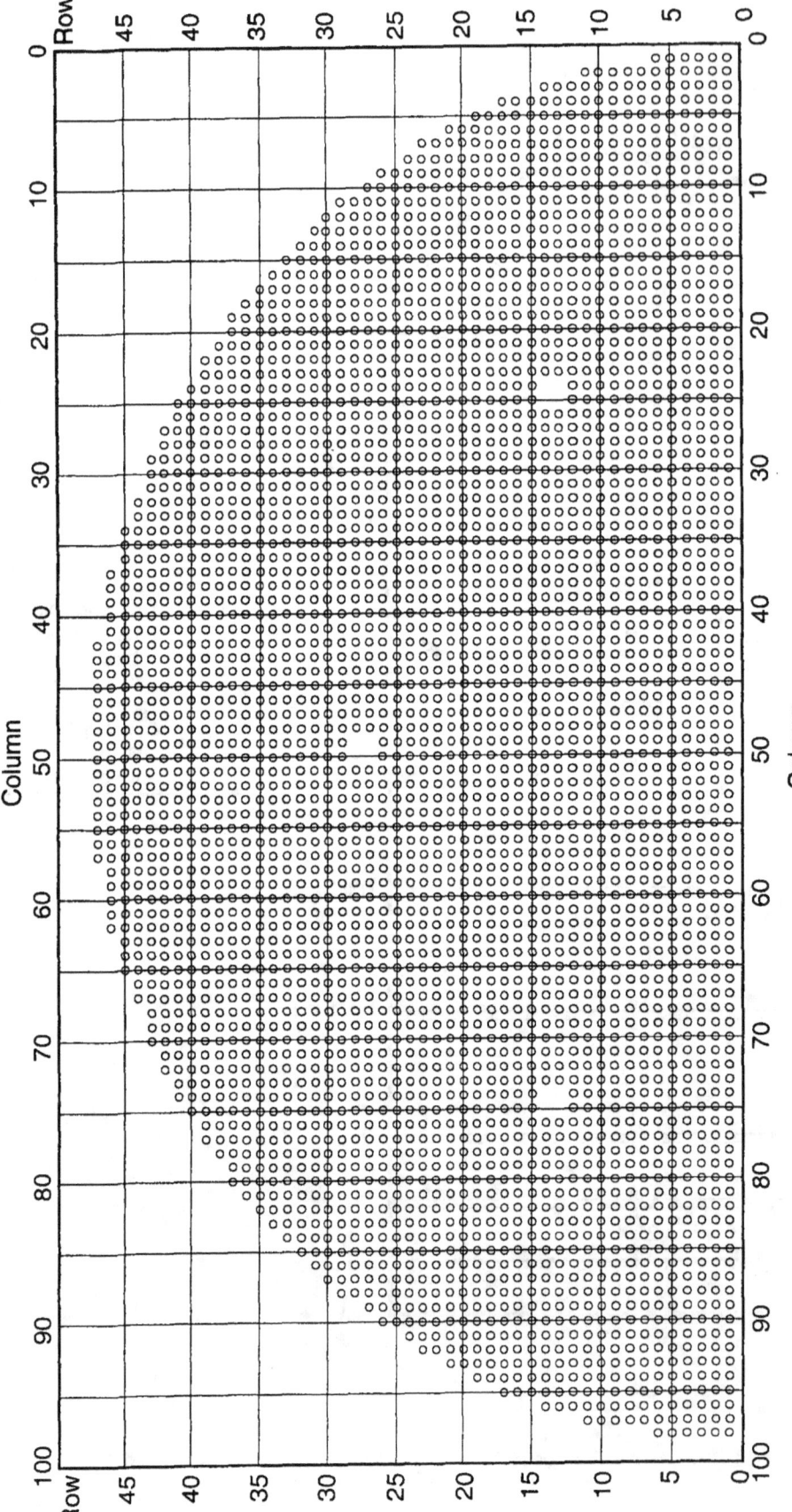

Figure 2-21: Tubesheet Map for Farley Units 1 and 2

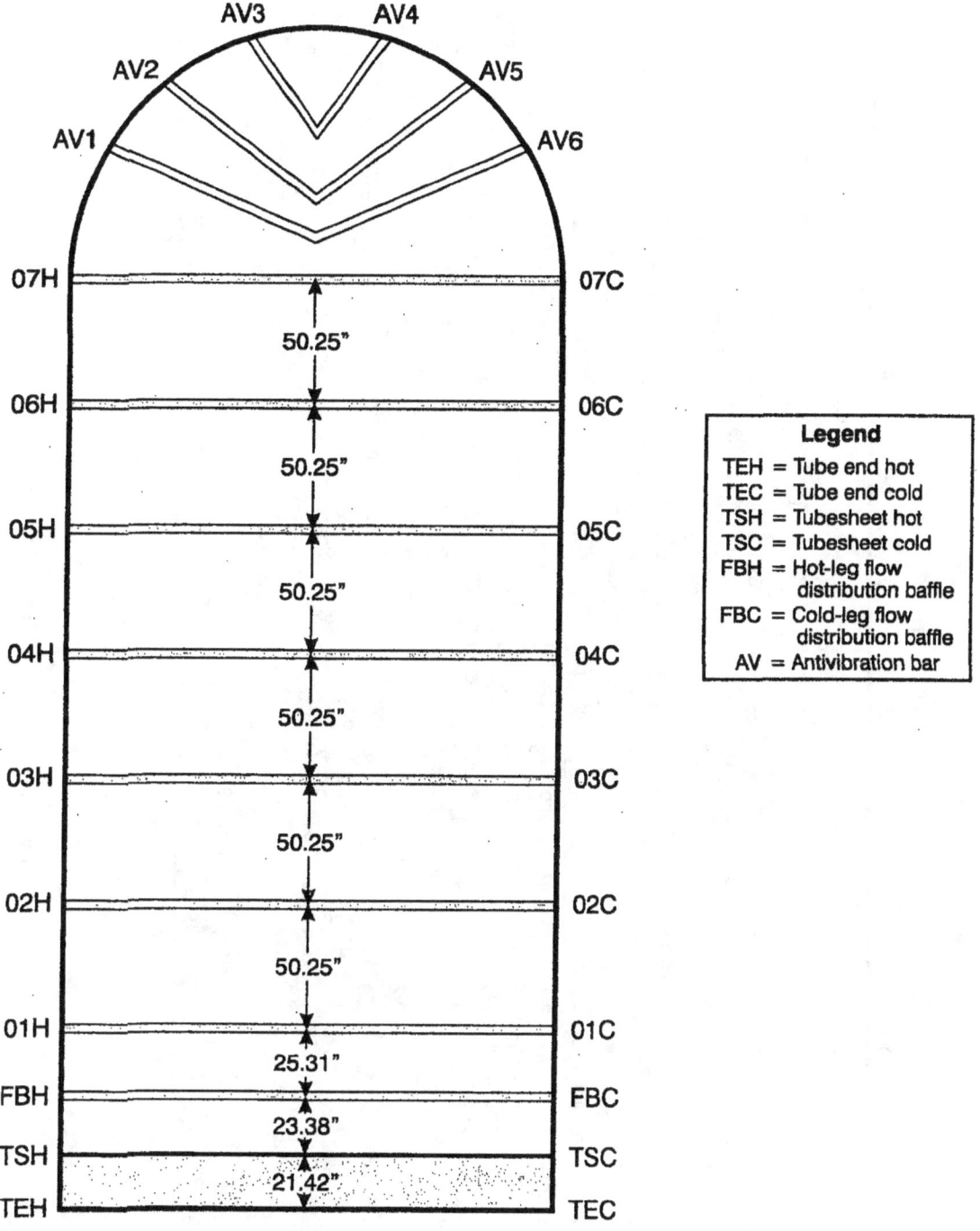

Figure 2-22: Tube Support Naming Convention at Farley Units 1 and 2

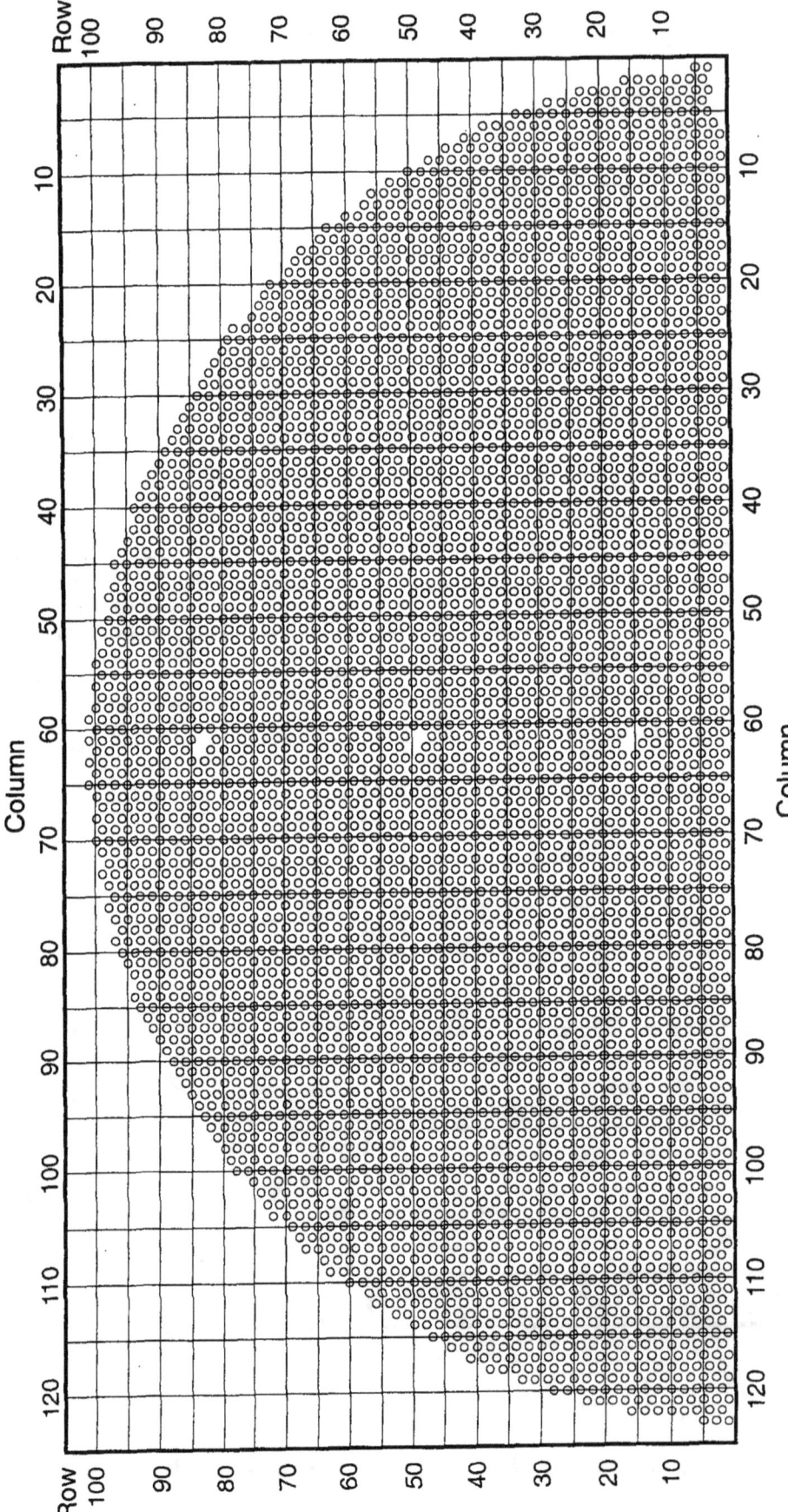

Figure 2-23: Tubesheet Map for Ginna

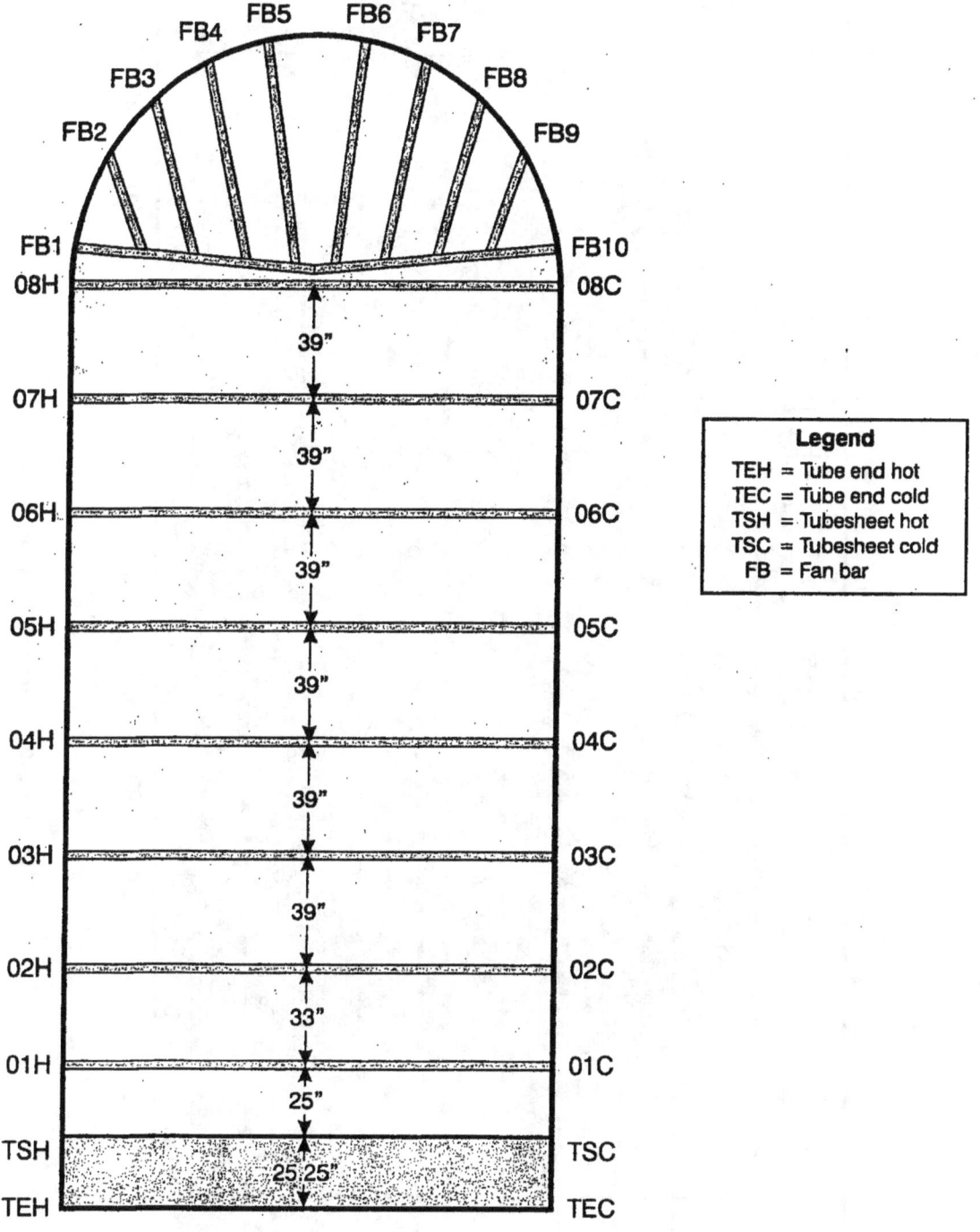

Figure 2-24: Tube Support Naming Convention at Ginna

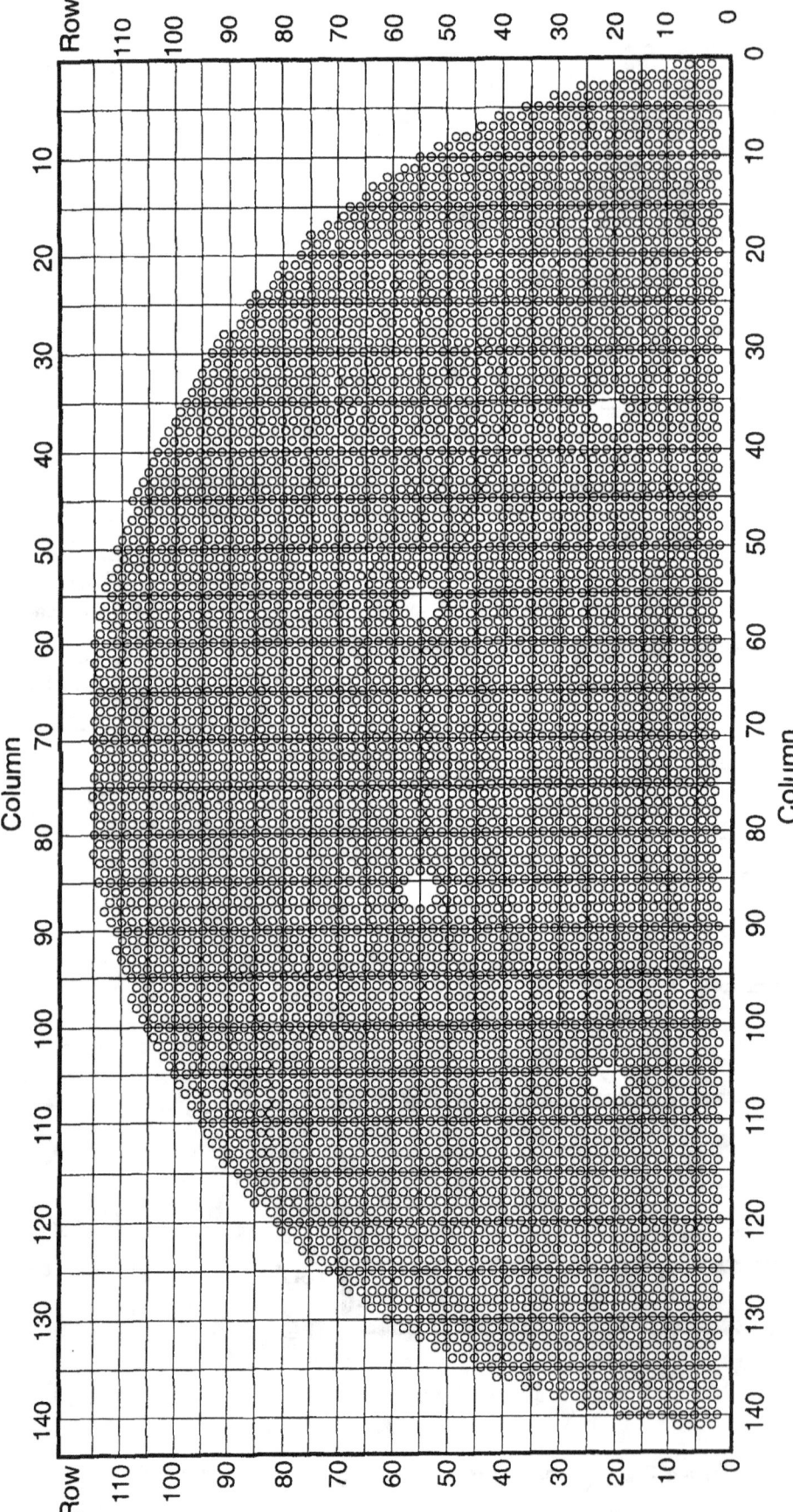

Figure 2-25: Tubesheet Map for Harris

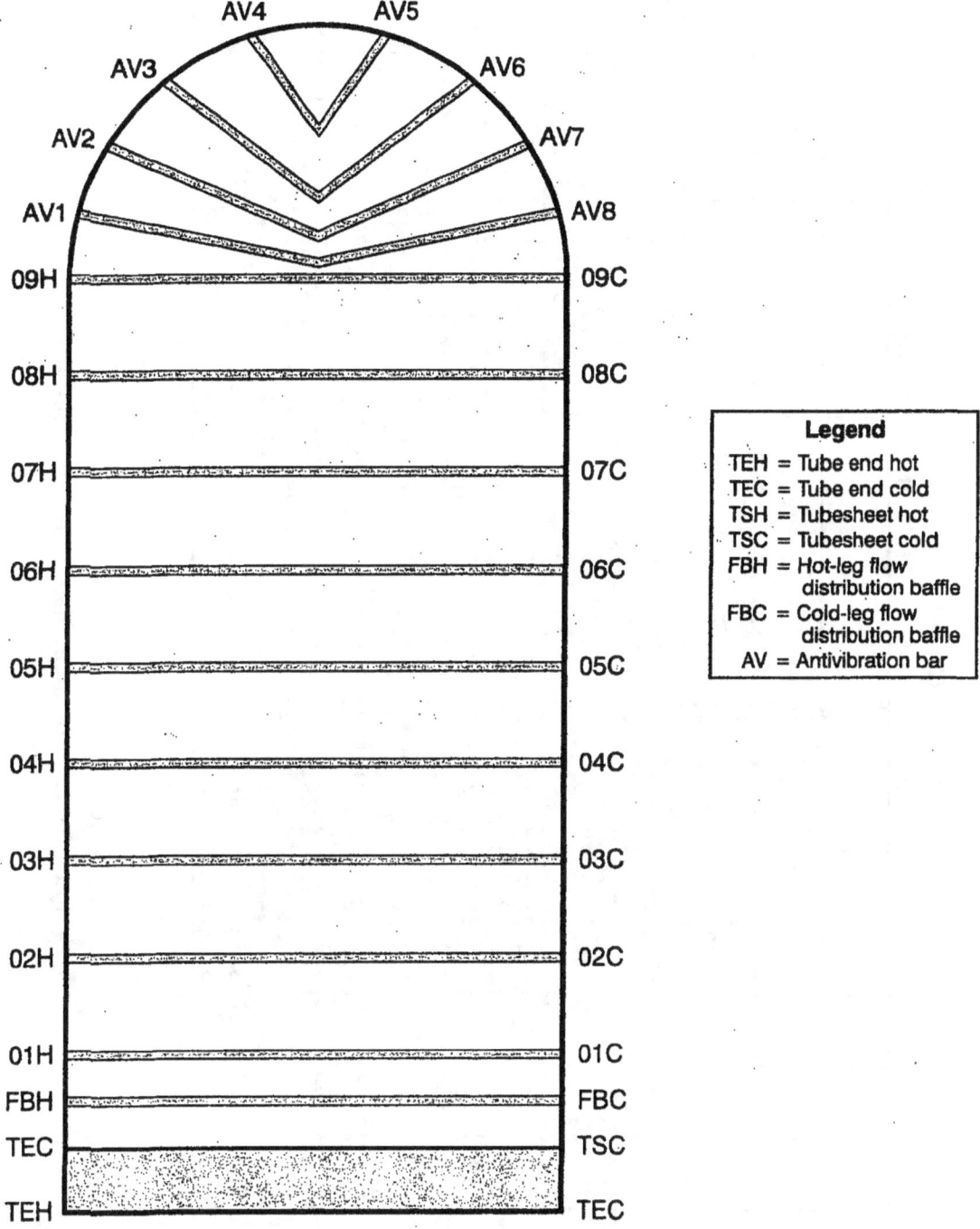

Figure 2-26: Tube Support Naming Convention at Harris

Legend

TEH	= Tube end hot
TEC	= Tube end cold
TSH	= Tubesheet hot
TSC	= Tubesheet cold
FBH	= Hot-leg flow distribution baffle
FBC	= Cold-leg flow distribution baffle
AV	= Antivibration bar

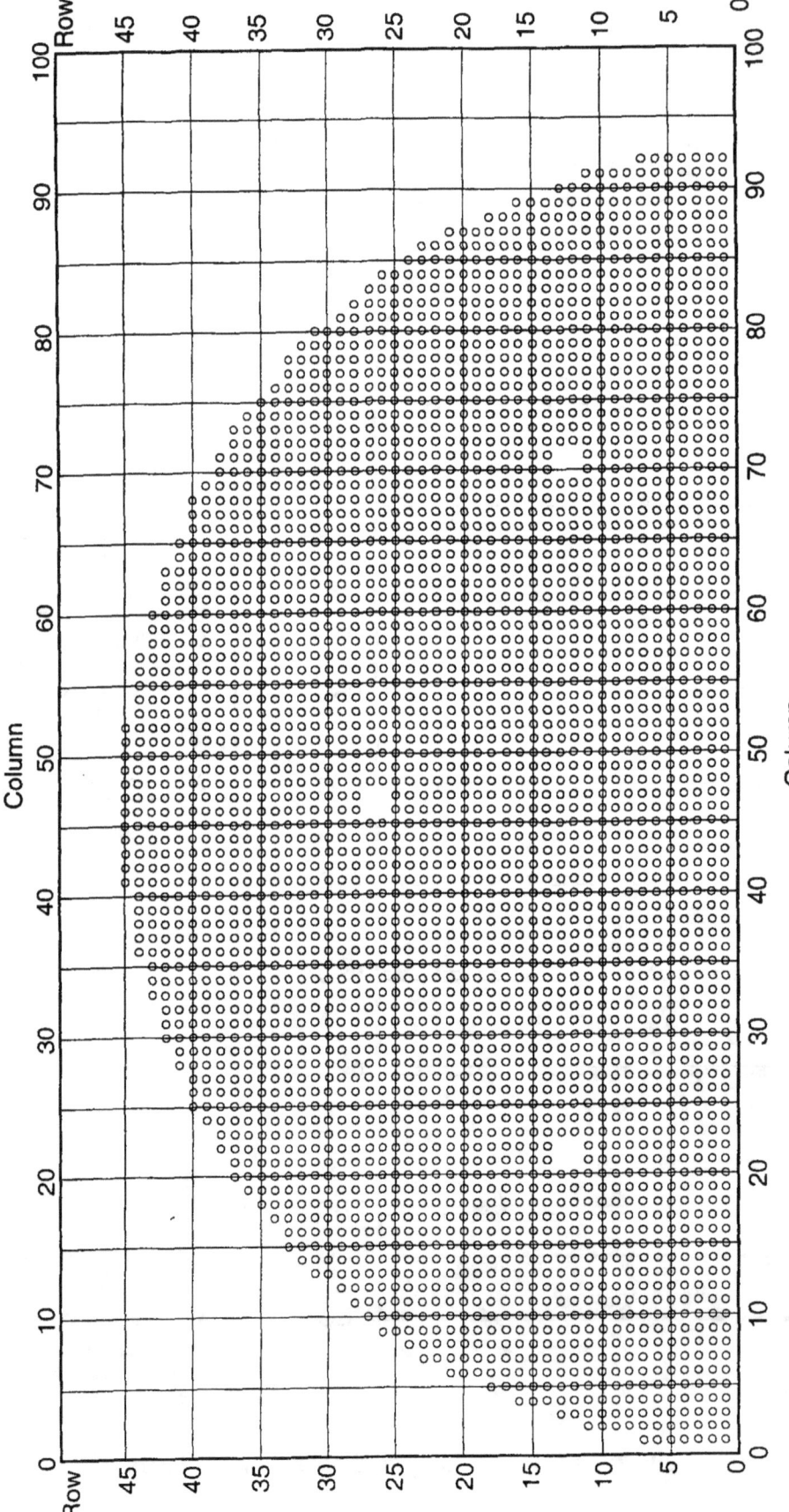

Figure 2-27: Tubesheet Map for Indian Point Unit 3

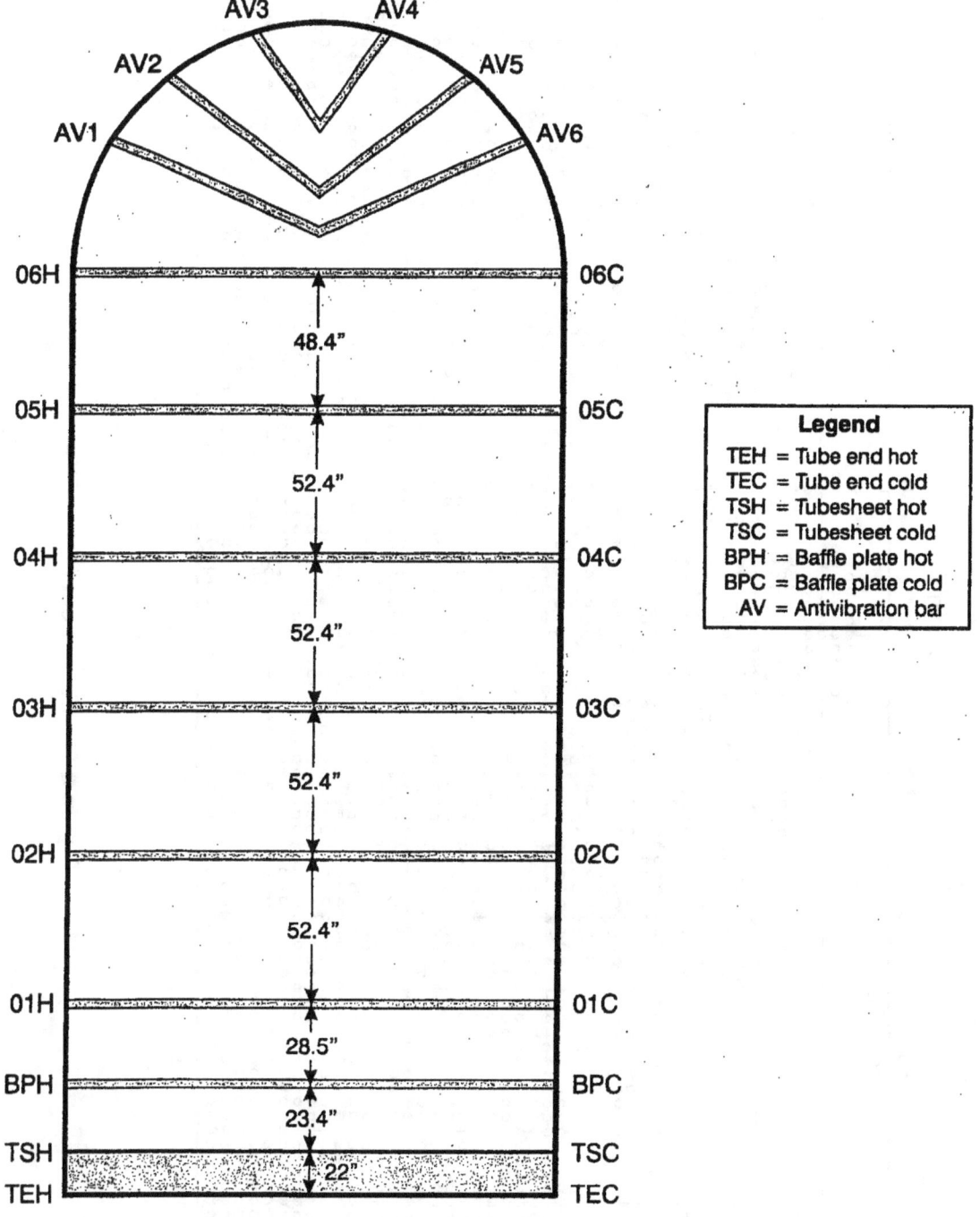

Figure 2-28: Tube Support Naming Convention at Indian Point Unit 3

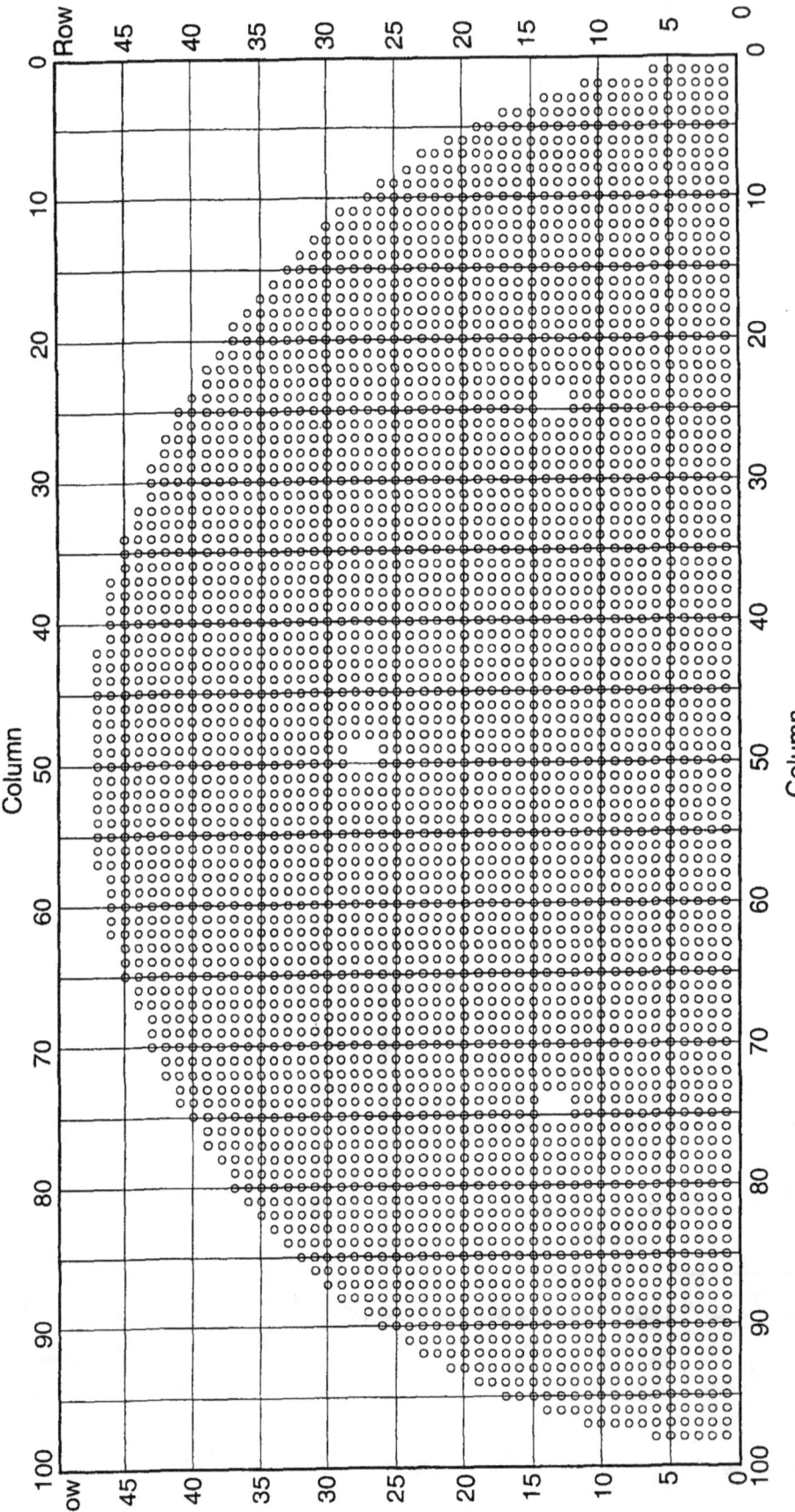

Figure 2-29: Tubesheet Map for Kewaunee

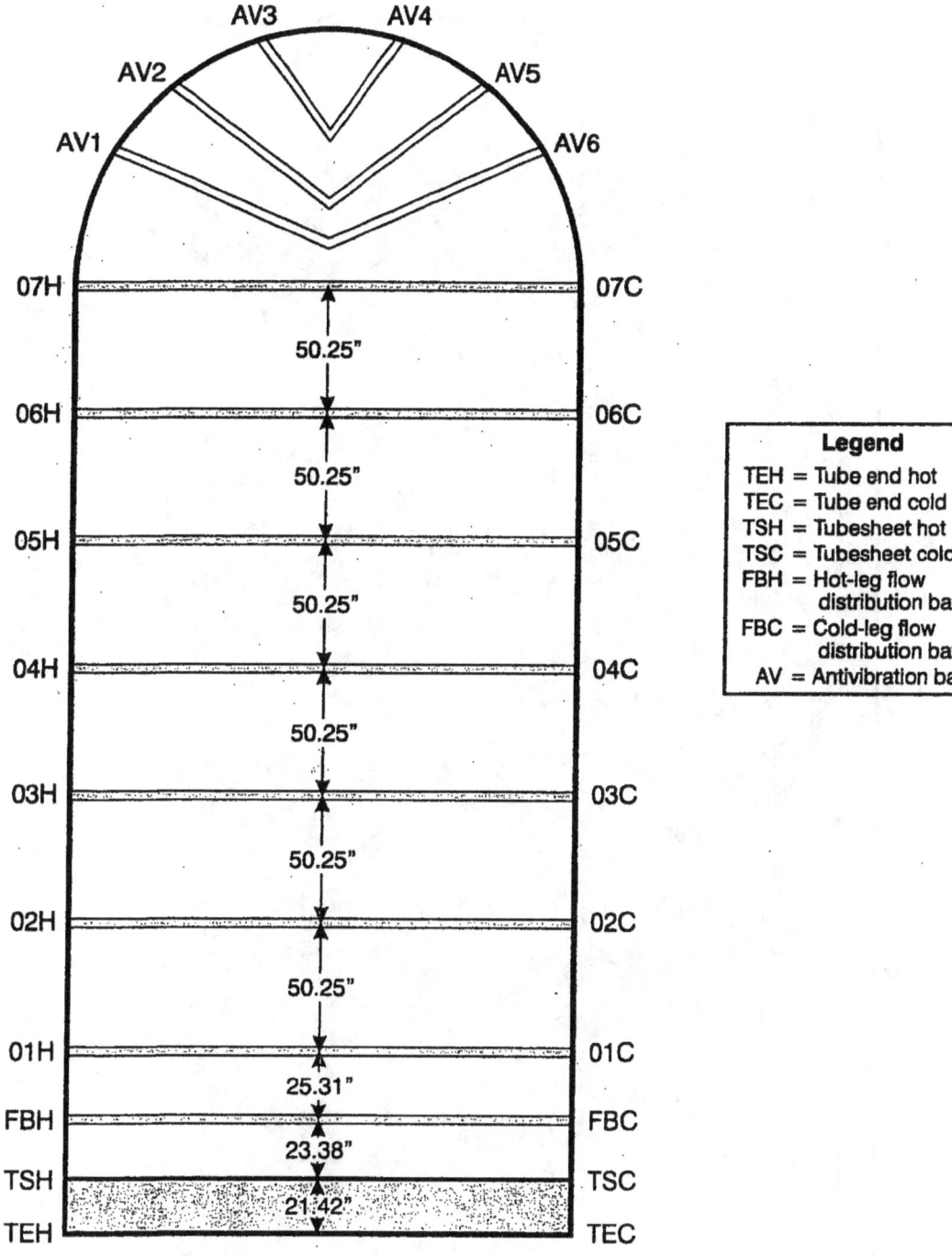

Figure 2-30: Tube Support Naming Convention at Kewaunee

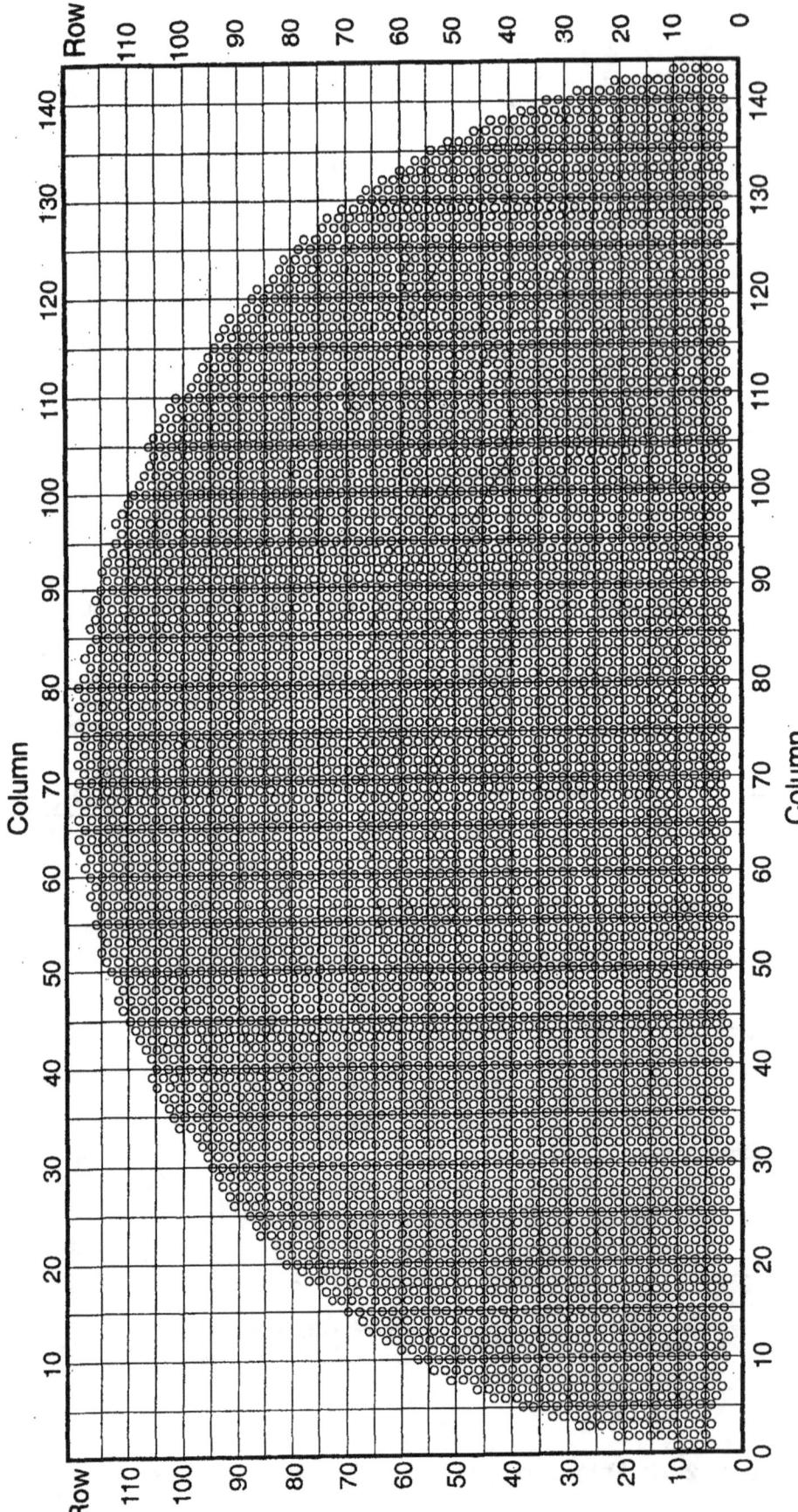

Figure 2-31: Tubesheet Map for McGuire Units 1 and 2

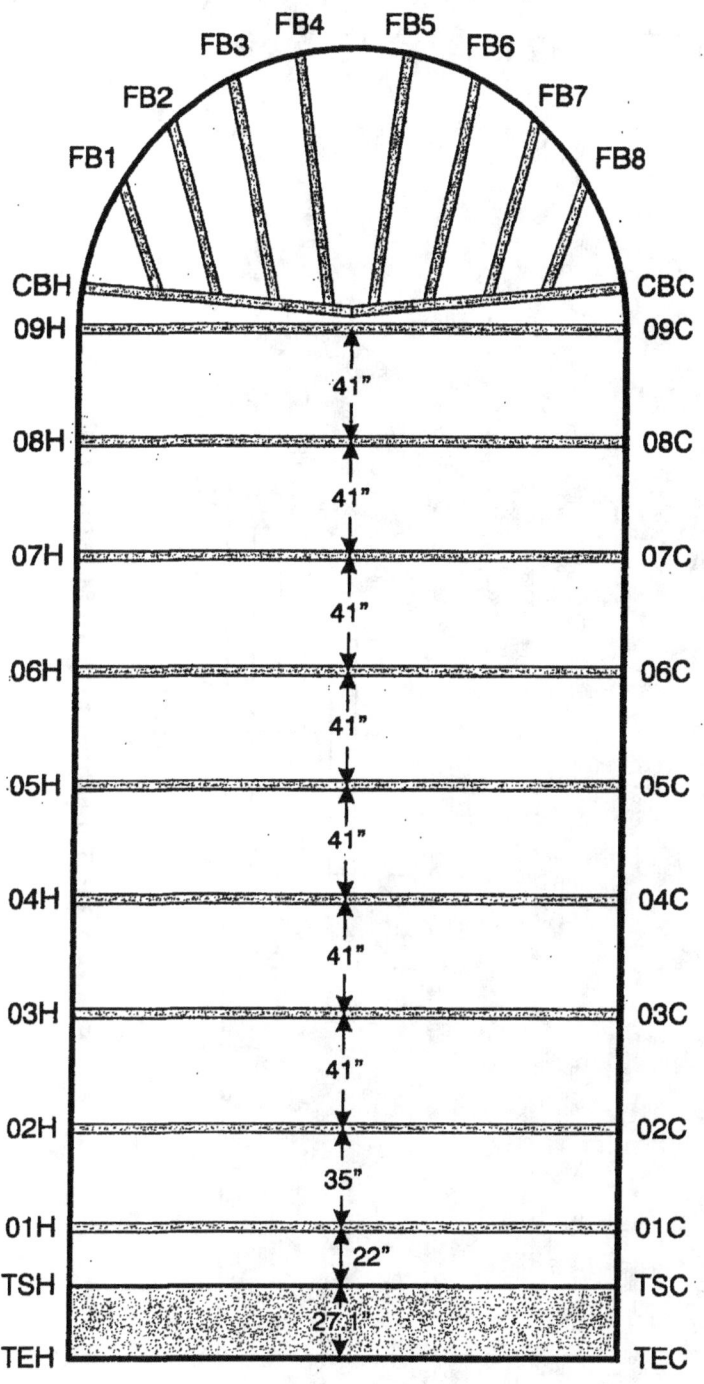

Legend

TEH = Tube end hot
TEC = Tube end cold
TSH = Tubesheet hot
TSC = Tubesheet cold
CBH = Connector bar hot
CBC = Connector bar cold
 FB = Fan bar

Figure 2-32: Tube Support Naming Convention at McGuire Unit 1 and 2

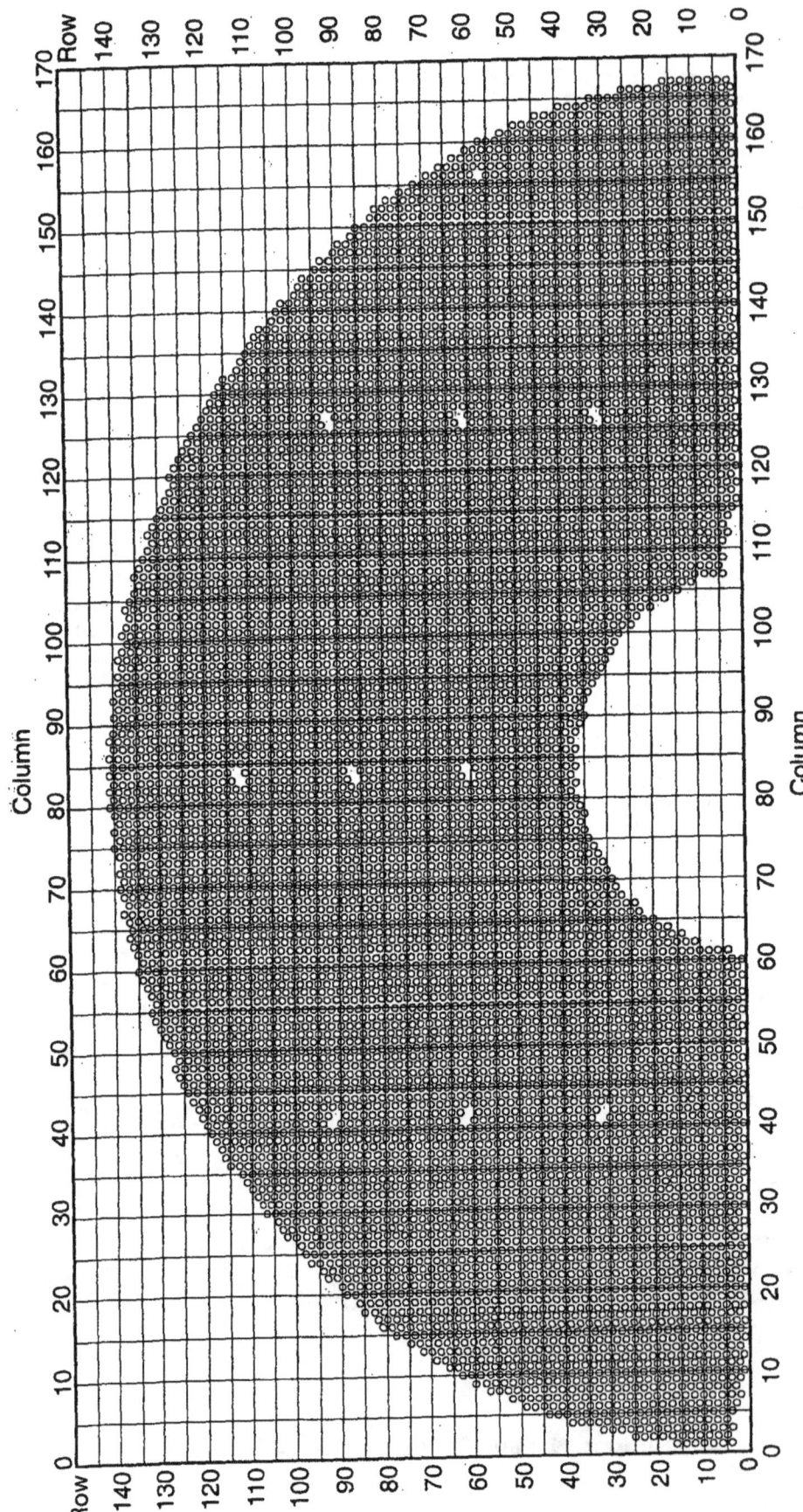

Figure 2-33: Tubesheet Map for Millstone Unit 2

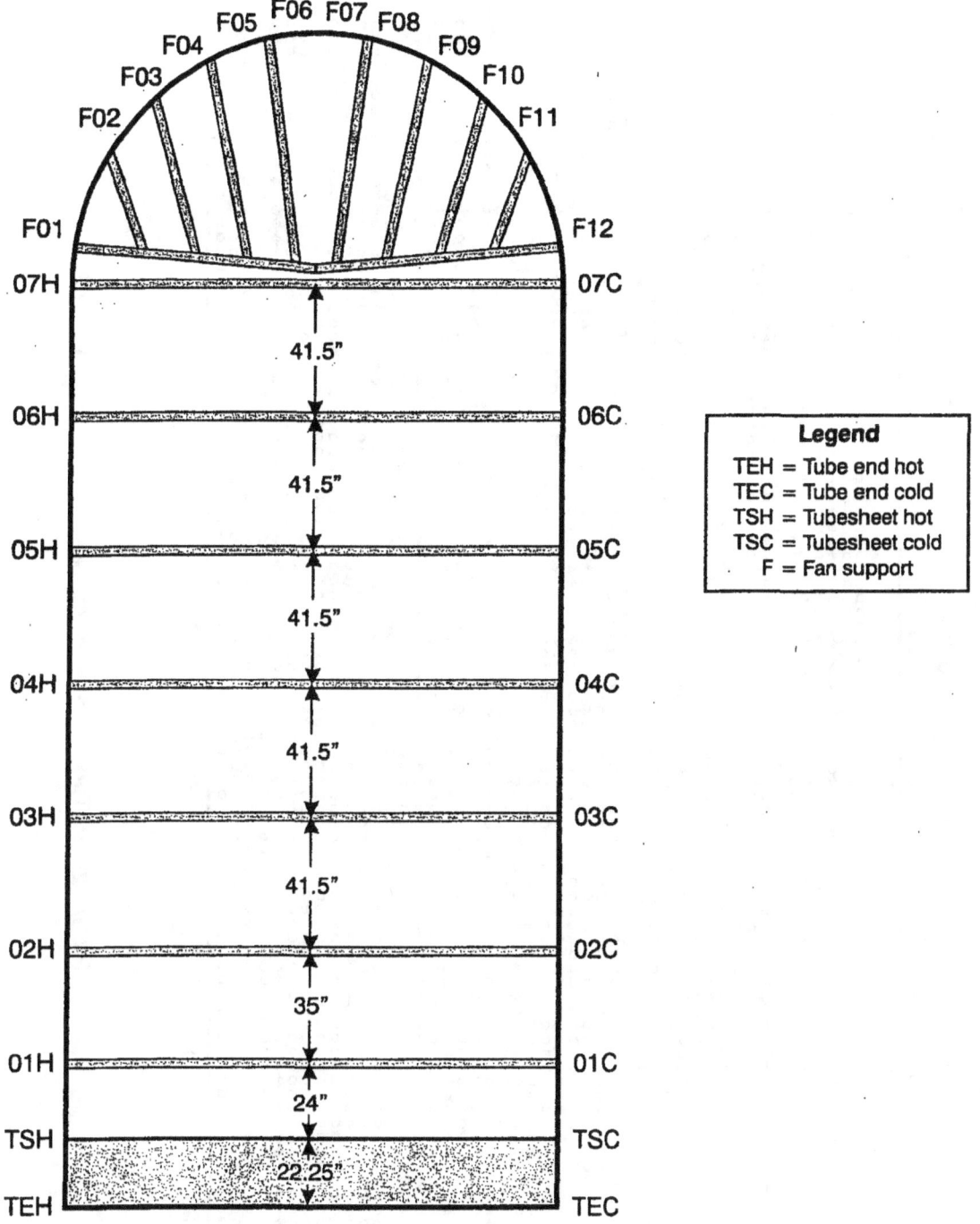

Figure 2-34: Tube Support Naming Convention at Millstone Unit 2

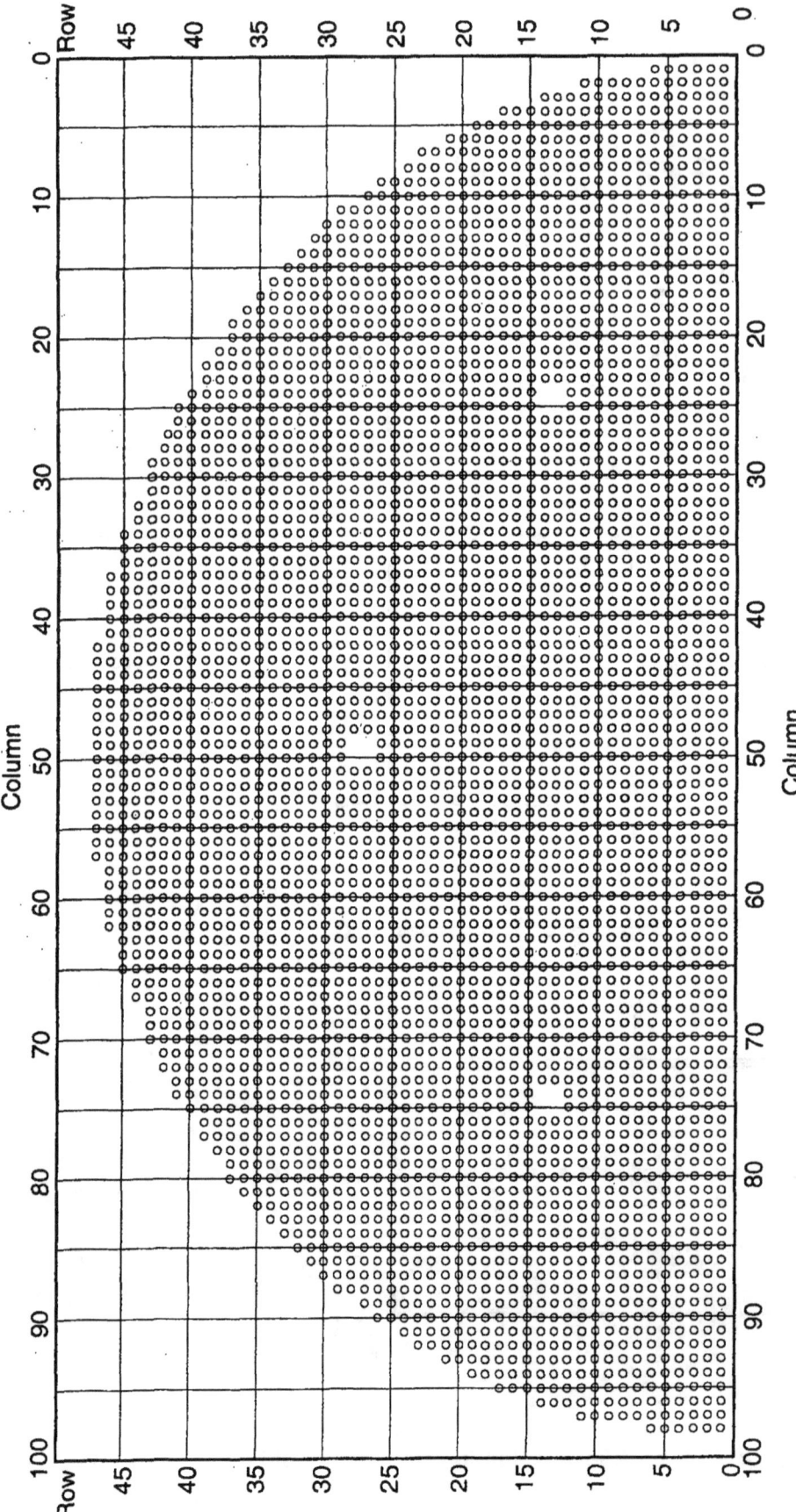

Figure 2-35: Tubesheet Map for North Anna Units 1 and 2

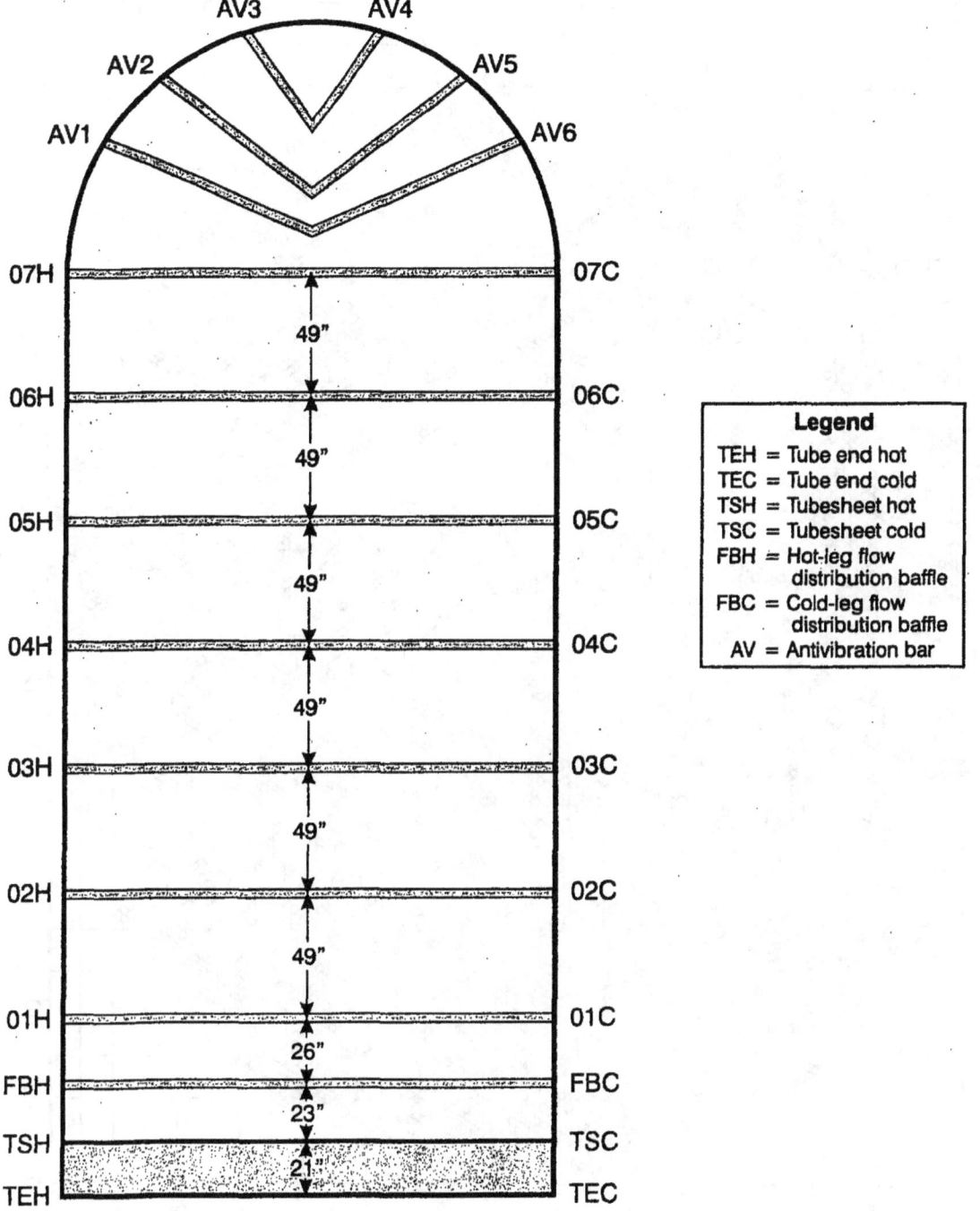

Figure 2-36: Tube Support Naming Convention at North Anna Units 1 and 2

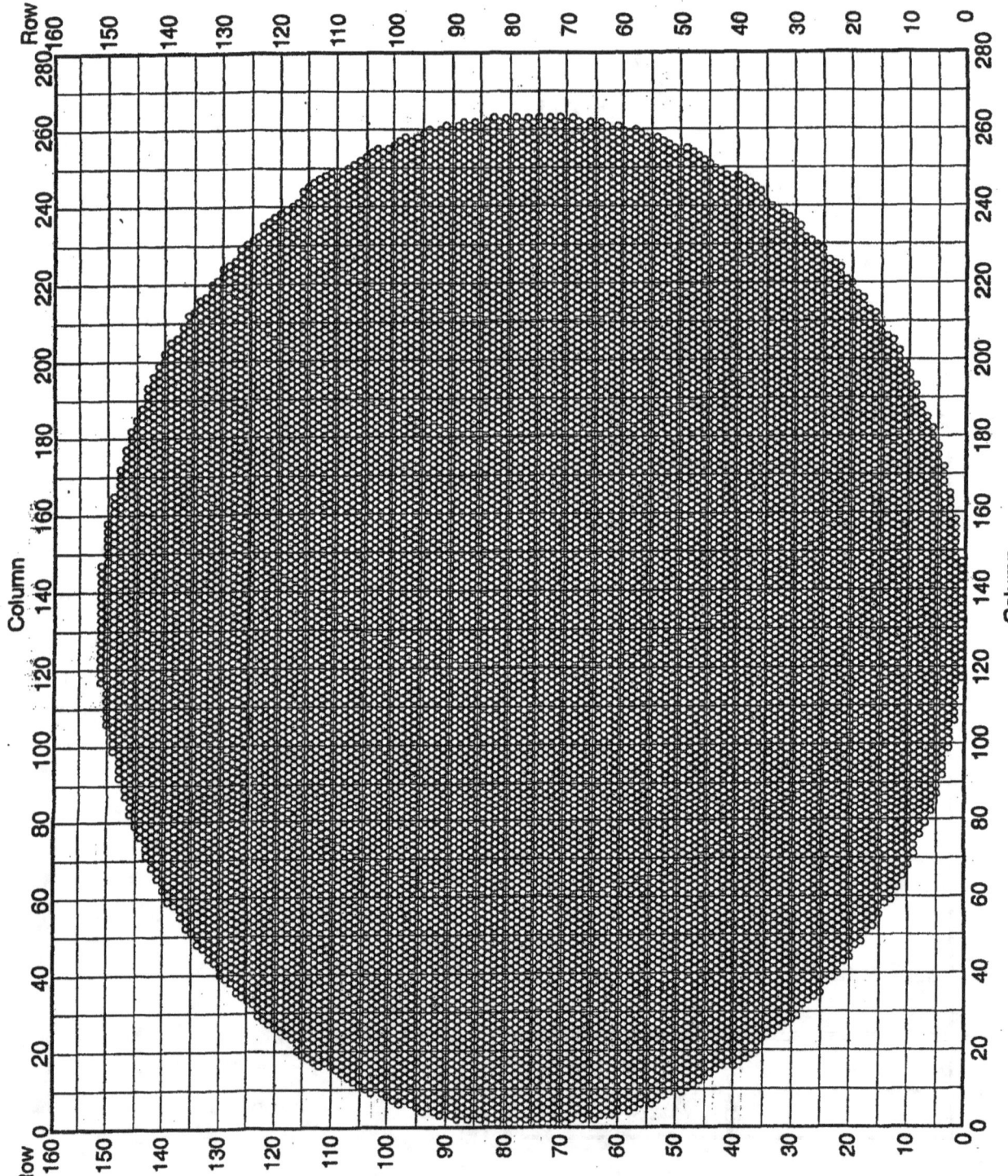

Figure 2-37: Tubesheet Map for Oconee Units 1, 2, and 3

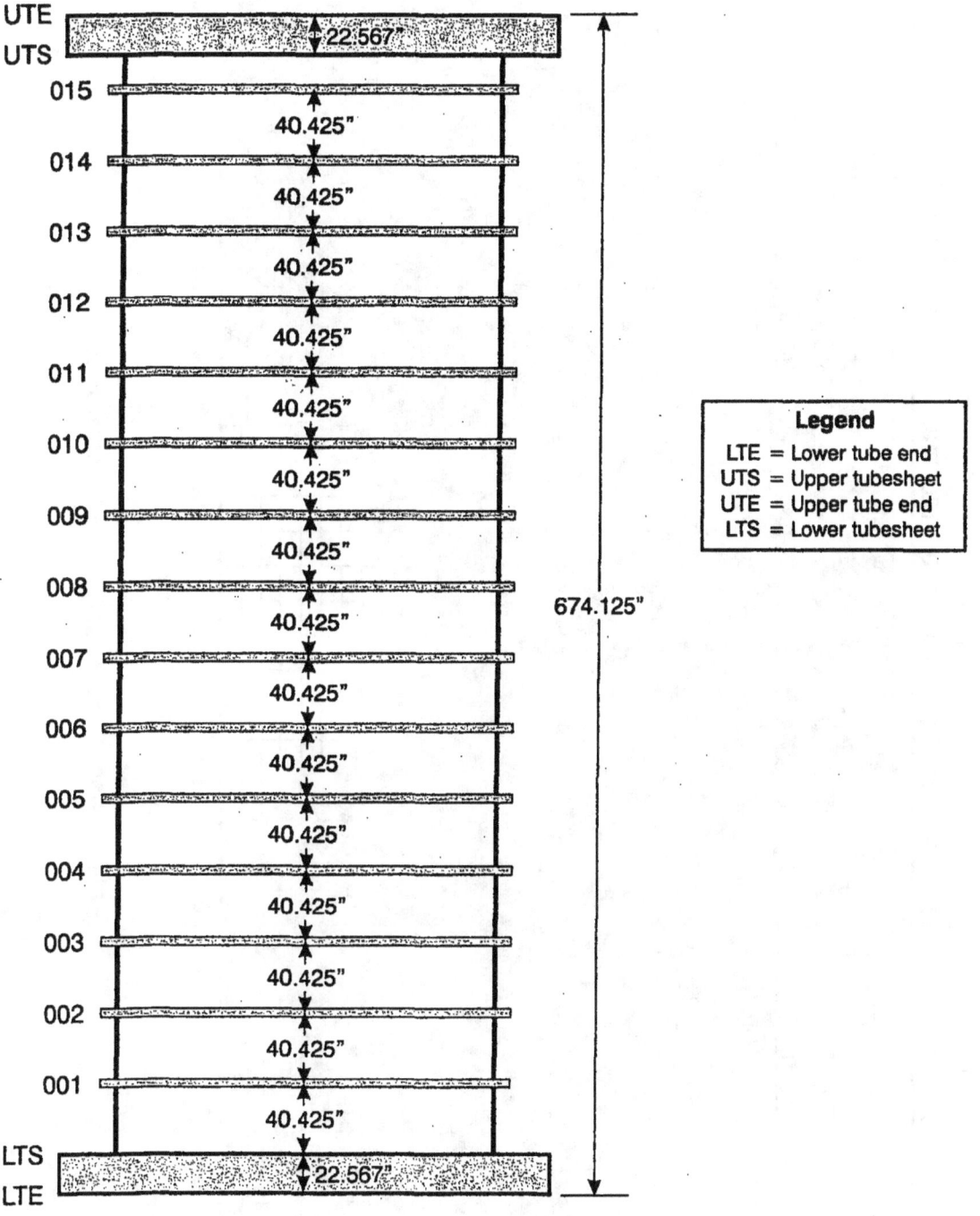

Figure 2-38: Tube Support Naming Convention at Oconee Unit 1, 2, and 3

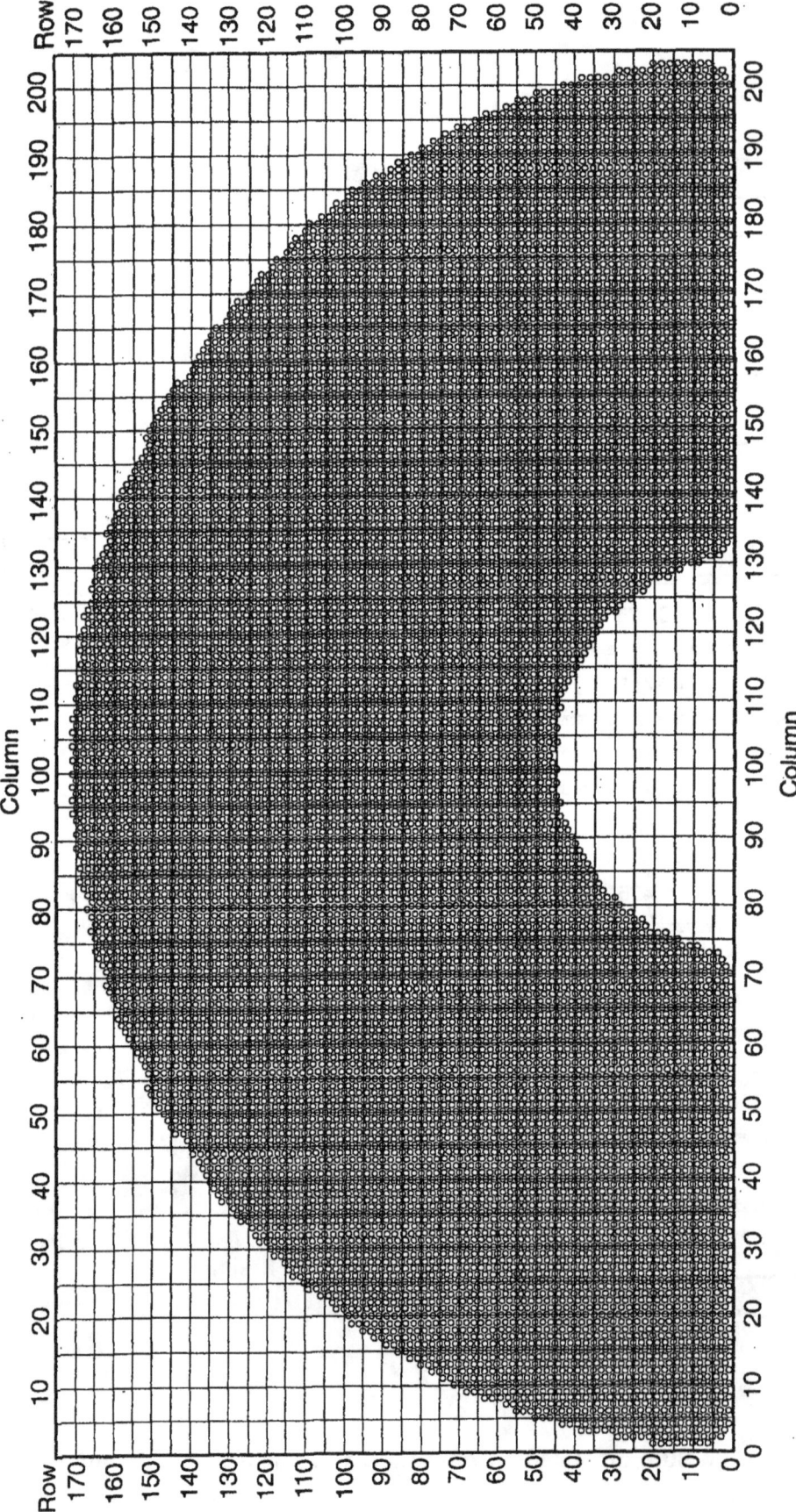

Figure 2-39: Tubesheet Map for Palo Verde Unit 2

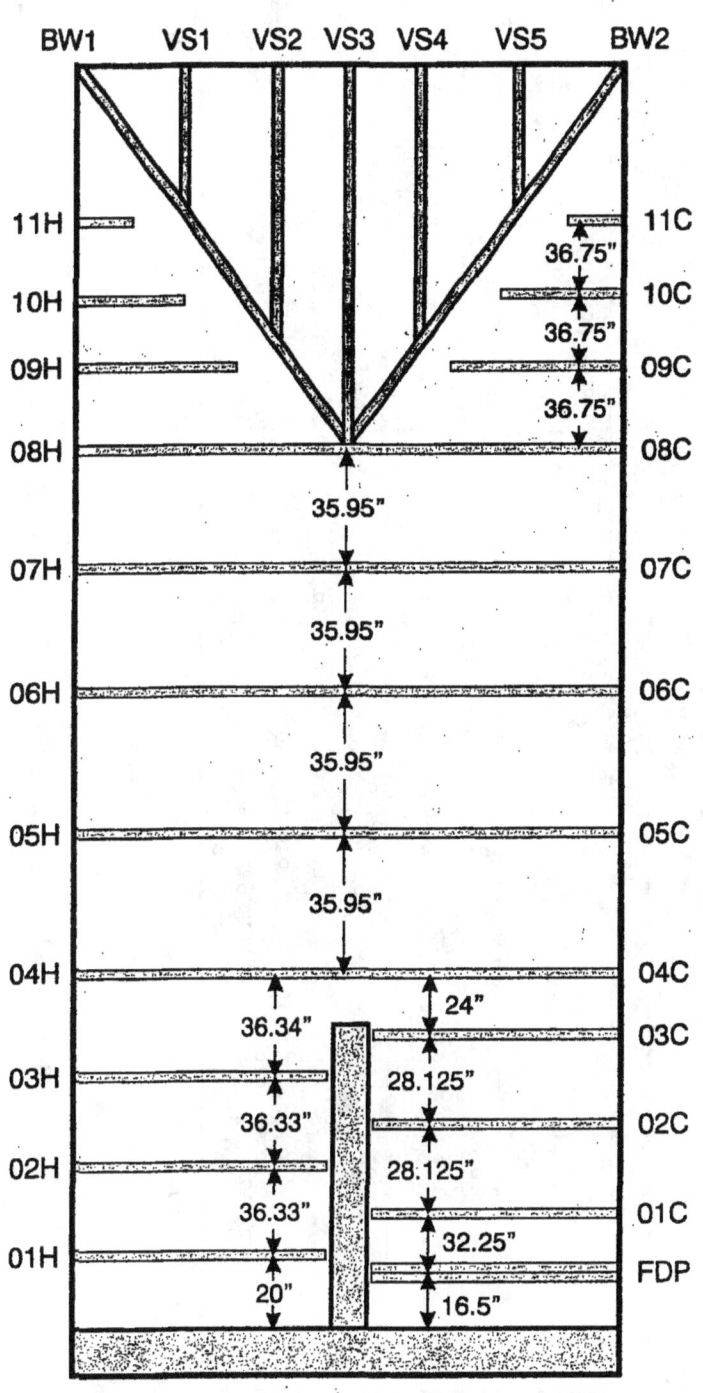

| | | | | | | |
|BW1|VS1|VS2|VS3|VS4|VS5|BW2|

11H — 11C

36.75"

10H — 10C

36.75"

09H — 09C

36.75"

08H — 08C

35.95"

07H — 07C

35.95"

06H — 06C

35.95"

05H — 05C

35.95"

04H — 04C

24"

36.34"

03H — 03C

28.125"

02H — 02C

28.125"

36.33"

01C

36.33"

32.25"

01H — FDP

20"   16.5"

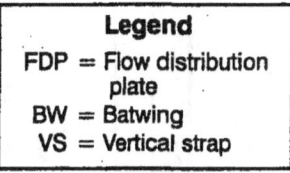

Legend

FDP = Flow distribution plate
BW = Batwing
VS = Vertical strap

Figure 2-40: Tube Support Naming Convention at Palo Verde Unit 2

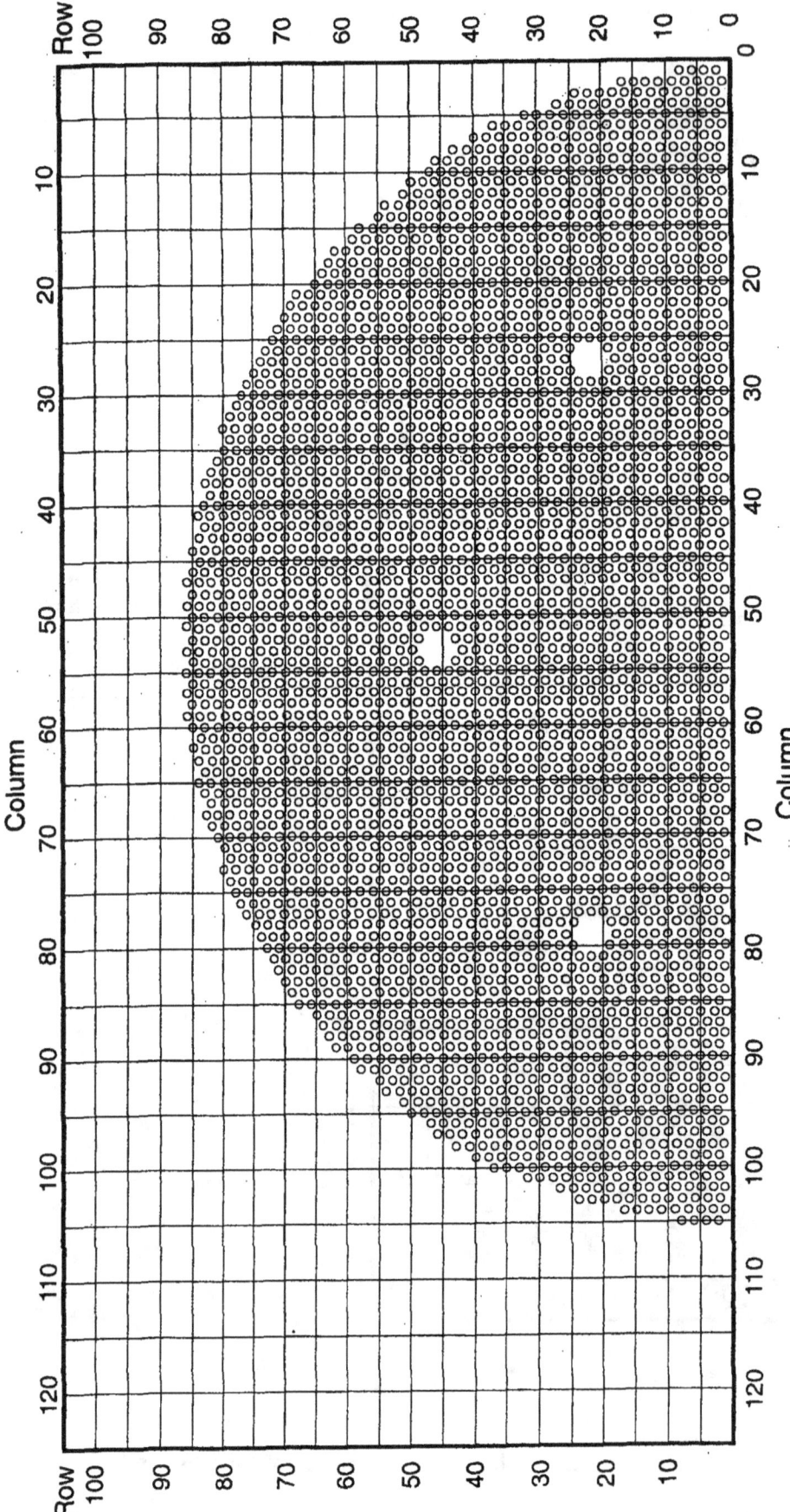

Figure 2-41: Tubesheet Map for Point Beach Unit 2

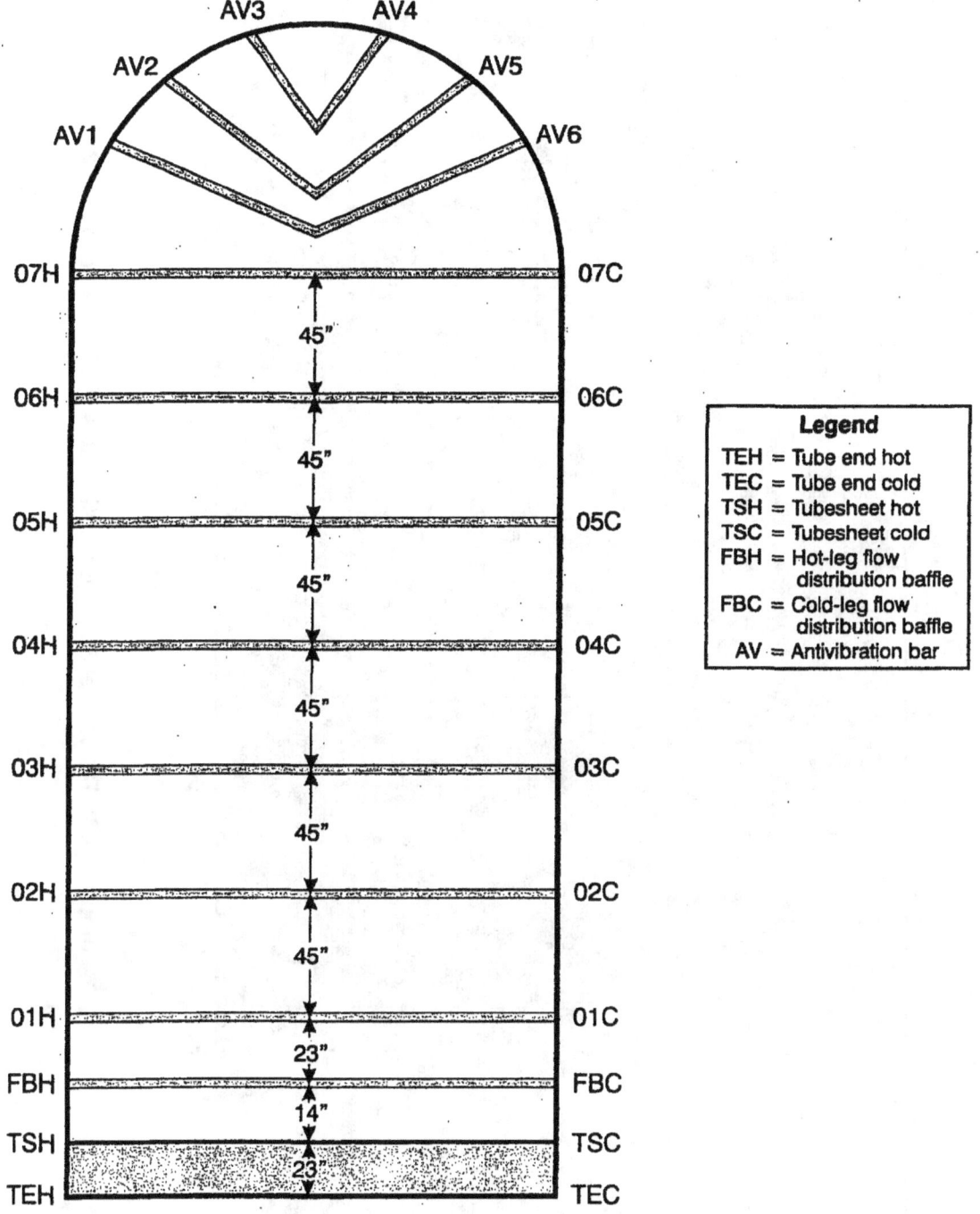

Figure 2-42: Tube Support Naming Convention at Point Beach Unit 2

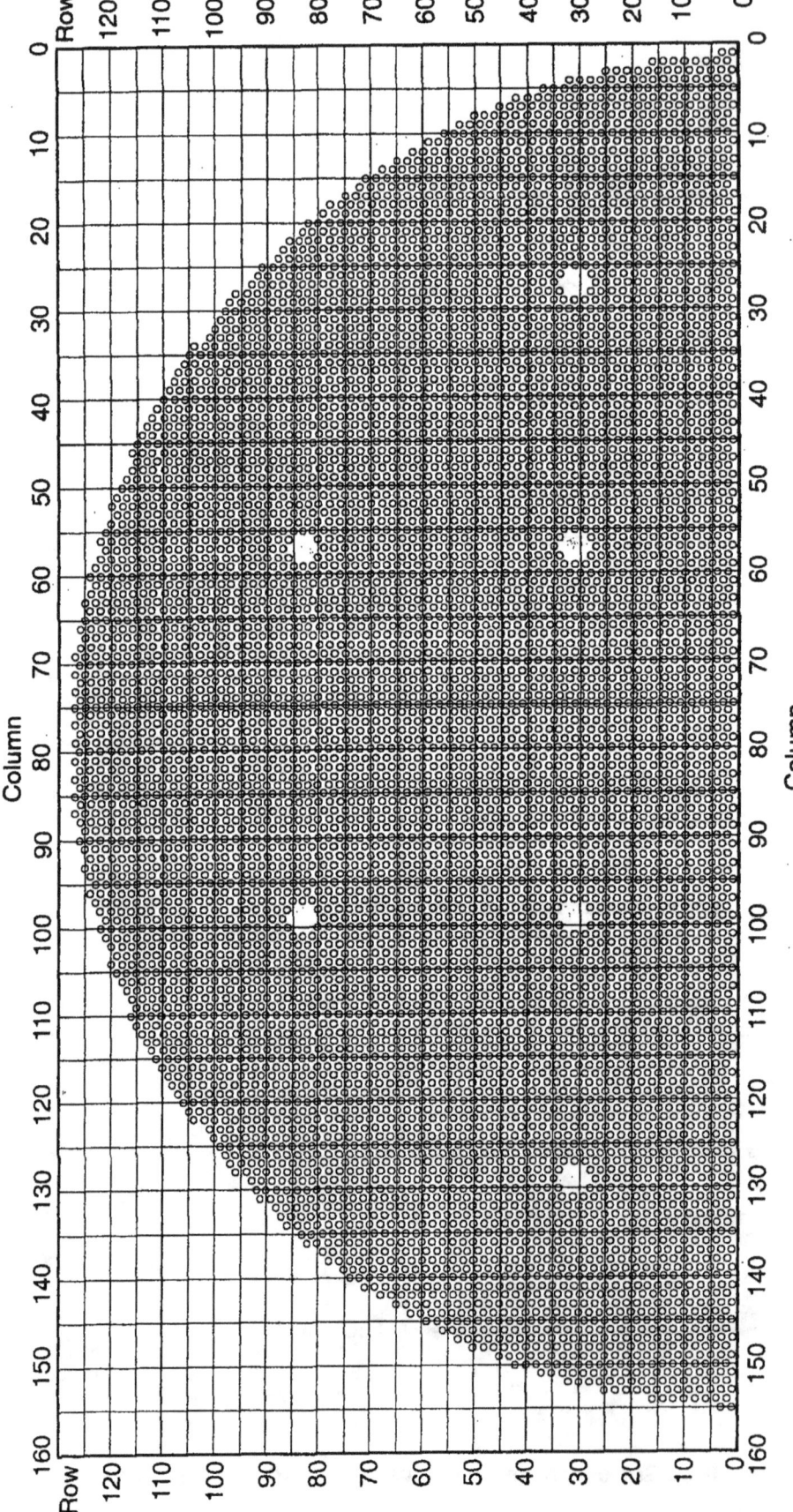

Figure 2-43: Tubesheet Map for South Texas Project Units 1 and 2

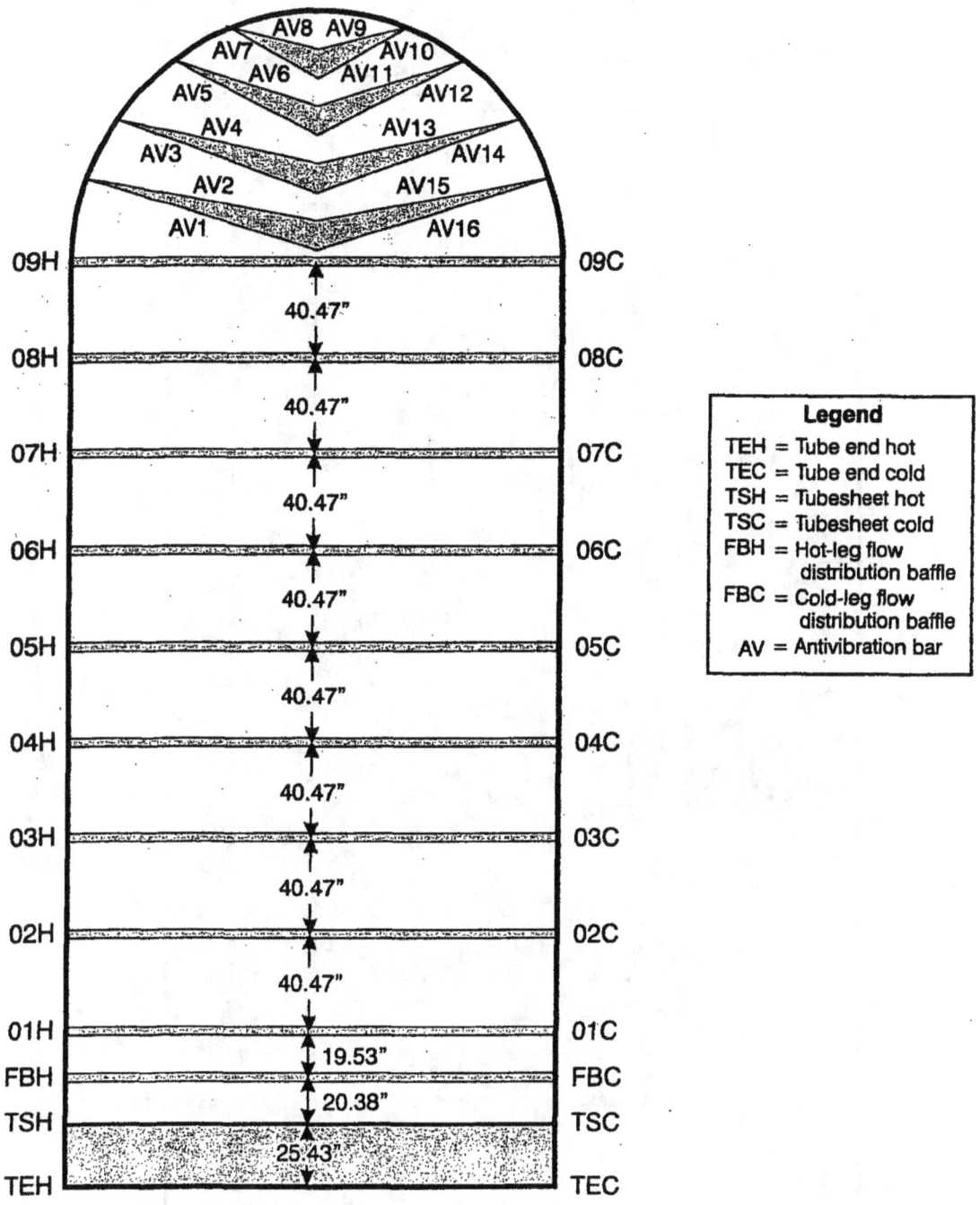

Figure 2-44: Tube Support Naming Convention at South Texas Project Units 1 and 2

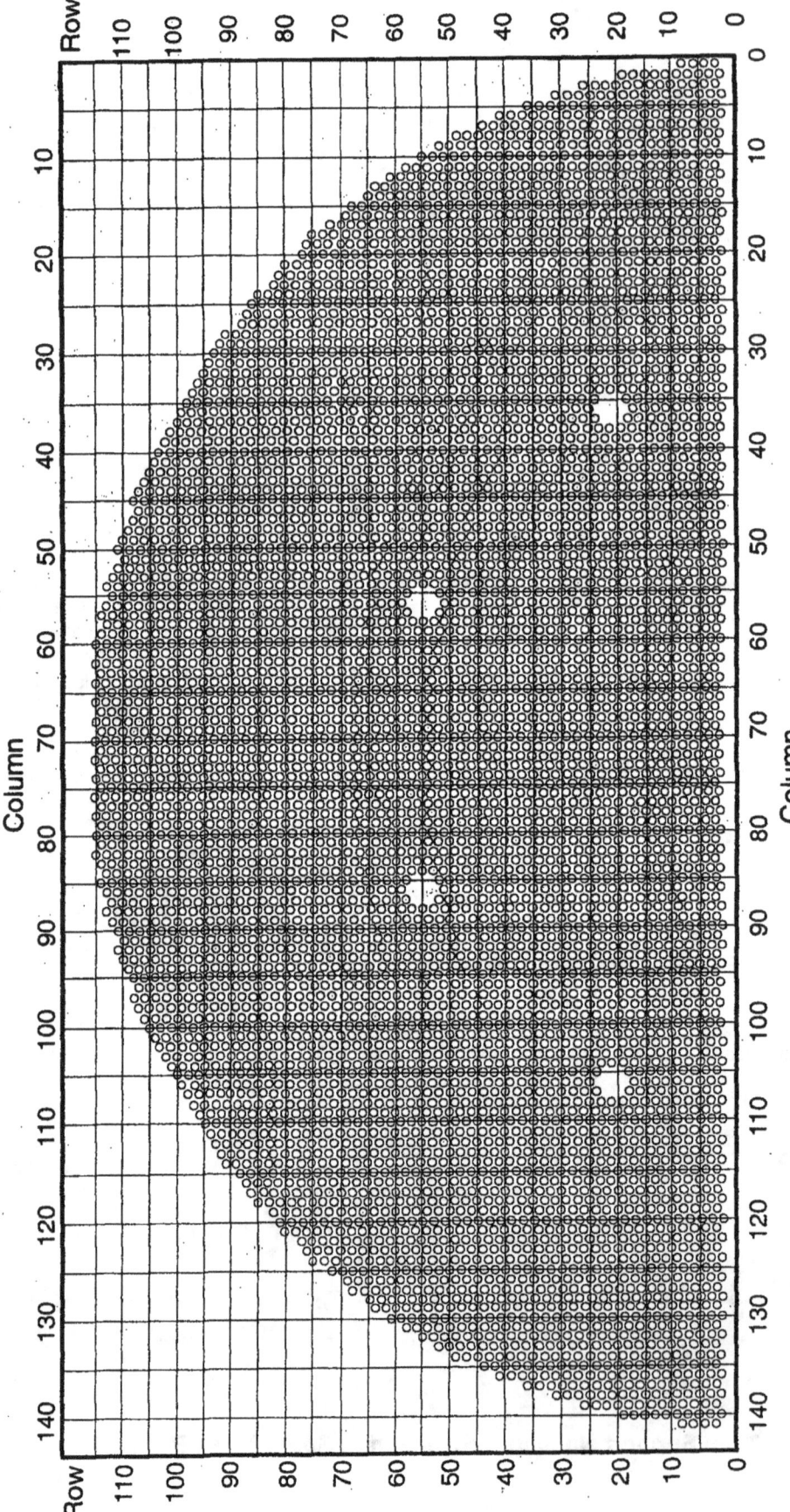

Figure 2-45: Tubesheet Map for Summer

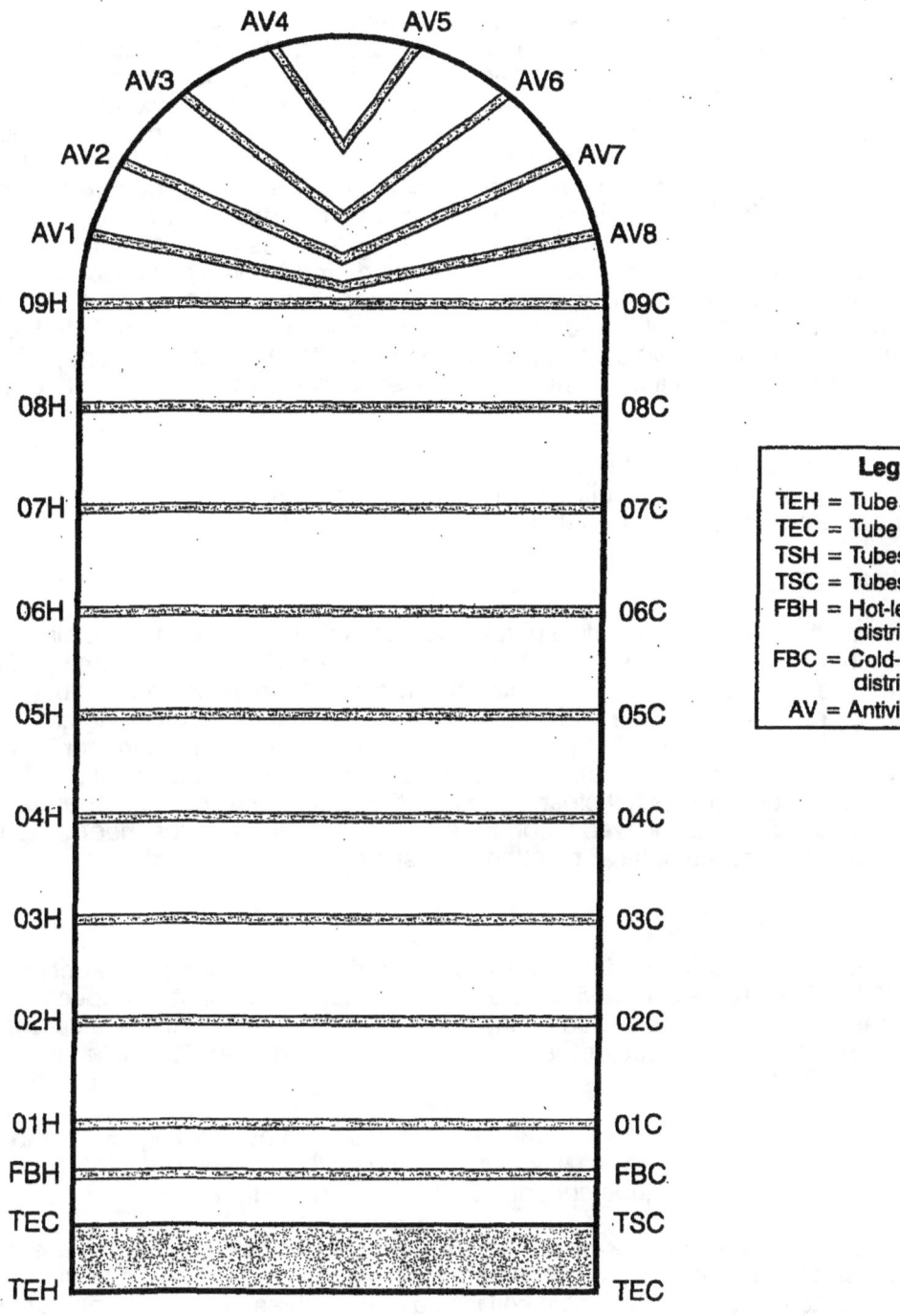

Figure 2-46: Tube Support Naming Convention at Summer

3 THERMALLY TREATED ALLOY 690 STEAM GENERATOR TUBE OPERATING EXPERIENCE

3.1 Data-Gathering Methodology and Introduction

This section summarizes inspection results for units with thermally treated Alloy 690 steam generator tubes through the 2003–2004 timeframe. The summary includes results for all units with Alloy 690 tubes as of December 31, 2004, regardless of whether an inservice inspection of the tubes was performed. The primary source of information was reports provided by licensees to the Nuclear Regulatory Commission (NRC) in accordance with plant technical specifications. These licensee reports typically discuss the number and extent of tubes inspected, the number and location of tubes plugged, and the location and percent of wall thickness penetration for each indication of an imperfection. The level of detail provided in these reports varies from unit to unit and frequently from tube inspection outage to outage. As a result, some units may not have reported all steam generator tube inspection activities during a given inspection outage and/or may not have provided all of their insights in their reports. In addition, the results and interpretation of the results represent the licensee's analysis and evaluation at the time the report was submitted. This may have changed over time. In spite of these limitations, this report provides useful insights into the extent of tube inspections and repairs, and the general conclusions of the report are valid.

Some inspection results were obtained through regional inspection reports, summaries of conference calls with licensees, and meeting summaries. A detailed review of regional inspection reports was not conducted, and that data was not compiled.

3.2 Unit Inspection Results

This section presents the inspection results alphabetically by unit for units with thermally treated Alloy 690 steam generator tubes. For each unit, the discussion provides (1) a summary of the inspections, (2) a table summarizing the full-length bobbin coil examinations and number of tubes plugged during each outage, (3) a table summarizing the reasons for plugging each tube, and (4) a table listing the tubes plugged. In the tables that summarize the reasons for tube plugging, the category "other" captures tubes that were plugged although the specific reason for plugging was not provided or was not clear. Tubes in this category were subdivided based on the location where the degradation was reported (e.g., at the top of the tubesheet). None of these indications were considered to have resulted from stress-corrosion cracking.

3.2.1 Arkansas Nuclear One Unit 2

Tables 3-1, 3-2, and 3-3 summarize the information discussed in this section for Arkansas Nuclear One (ANO) Unit 2. Table 3-1 provides the number of full-length bobbin inspections and the number of tubes plugged and deplugged during each outage for each of the two steam generators. Table 3-2 lists the reasons for the plugging of the tubes. Table 3-3 lists the plugged tubes.

ANO 2 has two recirculating steam generators which were designed by Westinghouse and fabricated at ENSA. The steam generators went into service in 2000 during refueling outage (RFO) 14. The licensee numbers its tube supports as depicted in Figure 2-8.

During the preservice inspection, 100 percent of the tubes in each of the two steam generators were inspected full length with a bobbin coil. In addition, a rotating probe equipped with a +Point™ coil was used to inspect select bobbin coil indications such as manufacturing burnish marks (MBMs) and dents. As a result of the bobbin inspections, 156 MBMs and 1738 dents (dings) were identified in steam generator A, and 314 MBMs and 871 dents (dings) were identified in steam generator B. Approximately 20 percent of the MBMs (33 in steam generator

A and 63 in steam generator B) and 20 percent of the dents (dings) greater than 7 volts (95 in steam generator A and 80 in steam generator B) were inspected with a rotating probe.

Before the steam generators went into service, one tube in steam generator B was plugged. This tube was plugged because of a failure of the mandrel during the hydraulic expansion process (i.e., an equipment failure). Welded Alloy 690 plugs were installed in both ends of this tube.

During RFO 15 in 2002, 100 percent of the tubes in both of the steam generators were inspected full length with a bobbin coil. In addition, a rotating probe equipped with a +Point™ coil was used to inspect the U-bend region of 100 percent of the row 1 and 2 tubes (181 tubes per steam generator) and selected bobbin indications (15 in steam generator A and 11 in steam generator B). In addition to the eddy current inspections, a visual inspection of the annulus and the tube lanes was performed.

As a result of these inspections, 139 MBMs and 1017 dings were identified in steam generator A and 216 MBMs and 596 dings were identified in steam generator B. In addition, one tube was reported as experiencing wear at an antivibration bar (AVB), and two tubes were affected by a loose part. The loose part, which was a small corkscrew-shaped metal shaving in steam generator B approximately 12 inches above the cold-leg tubesheet, was removed. No other loose parts were identified during the inspections. The maximum depth reported for the AVB wear indication was 12 percent through-wall, and the maximum depth reported for the loose part wear indications was 18 percent through-wall. No tubes were plugged and no in-situ pressure tests were performed during RFO 15.

As a result of receiving NRC approval to extend the interval between tube inspections from a maximum of 24 to 40 calendar months, no steam generator tube inspections were performed during RFO 16 in 2003.

On February 26, 2005, a 2-gallon-per-day (gpd) primary-to-secondary leak was observed in steam generator A. The leak rate gradually increased to approximately 35 gpd. On March 8, 2005, the unit was shut down just one week before a scheduled refueling outage.

A visual inspection revealed that the hot-leg portion of the tube in row 70, column 169, in steam generator A was dripping under the static head from the water on the secondary side of the steam generator. Eddy current testing indicated that this tube and two adjacent tubes had indications just above the tubesheet. A secondary-side visual inspection revealed a piece of metal wedged among these tubes. This piece was removed and examined. The piece measured 1.375 inches long, 0.7 inches wide, and 0.25 inches thick and was heavily cold worked, low-carbon steel. The source of this material could not be determined.

The tube in row 70, column 169 was in-situ pressure tested prior to being plugged and stabilized. The tube did not burst at three times the normal operating pressure (adjusted to account for the difference in the operating and the test temperature, this pressure is 4485 pounds per square inch (psi)). Leakage at normal operating pressure (adjusted to account for the difference in the operating and the test temperature) was measured to be approximately 53 gpd which is reasonably consistent with the 37 gpd measured before the unit shut down. Leakage at main steam line break differential pressures measured approximately 64 gpd, which is below the amount of leakage assumed in the design-/licensing- basis analysis for a main steam line break.

In addition to the foreign object near the tube in row 70, column 169, visual examinations on the secondary side of the steam generators led to the identification of additional foreign objects. In steam generator A, a 1-inch long piece of carbon steel, a piece of weld wire, and one machine screw were identified and removed. In steam generator B, a small machine winding was identified and removed. Additional small foreign objects were identified but could not be removed. These pieces were located in relatively low-flow areas of the tube bundle and were determined not to pose a tube integrity concern for the next operating cycle.

3.2.2 Braidwood 1

Tables 3-4, 3-5, and 3-6 summarize the information discussed in this section for Braidwood 1. Table 3-4 provides the number of full-length bobbin inspections and the number of tubes plugged and deplugged during each outage for each of the four steam generators. Table 3-5 lists the reasons why the tubes were plugged. Table 3-6 lists the plugged tubes.

Braidwood 1 has four recirculating steam generators designed and fabricated by Babcock and Wilcox International (BWI). The steam generators were put into service in 1998 during RFO 7. The licensee numbers its tube supports as depicted in Figure 2-10.

During the preservice inspection, 100 percent of the tubes in each of the four steam generators were inspected full length with a bobbin coil. In addition, a rotating probe equipped with a +Point™ coil was used to investigate bobbin coil indications classified as requiring additional diagnostic testing. A total of 54 indications were inspected with a rotating probe. All 54 locations were classified as "no defect found" or as MBMs. Tubesheet profilometry was also performed during the preservice inspection in the tubesheet region on both the hot- and cold-leg. Profilometry is a method for measuring the tube expansion within the tubesheet region and for determining the expansion transition region relative to the top of the secondary face of the tubesheet. No tubes were identified that were expanded beyond the secondary face of the tubesheet (i.e., no overexpanded tubes). A rotating probe equipped with a +Point™ coil was used to inspect the U-bend region of 100 percent of the tubes with the tightest radius (i.e., those in row 3).

Before the steam generators went into service, three tubes were plugged.

During cycle 8, the first cycle of operation with the replacement steam generators, no primary-to-secondary leakage was detected.

During RFO 8 in 2000, 100 percent of the tubes in each of the four steam generators were inspected full length with a bobbin coil. The bobbin coil eddy current data was analyzed with two independent automated data screening algorithms. In addition to the bobbin coil inspections, a rotating probe equipped with a +Point™ coil was used to investigate bobbin coil indications classified as requiring additional diagnostic testing. Seven indications were inspected with a rotating probe due to nonquantifiable signals detected during the bobbin coil inspection. No hot-leg dents or dings greater than 5 volts as measured with a bobbin coil were detected. As a result of these inspections, one tube was plugged. This tube was plugged as a result of wear at a fan bar. The maximum depth reported for this fan bar wear indication was less than 10 percent through-wall. A rotating probe examination was performed of this indication.

The tube-to-tube contact condition discussed in Sections 2.3.2 and 2.4.2 was monitored during RFO 8. A total of 85 tubes were found to be in contact (close proximity). These tubes showed no indications of degradation. The majority of the tubes in contact were located at the outer periphery of the tube bundle or within one tube of the outermost tube. A total of 508 tubes were identified as being in contact during the preservice inspection while the steam generators were in the horizontal position.

Sludge lancing was performed on the secondary side of the tubesheet in all four steam generators during RFO 8. Following sludge lancing, a video inspection of the secondary side of the tubesheet was performed to identify foreign objects. Minor debris was identified during the inspection; however, no foreign objects detrimental to the steam generator tubing were identified. All foreign objects identified (3 objects) were removed from the steam generators.

As a result of receiving NRC approval to extend the interval between tube inspections from a maximum of 24 to 40 calendar months, no steam generator tube inspections were performed during RFO 9 in 2001.

During RFO 10 in 2003, 100 percent of the tubes in steam generator A, and 54 percent of the tubes in steam generators B, C, and D were inspected full length with a bobbin coil. The bobbin coil eddy current data was analyzed with two independent automated data screening algorithms. In addition, a rotating probe equipped with a +Point™ coil was used to investigate bobbin coil indications classified as requiring additional diagnostic testing and to inspect 100 percent of the dents/dings greater than 5 volts in the hot-leg.

The bobbin inspections performed during RFO 10 included 100 percent of the peripheral tubes (two tubes in from the periphery) in all four steam generators. The intent of these inspections was to identify the potential presence of foreign objects and tube wear resulting from foreign objects. (Twenty-eight tubes were excluded from this inspection scope because of interference with the steam generator tube inspection equipment.) In addition to these inspections, 100 percent of the steam generator tube plugs (both previously installed and newly installed plugs) were inspected visually from the primary side, and foreign object search and retrieval (FOSAR) was performed in the periphery and free lanes of all four steam generators. In-bundle visual inspections were conducted in two steam generators.

As a result of these inspections, 21 tubes were plugged either because of wear associated with a loose part or because a loose part was in a location where it could not be retrieved. A total of 15 tubes were plugged because of wear associated with three loose parts. An additional six tubes, which did not show any signs of wear, were also plugged in order to surround foreign objects that could not be successfully removed from the steam generator. Of the 15 tubes with wear from foreign objects, the largest was 48 percent through-wall. Of the 15 tubes plugged with wear indications, eight were stabilized. Of the 21 tubes plugged either because of wear associated with a loose part or because of bound locations where a loose part could not be retrieved, 14 were stabilized.

The inspections identified seven tubes with indications of wear at the fan bar locations. The largest indication was 10 percent through-wall. The inspections detected five tubes with indications of wear at lattice grid locations. The largest indication was 12 percent through-wall. The wear at these locations (fan bar and lattice grid) is considered typical, as discussed in Section 2.3.2. The tubes with these indications remained in service. No hot-leg dents or dings greater than 5 volts, as measured with the bobbin coil, were detected. No anomalies were found during the visual inspection of the plugs.

Based on the FOSAR inspection, one foreign object was identified near the top of the tubesheet on the periphery of the cold leg side of steam generator A. It was determined to be a piece of flexitallic gasket and could not be retrieved because it was fixed in place between six tubes. The gasket was 2.3 inches long, 0.08 inches wide and 0.031 inches deep. The object was not detected in the eddy current data from the bobbin probe inspection. The lack of an eddy current response was attributed to the part being a small piece of flexitallic gasket which is stainless steel.

An inspection was performed with a rotating probe containing a +Point™ coil on 37 tubes surrounding the object. The tubes were inspected from 3 inches above to 3 inches below the top of the tubesheet. Seven tubes were determined to contain degradation due to wear, which ranged from 5 to 48 percent through-wall. Only the two deepest flaws (48 percent and 25 percent through-wall) were detected with the bobbin probe. A review of the earlier inspection data (bobbin coil data) for the location with the 48 percent through-wall indication indicated that a flaw was not detected at that time. All tubes with wear (regardless of depth) were plugged, and all tubes in contact with the object were plugged and stabilized. In steam generator A, eight tubes were plugged, of which six were stabilized. Because of the identification of one tube which exceeded the technical specification repair criteria of 40 percent through-wall, the scope of the bobbin probe inspections was expanded to include 100 percent of the tubes. No indications of possible loose parts (PLPs) or tube wear were detected in the additional tubes inspected.

In steam generator B, the FOSAR inspection detected two foreign objects. One foreign object identified near the top of the tubesheet on the periphery of the hot-leg side was determined to be a piece of flexitallic gasket, which could not be retrieved because it was fixed in place. The gasket was 1.6 inches long, 0.12 inches wide, and 0.12 inches deep. This object was not detected in the eddy current data from the bobbin probe inspection. The lack of an eddy current response was attributed to the part being a small piece of stainless steel flexitallic gasket. An inspection with a rotating probe containing a +Point™ coil was performed on 37 tubes surrounding the object. The tubes were inspected from 3 inches above to 3 inches below the top of the secondary face of the tubesheet. Three tubes were identified with degradation because of wear, which ranged from 7 to 17 percent through-wall. None of these flaws were detected with the bobbin probe. All tubes with wear (regardless of depth) were plugged, and all tubes in contact with the object were plugged and stabilized. Five tubes were plugged and stabilized in steam generator B because of the foreign object.

The second foreign object in steam generator B was identified near the top of the tubesheet on the periphery of the hot-leg side. It was determined to be a piece of weld spatter and was removed. The object was oval and was 2.0 inches long, 1.25 inches wide, and 0.062 inches deep. An inspection with a rotating probe containing a +Point™ coil was performed on 36 tubes surrounding the object. The tubes were inspected from 3 inches above to 3 inches below the top of the secondary face of the tubesheet. Five tubes were identified with degradation resulting from wear. None of these indications were deeper than 13 percent through-wall. One of these flaws, 11 percent through-wall, was detected with the bobbin probe. All tubes with wear were plugged (regardless of depth), and five tubes were plugged in steam generator B because of this foreign object.

In steam generator C, several foreign objects were identified. One foreign object identified near the top of the tubesheet on the periphery of the hot-leg side was determined to be a piece of flexitallic gasket which could not be retrieved. The foreign object was moved further into the tube bundle where the flow rate is low enough so that it would not be expected to cause wear. The gasket was 0.75 inches long, 0.1 inches wide, and 0.1 inches deep. A rotating probe containing a +Point™ coil was used to inspect 18 tubes surrounding the original location of the object. The tubes were inspected from 3 inches above to 3 inches below the top of the secondary face of the tubesheet. None of the tubes had detectable degradation. No tubes were plugged as a result of this foreign object. A second foreign object identified in-bundle near the top of the tubesheet on the hot-leg side of steam generator C was determined to be a piece of flexitallic gasket which was retrieved. The gasket was 1.25 inches long and 0.063 inches deep. Another foreign object was introduced into the steam generator during the current refueling outage when a tooling collet fell off the loose part retrieval equipment and became lodged between three tubes in-bundle on the hot-leg side of the steam generator. All three tubes surrounding the object were plugged and stabilized.

No foreign objects were identified in steam generator D.

A number of pieces of flexitallic gaskets were identified in the steam generators during RFO 10 and an investigation into the cause of this was performed. Flexitallic gaskets have stainless steel metal faces which are spot-welded to the gasket. In some cases, the stainless steel face on the gasket extrudes into the flow stream and pieces break off and become foreign objects or loose parts, possibly as a result of misapplication of the gasket. Flexitallic gaskets are used upstream and downstream from the feed pump strainers (which are upstream of the steam generators).

The majority of the objects identified in the secondary side of the steam generator during the RFO 10 inspections are believed to have resulted from a collapsed feedwater pump suction strainer. The collapsed suction strainer was identified during RFO 10 and repaired prior to returning the associated feedwater pump to service. The remaining feedwater suction strainers were inspected during the outage and were intact. A review of online feedwater iron transport data over the previous two cycles of operation indicated a spike in iron concentrations in August 2001. This is most likely when the suction strainer collapsed. The spike in August 2001 occurred late in cycle 9.

To address the potential for flexitallic gaskets downstream of the feedwater pump suction strainer to enter the steam generator, additional corrective actions considered were (1) identifying an acceptable alternative gasket, (2) performing an inventory of all gaskets in the feedwater system, and (3) investigating the need to revise work packages related to the installation of these gaskets.

As a result of the loose parts detected, an evaluation of loose parts at the top of the tubesheet region was conducted. This evaluation indicated that the outermost five tubes on the bundle periphery were most susceptible to foreign object wear. Inboard of the first five tubes, the flow velocity is reduced which lowers the potential for tube wear. The evaluation concluded that for an object similar to the bounding object identified in the 2003 inspection and assuming that the object was located in the limiting high-flow region of the periphery of the tube bundle, the expected wear morphology would not be projected to exceed the steam generator tube structural limit for two cycles of operation.

During RFO 10, all tubes previously identified as being in contact (close proximity) in RFO 8 were reinspected, and none showed any signs of wear. Of the 85 tubes identified to be in contact (close proximity) in RFO 8, only 67 were identified as still being in contact (close proximity) during RFO 10. However, 65 additional tubes were identified as being in contact (close proximity) during RFO 10.

During RFO 11 in 2004, 100 percent of the tubes in steam generator B were inspected full length with a bobbin coil. In addition, a rotating probe equipped with a +Point™ coil was used to investigate bobbin coil indications classified as requiring additional diagnostic testing and to inspect 25 percent of the dents/dings greater than 5 volts in the hot-leg in steam generator B (no tubes met this criterion). All 36 tubes previously identified to be in close proximity (tube-to-tube contact) in steam generator B were inspected. In response to the findings in steam generator B, the straight (i.e., non-bent) portion of the hot-leg of approximately 22 percent of the tubes in steam generators A, C, and D were inspected with a bobbin coil. In addition, the straight (i.e., non-bent) portion of the cold-leg of approximately 16 percent of the tubes (which includes all peripheral tubes) in steam generators A, C, and D were inspected with a bobbin coil. In some instances, the tubing in steam generators A, C, and D was inspected full length. Analysis of the bobbin coil eddy current data entailed the use of two independent automated data screening algorithms. No secondary-side inspections or tubesheet sludge lancing were performed. Steam generator tube plugs (both previously installed and newly installed plugs) were inspected visually from the primary side during the outage.

As a result of the inspections in steam generator B, four indications of wear at the fan bars were detected and two indications of wear at a lattice grid was detected. The depth of the fan bar wear indications measured 3 to 7 percent through-wall, and the depth of the lattice grid wear indication measured 3 to 10 percent through-wall. In addition, two volumetric indications were detected. One of the volumetric indications, which measured 42 percent through-wall, was located at the sixth cold-leg lattice grid tube support in the tube at row 1, column 116. This tube borders the tube-free lane. The second volumetric indication measured 27 percent through-wall and was located at the sixth hot-leg lattice grid tube support in the tube at row 48, column 89. This tube is in the interior of the tube bundle. These two indications had depths less than the structural limit of 64.7 percent through-wall; therefore, these two tubes had adequate structural and leakage integrity.

The volumetric indications in these two tubes were attributed to wear from loose parts rather than to indications of lattice grid wear. The indications were not located at the tube-to-lattice grid contact points but in the high-flow region of the tube bundle. These areas are currently inaccessible for visual inspection. The bobbin and rotating probe did not provide any indication that the loose part was still present. The indications were not present in the 2003 inspection data.

Based on these results, the scope of the examination was expanded to include selected tubes in steam generators A, C, and D. The inspections in these generators focused on the two regions of high flow in the BWI steam generators as determined using the ATHOS computer code (a thermal hydraulic computer code). These two regions of high flow included tubes in the periphery of the tube bundle on both the hot- and cold-leg side of the steam generator and tubes in the interior of the tube bundle on the hot-leg side. The inspections in steam generators A, C, and D included 22 percent of the tubes in the steam generator. The 22-percent sample included 100 percent of the tubes in the periphery of the tube bundle and 20 percent of the tubes in the high-flow region of the interior of the tube bundle.

As a result of the inspections in steam generator A, no wear due to foreign objects was detected; however, two indications of wear at lattice grid supports were detected. These indications ranged from 3 to 6 percent through-wall. The inspections in steam generator C detected two indications, both attributable to wear from foreign objects. One of the indications, which measured 17 percent through-wall, was located at the first hot-leg lattice grid tube support in the tube at row 66, column 13, a peripheral tube. This indication was not present in the 2003 data. An inspection of the adjacent tubes was performed, and these tubes did not have any indications nor were any foreign objects detected. This tube was plugged and stabilized. This tube was plugged and stabilized. The second indication in steam generator C was located in an interior tube at row 69, column 50. The indication measured 9 percent through-wall and was located at the fourth hot-leg lattice grid tube support. This indication was present in the 2003 inspection data and was classified as lattice grid wear. The signal did not change from 2003 to 2004. This tube was plugged in 2004. Bobbin and rotating probe examinations of adjacent tubes did not result in identifying any additional indications or foreign objects. This indication was in the high-flow region of the tube bundle.

As a result of the inspections in steam generator D, no wear due to foreign objects was found in any of the peripheral tubes inspected; however, one indication attributed to wear from a foreign object was found in an interior tube. This indication, which measured 25 percent through-wall, was located at the second hot-leg lattice grid tube support in the tube at row 70, column 41. The indication at this location in the high-flow region of the tube bundle was present during the 2003 inspection and was attributed to lattice grid wear. The signal did not change between 2003 and 2004. Bobbin and rotating probe examinations of adjacent tubes did not result in identifying any additional indications or foreign objects. This tube was plugged. In addition to this indication, one indication of wear at a fan bar was detected. This indication measured 7 percent through-wall.

As a result of the 2004 inspections, a total of five tubes were plugged (two in steam generator B, two in steam generator C, and one in steam generator D). The five tubes were removed from service because of wear associated with secondary-side foreign objects located at the lattice grid supports. The foreign objects causing the wear could be spiral-wound gasket material (e.g., the stainless steel windings of flexitallic gaskets). This material may also subsequently degrade (e.g., by fatigue) with time such that it could break up and move to another location in the steam generator and not be present during the steam generator tube inspections. No indications of possible loose part signals were detected during the eddy current inspections; however, one possible loose part indication was initially identified with a bobbin coil, but based on rotating probe examinations was reclassified as sludge. No anomalies were found during the visual inspection of the plugs.

During RFO 11, tube-to-tube proximity was monitored as part of the bobbin coil inspection program in steam generator B. No tube degradation was identified in any of the tubes considered to be in close proximity. Of the 36 tubes identified as being in close proximity in steam generator B during RFO 10, 34 were still in close proximity and 2 were no longer in close proximity in RFO 11. In addition, 5 new tubes were identified as being in close proximity (resulting in a total of 39 tubes being in close proximity).

3.2.3 Byron 1

Tables 3-7, 3-8, and 3-9 summarize the information discussed in this section for Byron 1. Table 3-7 provides the number of full-length bobbin inspections and the number of tubes plugged and deplugged during each outage for each of the four steam generators. Table 3-8 lists the reasons why the tubes were plugged. Table 3-9 lists the plugged tubes.

Byron 1 has four recirculating steam generators designed and fabricated by BWI. The steam generators were put into service in 1998 during RFO 8. The licensee numbers its tube supports as depicted in Figure 2-12.

Before the steam generators went into service, one tube was plugged. This tube was plugged with welded plugs at each end.

During cycle 9, the first cycle of operation with the replacement steam generators, no primary-to-secondary leakage was detected.

During RFO 9 in 1999, 100 percent of the tubes in steam generators B, C, and D were inspected full length with a bobbin coil. In addition, a rotating probe equipped with a +Point™ coil was used to investigate bobbin coil indications classified as requiring additional diagnostic testing. A total of 7 indications were inspected with a rotating probe. No eddy current inspections were performed in steam generator A. In addition to these inspections, a visual inspection was performed on (1) all installed plugs, (2) the secondary side of the tubesheet in each of the four steam generators, and (3) the upper bundle (including the top lattice grid) and feedring region (including the J-tube region) of steam generator D.

As a result of the eddy current and visual inspections, no degradation was found in the tubing or internal components. As a result, no tubes were plugged.

The top of the tubesheet visual inspections encompassed the outer tube annulus, the tube-free divider lane, and two inner bundle passes on each of the hot- and cold-leg sides of the steam generator. The top lattice grid/upper bundle inspection encompassed inspection of the lattice grid, lattice grid support rim, lattice grid acorn nuts, acorn nut tack welds, U-bend region tube surfaces, and the general condition of the area surrounding the handhole where access was gained. The feedring and J-tube inspection included inspection of the feedwater header outside and inside surfaces, two J-tubes, J-tube-to-header welds, and the general condition of the area surrounding the handhole where access was gained. No observations of degradation were found during the visual inspections described above.

During cycle 10, no primary-to-secondary leakage was detected.

During RFO 10 in 2000, 100 percent of the tubes in steam generator A were inspected full length with a bobbin coil. The bobbin coil eddy current data was analyzed with two independent automated data screening algorithms. In addition to the bobbin coil inspections, a rotating probe equipped with a +Point™ coil was used to investigate bobbin coil indications classified as requiring additional diagnostic testing. A total of four indications were inspected with a rotating probe. No eddy current inspections were performed in steam generators B, C, or D.

As a result of these inspections, no degradation was found and no tubes were plugged. No hot-leg dents or dings greater than 5 volts, as measured with the bobbin coil, were detected.

During RFO 11 in 2002, 54 percent of the tubes in each of the four steam generators were inspected full length with a bobbin coil. This included inspection of 100 percent of the peripheral tubes (except those under the eddy current manipulator base plate), 50 percent of the interior tubes, and all tubes identified as being in contact based on previous inspection results (seven tubes). The bobbin coil eddy current data was analyzed with two independent automated data-screening algorithms. In addition to the bobbin coil inspections, a rotating probe equipped with a +Point™ coil was used to inspect the hot-leg expansion transition region (from 3 inches above to 3 inches below the top of the secondary face of the tubesheet) of 25 percent of the tubes in each of the four steam generators; to inspect 25 percent of the dents and dings greater than 5 volts in the hot-leg (no tubes met this criterion); and to investigate bobbin coil indications classified as requiring additional diagnostic testing. Approximately 90 locations with nonquantifiable indications were inspected with rotating probes. Approximately 85 locations were inspected with a bobbin probe to bound locations of foreign object signals and fan bar wear indications. In addition, a visual inspection was performed on all plugs.

As a result of these inspections, no tubes were plugged. Twelve tubes were found that contained indications of fan bar wear. The depths of these indications ranged from 2 to 8 percent through-wall. No other tubes exhibited degradation. No hot-leg dents or dings greater than 5 volts by the bobbin coil technique were detected. No evidence of degradation or leakage was found during the visual inspection of the plugs.

During RFO 12 in 2003, no steam generator tube inspections were performed.

During RFO 13 in 2005, 57 of the 67 row 1 tubes on the hot-leg side of the bundle in steam generator B were found to be disengaged from the collector bar. The 10 row 1 tubes on the hot-leg side that remained engaged, or partially engaged, with the collector bar were in close proximity to one another and were located near the periphery of the tube bundle. The collector bar engaged all row 1 tubes on the cold-leg side. There were no indications of tube wear in the affected row 1 tubes. All other steam generators had fully engaged collector bars, as confirmed by eddy current.

Eddy current analysis software designed to look for support structures as landmarks identified the absence of the collector bar from the row 1 hot-leg tubes. When the software identified the absence of the collector bar, past inspection data were reviewed, confirming that this condition had existed since the preservice inspection. Analysts had not noticed the condition in past inspections since they were looking only for changes in landmarks.

Since the "as found" condition was different than the condition analyzed during the design of the steam generator, an analysis was performed to verify that the Row 1 tubes will remain stable despite the absence of the collector bar (i.e., the tubes will remain fluid elastically stable, and there is no risk of high-cycle fatigue). This analysis concluded that no fluid elastic instability issue exists for the as found condition and that this condition is acceptable. Steam generator operating experience supports this conclusion since this condition existed for five cycles with no indication of wear in the affected tubes at this location or at the lattice grid supports. The greatest increase in unsupported tube length is in the row 1 tubes that are no longer in contact with the collector bar. An analysis by BWI showed that the greatest movement of the fan bars from their normal position is equal to one-half of a tube pitch (0.456 inches). A change of this magnitude does not significantly change the distance between supports for tubes in rows 2 and higher. For this reason, the vibration analysis performed for the row 1 tubes bounds all other tubes.

Since the movement of the collector bar in the 57 tubes could have affected the positioning of the collector bars and fan bars, the location of the collector bar in rows 116–119 on the cold-leg side of the steam generator was confirmed. These locations (i.e., rows 116–119) are the most susceptible to flow-induced vibration fatigue.

The condition of the disengaged collector bar is believed to have originated during fabrication of the steam generator. In order to correct the potential for tubes to come into contact (or close proximity) with each other, the fan bar support mechanisms (J-tabs) were repositioned during the fabrication process. The curvature of the tube bundle is believed to have interacted with the J-tabs in a way that caused the collector bar to shift out of contact with the row 1 tubes on the hot-leg side. The 10 row 1 tubes that remain in contact with the collector bar are located at the periphery of the tube bundle where the outermost tubes have a smaller U-bend radius. The smaller radius tubes are believed to have resulted in less rotation of the fan bar assembly during the J-tab repositioning process and, therefore, allowed the collector bar to remain engaged at those locations. Analysis predicts no additional stress has been added to the fan bar assemblies as a result of the shift in position. The assembly rotated approximately one-half of a degree. The fan bar assembly can tolerate movements of this magnitude without adding strain to fan bars or the tubes.

BWI stated that the J-tab repositioning procedure has not been performed at any other plant. This procedure was performed for all four steam generators at Byron 1; however, shifting of the collector bar was observed only in steam generator B.

3.2.4 Calvert Cliffs 1

Tables 3-10, 3-11, and 3-12 summarize the information discussed in this section for Calvert Cliffs Unit 1. Table 3-10 provides the number of full-length bobbin inspections and the number of tubes plugged and deplugged during each outage for each of the two steam generators. Table 3-11 lists the reasons why the tubes were plugged. Table 3-12 lists the plugged tubes.

Calvert Cliffs 1 has two recirculating steam generators which were designed and fabricated by BWI. The steam generators were put into service in 2002 during RFO 15. The licensee numbers its tube supports as depicted in Figure 2-14.

Before the steam generators went into service, no tubes were plugged in either steam generator. Several dings were detected during the preservice inspection (14 indications in 14 tubes in steam generator A and 16 indications in 15 tubes in steam generator B). Most of the dings were inspected with a rotating probe during the preservice inspection and no anomalous signals were noted. No tubes were overexpanded in the tubesheet region in either of the steam generators.

During the first cycle of operation with the replacement steam generators, no primary-to-secondary leakage was detected.

During RFO 16 in 2004, 100 percent of the tubes in both of the steam generators were inspected full length with a bobbin coil. In addition, a rotating probe equipped with a +Point™ coil was used to inspect all bobbin nonquantifiable indications (approximately 98 locations in steam generator A and 101 locations in steam generator B). As a result of these inspections, no tubes were plugged. In addition, no loose parts or any damage from loose parts were identified during the inspections, and no tubes were identified as being in close proximity to other tubes or have come in contact with adjacent tubes as a result of the first cycle of operation. No new dents were identified during this outage, and the existing ding signals did not exhibit any change since the preservice inspection.

The number of indications of fan bar wear detected during the first inservice inspection was more than that typically identified at other similarly designed and operated units; however, the locations and size of the wear are consistent with those found at other similar units. The wear indications are considered typical fan bar wear (i.e., caused by the thermal hydraulic conditions and tube-to-support clearances which can vary because of manufacturing tolerances). All fan bar wear indications were inspected with a rotating probe during RFO 16.

3.2.5 Calvert Cliffs 2

Tables 3-13, 3-14, and 3-15 summarize the information discussed in this section for Calvert Cliffs Unit 2. Table 3-13 provides the number of full-length bobbin inspections and the number of tubes plugged and deplugged during each outage for each of the two steam generators. Table 3-14 lists the reasons why the tubes were plugged. Table 3-15 lists the plugged tubes.

Calvert Cliffs 2 has two recirculating steam generators which were designed and fabricated by BWI. The steam generators were put into service in 2003 during RFO 14. The licensee numbers its tube supports as depicted in Figure 2-14.

Before the steam generators went into service, three tubes were plugged. Two of these three tubes were preventively plugged since they were identified as potentially being in contact during the preservice inspection. Several dings were detected during the preservice inspection (18 indications in 7 tubes in steam generator A and 2 indications in 2 tubes in steam generator B). Most of the dings were inspected with a rotating probe during the preservice inspection and no anomalous signals were noted. No tubes were overexpanded in the tubesheet region in either of the steam generators.

During the first cycle of operation with the replacement steam generators, no primary-to-secondary leakage was detected.

The first inservice inspection of the steam generators was scheduled for the spring of 2005.

3.2.6 Catawba 1

Tables 3-16, 3-17, and 3-18 summarize the information discussed in this section for Catawba Unit 1. Table 3-16 provides the number of full-length bobbin inspections and the number of tubes plugged and deplugged during each outage for each of the four steam generators. Table 3-17 lists the reasons why the tubes were plugged. Table 3-18 lists the plugged tubes.

Catawba 1 has four recirculating steam generators which were designed and fabricated by BWI. The steam generators were put into service in 1996 during RFO 9. The licensee numbers its tube supports as depicted in Figure 2-16.

During the preservice inspection, 100 percent of the tubes in each of the four steam generators were inspected full length with a bobbin coil. Profilometry was also performed on all inservice tubes. As a result of the preservice inspection, 144 tubes were repaired to reduce the crevice depth at the top of the tubesheet to less than 0.25 inches, and an additional 9 tubes were repaired because of missed tube-to-tubesheet expansions. These nine missed expansions were of two types—missed expansions and partial expansions (i.e., an expansion of approximately 12 inches). A partial expansion can occur in tubes expanded with a shorter than normal (28-inch) length mandrel since these tubes are expanded in a two-stage expansion process. The two-stage expansion process (i.e., the shorter mandrel) is used in tubes near the periphery of the steam generator because the normal length mandrel interferes with the steam generator bowl in this area. One leg of four tubes was only partially expanded, and one leg of five tubes was not expanded at all. The repairs were accomplished by re-expanding the tubes with an expansion mandrel. All repaired tubes were subsequently reinspected with eddy current techniques before commercial operation.

As a result of the preservice inspection, one tube was plugged with a thimble plug because of shallow scratches on the cold-leg tube end from a stuck expansion mandrel used to correct a missed expansion in this tube. In addition, four tubes were identified in steam generator A which were over "rolled" at the top of the tubesheet, and one tube was identified in steam generator D with a dent (0.002 inches or less). A total of 79 over expanded tubes were identified (32 in steam generator A, 2 in steam generator B, 41 in steam generator C, and 4 in steam generator D). An overexpansion is a tube whose diameter varies by more than 6 mils from the nominal tube diameter after expansion. A total of 35 locations were identified which had eddy current crevice depths (measured from the top of the tubesheet to the fully expanded portion of the tube) greater than 0.25 inches, but whose mechanically measured depth was less than 0.25-inches. A total of 141 MBMs were reported (63 in steam generator A, 37 in steam generator B, 14 in steam generator C, and 27 in steam generator D).

Before the steam generators went into service during RFO 9, the fabricator plugged 19 tubes because of manufacturing flaws. One of the plugged tubes in steam generator A was stabilized on the hot-leg side because a section of the tube was expanded above the tubesheet.

During RFO 10 in 1997, 100 percent of the tubes in each of the four steam generators were inspected full length with a bobbin coil. As a result of these inspections, no tubes were plugged; however, some peripheral tubes were noticed to be in close proximity based on the eddy current inspection (refer to Section 2.4.6). There was no degradation associated with these tubes.

During RFO 10, the upper steam drum of steam generator A was entered and visually inspected. The following areas were inspected, in part: primary and secondary decks, supports, hatches, primary and secondary moisture separators, feedring and supports, feedwater header "J" tubes, top of the tube bundle, downcomer components, seismic pins, wrapper lug, and a section of the top of the tubesheet. No erosion or corrosion was observed. The separator drains were open, and there was no meaningful scale buildup on the primary and secondary moisture separators. Nothing was adrift. Welds appeared normal. No debris was found on the primary or intermediate decks. The results of a sample of a light dusting of corrosion products taken on the primary deck near the access hatch indicated the material consisted principally of magnetite. There was no steam generator wrapper to shell misalignment nor was any wrapper drop noted in steam generator A.

During RFO 11 in 1999, approximately 21 percent of the tubes in each of the four steam generators were inspected full length with a bobbin coil. In addition to the full-length inspections, partial-length inspections were conducted on 7 percent of the tubes in steam generator A, 2 percent of the tubes in steam generator C, and 7 percent of the tubes in steam generator D. As a result of these inspections, no tubes were plugged.

During RFO 12 in 2000, approximately 80 percent of the tubes in steam generators B and C were inspected full length with a bobbin coil. No tube inspections were performed in steam generators A or D during this outage. As a result of these inspections, no tubes were plugged.

During RFO 13 in 2002, no steam generator tube inspections were performed.

During RFO 14 in 2003, 100 percent of the tubes in steam generators A and D were inspected full length with a bobbin coil, and 100 percent of the tubes in steam generators B and C were inspected with a bobbin coil from the tube end through the second lattice grid support on both the hot- and cold-leg side of the steam generator. In addition to the bobbin coil inspections, a rotating probe was used to inspect the expansion transition region (from 2 inches above to 8 inches below the top of the secondary face of the tubesheet) of 20 percent of the tubes in steam generators A and D. As a result of these inspections, seven tubes were plugged and stabilized in steam generator C because of a loose part on the hot-leg side. Attempts to retrieve the loose part were unsuccessful. The part was characterized as an S-hook from an unknown source. Two of the tubes showed some wear. Five sets of tubes were identified as being in proximity and were inspected with a rotating probe.

On January 13, 2005, Catawba 1 and 2 revised the steam generator portion of their technical specifications making them performance-based consistent with Technical Specification Task Force (TSTF) Improved Standard Technical Specifications Change Traveler TSTF-449 (see ADAMS Accession No. ML050110258).

3.2.7 Cook 1

Tables 3-19, 3-20, and 3-21 summarize the information discussed in this section for Cook Unit 1. Table 3-19 provides the number of full-length bobbin inspections and the number of tubes plugged and deplugged during each outage for each of the four steam generators. Table 3-20 lists the reasons why the tubes were plugged. Table 3-21 lists the plugged tubes.

Cook 1 has four recirculating steam generators which were designed and fabricated by BWI. The steam generators were put into service in 2000 during RFO 17. The licensee numbers its tube supports as depicted in Figure 2-18.

Before the steam generators went into service, no tubes were plugged.

During RFO 18 in 2002, 100 percent of the tubes in each of the four steam generators were inspected full length with a bobbin coil. In addition, a rotating probe was used to inspect selected bobbin coil signals. Approximately 90 locations (in approximately 80 tubes) were inspected with a rotating probe. The locations inspected included a dent, possible loose parts, and MBMs. The inspection identified one tube with wear at a fan bar. The depth of the wear scar measured 8 percent through-wall. A total of one dent (tubing diameter less than nominal at a support) and five dings (tubing diameter less than nominal in the free span) were reported using a two-volt criteria during the bobbin coil inspection. The dents and dings ranged from 2 to 8 volts, and all but one were present in the preservice inspection. In addition, only one was not in the U-bend region. The ding that was not present in the preservice inspection measured 2.15 volts and was located in the U-bend region. The indication was attributed (during RFO 18) either to probe interaction when traversing the U-bend or to steam generator movement during shipping. During an inspection of this location during RFO 19, this ding was not found leading analysts to theorize that foreign material on the outside surface of the tube (since removed) contributed to the false indication during RFO 18.

As a result of the RFO 18 inspections, four tubes were plugged because of changes in the bobbin coil voltage amplitude at five MBM indications (tube buffing operations following thermal treatment during the manufacturing process). The signal change was attributed to a microstructural change induced by the heat of the first operating cycle. The rotating probe detected no degradation at these locations.

During RFO 19 in 2003, approximately 20 percent of the tubes in steam generator D were inspected full length with a bobbin coil. In addition, a rotating probe was used to inspect approximately 20 locations to characterize dings, a freespan indication (which had not changed since the preservice inspection), MBMs (which had not changed since the preservice inspection), and a fan bar wear indication.

As a result of these inspections, no defective or degraded tubes were identified, and no tubes were plugged. However, four ding signals were reported using a 2-volt reporting threshold. All dings except one were present in the preservice inspection. The ding in this tube measured approximately 5 volts and was theorized to be the result of a loose part. One indication of wear at a fan bar was identified. This indication was present during RFO 18, and the depth of the wear scar increased from 8 percent in RFO 18 to 11 percent through-wall.

3.2.8 Cook 2

Tables 3-22, 3-23, and 3-24 summarize the information discussed in this section for Cook Unit 2. Table 3-22 provides the number of full-length bobbin inspections and the number of tubes plugged and deplugged during each outage for each of the four steam generators. Table 3-23 lists the reasons why the tubes were plugged. Table 3-24 lists the plugged tubes.

Cook 2 has four recirculating steam generators which were designed and fabricated by Westinghouse. The steam generators were put into service in 1989 during RFO 6. The licensee numbers its tube supports as depicted in Figure 2-20.

As of 2000, 10 percent of the steam generator tubes could be plugged, with a maximum of 15 percent in one steam generator.

Before the steam generators went into service, one tube was plugged with welded plugs. However, a total of 97 bulges within the tubesheet region were identified during the preoperational insepection (36 in steam generator A, 34 in steam generator B, 7 in steam generator C, and 20 in steam generator D). These bulges are attributed to hydraulic expansion of the tubes into drilling irregularities in the tubesheet holes.

During RFO 7 in 1990, approximately 2 percent of the tubes in steam generators B and C were inspected full length with a bobbin coil. In addition, partial-length inspections were conducted on approximately 5 percent of the tubes in steam generators B and C. These partial-length inspections were performed with a bobbin coil from the hot-leg tube end to the seventh cold-leg (7C) tube support. These inspections detected no degradation, and thus no tubes were plugged.

During RFO 8 in 1992, approximately 2 percent of the tubes in steam generators A and D were inspected full length with a bobbin coil. In addition, partial-length inspections were conducted on approximately 5 percent of the tubes in steam generators A and D. These partial-length inspections were performed with a bobbin coil from the hot-leg tube end to the seventh cold-leg (7C) tube support, with a few of these tubes being inspected to lower cold-leg tube supports (e.g., 6C and 5C). As a result of these inspections, no degradation was detected and no tubes were plugged.

During RFO 9 in 1994, the only planned steam generator activities were pressure pulse cleaning and sludge lancing in all four steam generators. Following the completion of pressure pulse cleaning in the first steam generator, a visual inspection on the secondary side of the steam generator was performed to judge the effectiveness of the cleaning operations. During this inspection, mechanical damage of the steam generator tubes near the hand holes was noted. An investigation revealed that the pressure pulse cleaning nozzles, which consist of flexible stainless steel hoses tipped with solid stainless steel rings, were damaged. These nozzles were determined to be the cause of the tube damage. The pressure pulse cleaning operations on the second steam generator then in progress were immediately stopped, and the pressure pulse cleaning of the remaining two steam generators was cancelled.

As a result of the visual evidence of tube damage, an eddy current inspection was conducted to investigate those portions of tubes necessary to bound the damage inflicted by the pressure pulse cleaning nozzles. The areas that the pressure pulse cleaning nozzles could have reached were limited to an area above the tubesheet and below the flow distribution baffle. All (except for one) of the potentially affected tubes and the row 1 tubes were inspected from 3 inches above the flow distribution baffle to the hot- and cold-leg tube end with a bobbin coil. One of the potentially affected tubes in steam generator C was damaged to the extent that it could not be inspected with a qualified technique.

Since inspection personnel were available as a result of the pressure pulse cleaning damage, they conducted an inspection pursuant to the technical specification requirements. During this inspection, approximately 2 percent of the tubes in steam generators B and C were inspected full length with a bobbin coil. In addition, partial-length inspections were conducted on approximately 5 percent of the tubes in steam generators B and C. These partial-length inspections were performed with a bobbin coil from the hot-leg tube end to the seventh cold-leg (7C) tube support. As discussed above, approximately 3 percent of the tubes were inspected from the tube end (both hot- and cold-leg) to 3 inches above the flow distribution baffle with a bobbin coil to investigate damage observed in the tube lanes on the secondary side of the steam generators as a result of pressure pulse cleaning on the secondary side of the steam generators. As a result of these inspections, nine tubes were plugged as a result of this tube damage.

Selected areas on the secondary side of one of the steam generators were inspected during RFO 9. Inspections were performed on ten J-nozzles and the feedring tee section. No erosion was noted.

During RFO 10 in 1996, no steam generator tube inspections were performed.

During cycle 11, there was a brief period of elevated chloride levels in the steam generators following unit startup. In addition, for a period of time in the middle of the cycle, the dissolved oxygen in the condensate was elevated because of unit testing for condenser in-leakage. During cycle 11, no primary-to-secondary leakage was detected.

During RFO 11 in 1997, approximately 50 percent of the tubes in each of the four steam generators were inspected full length with a bobbin coil. In addition, a rotating probe was used to inspect the hot-leg expansion transition region of 20 percent of the tubes in steam generator D.

As a result of these inspections, five tubes were plugged. One tube was plugged in steam generator A because of a 28 percent through-wall indication attributed to a burr or small foreign object near the center of the first cold-leg support plate. No detectable degradation was recorded for this tube during its last inspection in 1988 (i.e., the preoperational inspection). Bobbin and rotating probe inspections of surrounding tubes detected no degradation. Four tubes were plugged because of six single volumetric indications characteristic of wear from a loose part near the top of the tubesheet on the hot-leg side of the steam generator. The estimated depths of these indications ranged from 7 to 20 percent through-wall. These indications were detected during the rotating probe inspection of the hot-leg expansion transition region in steam generator D (these tubes were inspected with a bobbin coil, but the indications were not present in the bobbin coil data). No detectable degradation was recorded for these tubes during their last inspection in 1988 (i.e., the preoperational inspection). As a result of the findings from the rotating probe inspection, the eddy current inspection in steam generator D was expanded to include the tubes surrounding these six indications and approximately 100 percent of the tubes in the periphery of the tube bundle (i.e., the last two tubes in each column). The inspections in the periphery were performed with a bobbin coil, and the inspections surrounding the tubes with indications were performed with a rotating probe. No indications of a possible loose part were present in the eddy current data for these four tubes that were plugged; however, a PLP signal was detected in three tubes neighboring one of the plugged tubes. During a secondary- side visual inspection, a weld wire was removed from this area of the tube bundle. No other loose parts were identified in the steam generators. No additional indications were detected during the bobbin coil inspection of the peripheral tubes.

In addition to the above, four indications of hot-leg support-plate wear were found in two tubes in steam generator A during RFO 11. The depth of the wear scars ranged from 4 to 11 percent through-wall. A comparison with the 1988 preoperational inspection data found the inspection data for all tubes with MBMs to be unchanged. Dent indications (ranging from 3.0 to 16.24

volts) were also compared to the 1988 preoperational inspection data. Of the 90 dent indications, 87 were matched to signals in the preservice data. The preoperational data for the three remaining dent indications (all less than 4 volts) could not be retrieved from the digital tapes. The vast majority of the dents are in the 3 to 6 volt range with only two dents exceeding 10 volts.

During RFO 11 in 1997, the secondary side of each steam generators was inspected. In each of the steam generators, a visual inspection was performed before sludge lancing to identify the tube and sludge conditions at the tubesheet. This inspection included the annulus and the divider lane. Following sludge lancing, a visual inspection was performed to verify the effectiveness of the sludge removal process and to observe the general condition of the tubesheet and the tubes at the top of the tubesheet. Areas inspected included the annulus, divider lane, and an inner bundle pass on both the hot- and cold-leg sides of the steam generator. In one steam generator, a visual inspection of the sixth and seventh tube support plates was performed to assess the sludge and fouling conditions in the upper region of the steam generator. No degradation or signs of abnormalities were noted during these inspections.

To resolve design-basis concerns identified during an NRC Architect and Engineering Inspection, Unit 2 did not operate from the end of 1997 to June 2000. These design-basis concerns were not related to the steam generator tube inspections. As a result of this extended shutdown, the next steam generator tube inspections were delayed, with NRC approval, until 2002.

During RFO 12 in 2002, approximately 50 percent of the tubes in each of the four steam generators were inspected full length with a bobbin coil. In addition, a rotating probe was used to examine the hot-leg expansion transition region of 20 percent of the tubes in steam generator A; the U-bend region of 100 percent of the row 1 and 2 tubes in steam generator C; 20 percent of all dents and dings with voltage amplitudes of 2 volts or greater (as determined from the bobbin coil probe); and a 20-percent sample of the freespan and MBMs which revealed no change since the preoperational inspection. The bobbin coil was not used to inspect the U-bend region of the tubes in rows 1 and 2 because of potential difficulties in traversing the bend region.

In steam generator A, three tubes were identified with possible loose part indications at the top of the hot-leg tubesheet during the rotating probe inspections. As a result, additional rotating probe inspections in these areas identified two additional tubes as having PLP indications. A total of 15 additional tubes were inspected with a rotating probe. Visual inspections in these areas identified a small wire and sludge rock at these locations. The wire was removed from the steam generator. No tube degradation was associated with these PLP indications.

As a result of these inspections, no tubes were plugged; however, several bulges, dents, dings, nonexpanded tubes, and MBMs were identified. In steam generator A, one bulge was detected, and in steam generator D, two bulges were detected. These bulges, located in the tubesheet, are attributed to expansion of the tube into a region of the tubesheet which was not perfectly round (as a result of drill-bit wobble). Approximately 220 dents and dings (greater than 2 volts) were detected in the steam generators. All of the dents and dings were less than 9 volts. A comparison of these dents and dings to the preoperational inspection indicated no change in the data (however, the historical data for four dents and dings could not be reviewed because of problems encountered when duplicating the data from the obsolete cartridge tapes used during the preoperational inspection to the current industry standard data media (optical disks)). Rotating probe inspection of a 20-percent sample of the dents and dings revealed no degradation. None of the dents and dings inspected was in the U-bend (i.e., all were in the straight portion of the tube). In steam generator B, two tubes were reported as not having been hydraulically expanded into the cold-leg tubesheet. Similarly, in steam generator C, two tubes were reported as not having been hydraulically expanded into the hot-leg tubesheet. These four tubes were inspected the full length of the tubesheet using a rotating coil probe and no degradation was detected. Several freespan and MBM indications were detected. All but five of these indications could be traced back to the pre-operational inspection. These five indications could not be traced back to the pre-operational inspection because of problems in recovering the data from the obsolete cartridge tapes. A rotating probe inspection of these five indications revealed no degradation. In addition, rotating probe inspections of a 20-percent sample of the freespan and MBMs which revealed no change since the pre-operational inspection which had revealed no degradation.

One tube in steam generator A was reported as having tube wear at hot-leg tube support 06H. The wear was estimated to be less than 10 percent through-wall. This indication was present during the RFO 11 inspections. No indication of wear was reported at three other locations in steam generator A which were identified as having wear during RFO 11, which measured less than 10 percent through-wall.

Following RFO 12, data were reviewed to determine if any additional tubes were not expanded into the tubesheet. This data review identified three additional tubes (for a total of seven tubes) that were not hydraulically expanded in the tubesheet. Three of these tubes are in steam generator B and were not expanded into the tubesheet on the cold-leg side of the steam generator. The other four tubes are in steam generator C and were not expanded into the tubesheet on the hot-leg side of the steam generator.

As of RFO 12 (2002), no primary-to-secondary leakage had been detected since the steam generators were put into service in 1989.

During RFO 13 in 2003, no steam generator tube inspections were performed.

During RFO 14 in 2004, approximately 25 percent of the tubes in each of the four steam generators were inspected full length (except for the U-bend region of row 1 and 2 tubes) with a bobbin coil. In addition, a rotating probe was used to examine the hot-leg expansion transition region of 20 percent of the tubes (from 3 inches above to 3 inches below the top of the secondary face of the tubesheet) in steam generators B and C; the U-bend region of 20 percent of the row 1 and 2 tubes in steam generator B; 20 percent of all dents and dings with voltage amplitudes of 2 volts or greater (as determined from the bobbin coil probe); a 20-percent sample of the freespan indications greater than or equal to 0.50 volts; the tubesheet region in all tubes that were not hydraulically expanded into the tubesheet; 100 percent of all hot-leg tubesheet bulges which were in the planned bobbin coil inspection population; 100 percent of the PLP indications and the associated surrounding tubes; and areas where loose parts were identified to ensure that no tube damage was present.

A total of eleven indications were identified with the bobbin coil. Four of these indications (on two tubes) were attributed to wear at the tube supports that had been identified during previous inspections. These indications were shallow and essentially had not changed. The remaining seven indications were characterized with a rotating probe and included a distorted indication in the tubesheet, a permeability variation, a distorted tube support indication, and four MBMs. With the exception of the distorted tube support indication, no degradation was detected at these locations. At the distorted tube support indication, the rotating probe indicated the presence of a loose part. The wear was estimated to be 33 percent through-wall from the rotating probe data (25 percent through-wall based on the bobbin data). A visual inspection revealed a small metallic part lodged between the flow distribution baffle and the tube at this location. Since the part could not be removed, this tube was subsequently plugged.

In addition to the above-mentioned loose part, a wire fragment was found during the secondary-side visual inspection in steam generator B. The wire was adhering to the base of the tube in row 3, column 55, at the top of the tubesheet on the cold-leg side of the steam generator. There was no degradation at this location. Attempts to retrieve the part were unsuccessful. In steam generator C, all loose parts, including three wire fragments and a 3/8-inch square nut, were removed.

In general, the RFO 14 inspections revealed only minor support plate and foreign-object-induced wear on a total of three tubes. As a result of these inspections, one tube was plugged for wear associated with a loose part.

3.2.9 Farley 1

Tables 3-25, 3-26, and 3-27 summarize the information discussed below for Farley Unit 1. Table 3-25 provides the number of full-length bobbin inspections and the number of tubes plugged and deplugged during each outage for each of the three steam generators. Table 3-26 lists the reasons why the tubes were plugged. Table 3-27 lists the plugged tubes.

Farley 1 has three recirculating steam generators which were designed by Westinghouse and fabricated by ENSA. The steam generators were put into service in 2000 during RFO 16. The licensee numbers its tube supports as depicted in Figure 2-22.

Before the steam generators went into service, no tubes were plugged.

During RFO 17 in 2001, 100 percent of the tubes in each of the steam generators were inspected full length with a bobbin coil except for the U-bend of the low-row tubes (i.e., those in rows 1 and 2). In addition, a rotating probe equipped with a +Point™ coil was used to inspect the hot-leg expansion transition region of 20 percent of the tubes (from 3 inches above to 3 inches below the top of the secondary face of the tubesheet); the U-bend region of 100 percent of the low-row tubes; 100 percent of hot-leg straight section dents and dings with voltage amplitudes greater than or equal to 5 volts (as determined from the bobbin coil probe); and bobbin signals that were not observed during the preservice inspection or which had changed since the preservice inspection. A total of 37 bobbin indications were inspected with a rotating probe to investigate signals that were either not observed during the preservice inspection or had changed since the preservice inspection. As a result of these inspections, no indications of degradation were observed, and no tubes were plugged. However, these inspections revealed 2 bulge indications, 5 dings (in the freespan), 2 dents (at tube supports), 10 expansion anomalies within the tubesheet, and 2 PLP indications. Visual inspections did not confirm the existence of a loose part at the location of the PLP indications, and no tube wear was associated with these PLP indications.

Sludge lancing was performed on the secondary side of the tubesheet in all three steam generators during RFO 17. Upon completion of the sludge lancing, a video inspection was performed in this region to identify any foreign objects. No foreign objects detrimental to the steam generator tubing were identified. Minor debris was removed from the steam generators as part of the inspection, including 13 small pieces of nonmetallic material resembling flexitallic gasket-like material, 1 metal shaving approximately 1/16 of an inch long, 1 piece of wire approximately 3/4 of an inch long, 1 1.25-inch nail, and 2 pieces of flexitallic gasket-like material (approximately 1 inch long). At the end of the inspection, no foreign objects were known to be present in the steam generators.

As a result of receiving NRC approval to extend the interval between tube inspections from a maximum of 24 to 40 calendar months, no steam generator tube inspections were performed during RFO 18 in 2003.

On September 10, 2004, Farley Units 1 and 2 revised the steam generator portion of their technical specifications making them performance-based and consistent with Improved Standard Technical Specifications Change Traveler TSTF-449 (see ADAMS Accession No. ML042570427). As a result of this revision, no steam generator tube inspections were performed during RFO 19 in 2004.

3.2.10 Farley 2

Tables 3-28, 3-29, and 3-30 summarize the information discussed below for Farley Unit 2. Table 3-28 provides the number of full-length bobbin inspections and the number of tubes plugged and deplugged during each outage for each of the three steam generators. Table 3-29 lists the reasons why the tubes were plugged. Table 3-30 lists the plugged tubes.

Farley 2 has three recirculating steam generators which were designed by Westinghouse and fabricated by ENSA. The steam generators were put into service in 2001 during RFO 14. The licensee numbers its tube supports as depicted in Figure 2-22.

Before the steam generators went into service, no tubes were plugged in any of the generators. However, during the preservice inspection, many dents were reported at the seventh tube support plate on the cold-leg side in steam generator A. These dents were formed during the heat treatment process of welds. All dents/dings in steam generator A were inspected with a rotating probe equipped with a +Point™ coil during the preservice inspection.

During cycle 15, the first cycle of operation with the replacement steam generators, no primary-to-secondary leakage was detected.

During RFO 15 in 2002, 100 percent of the tubes in each of the steam generators were inspected full length with a bobbin coil, except for the U-bend of the low-row tubes (i.e., those in rows 1 and 2). In addition, a rotating probe equipped with a +Point™ coil was used to inspect the hot-leg expansion transition region of 20 percent of the tubes (from 3 inches above to 3 inches below the top of the secondary face of the tubesheet); the U-bend region of 100 percent of the low-row tubes; 100 percent of the hot-leg straight section dents and dings with voltage amplitudes greater than or equal to 5 volts (as determined from the bobbin coil probe); and bobbin signals which were not observed during the preservice inspection or which had changed since the preservice inspection. As a result of these inspections, no indications of degradation were observed and no tubes were plugged.

As a result of the RFO 15 inspections, 267 dents (265 in steam generator A, 1 in steam generator B, and 1 in steam generator C) and 5 dings (3 in steam generator A and 2 in steam generator C) were reported. The dents in steam generator A are from the heat treatment process of the welds in the steam generator (see above). The dents and dings in steam generators B and C all measured less than 5 volts by the bobbin coil. No rotating probe inspections were performed on dents and dings during RFO 15 since no new dents, dings, or signal changes at the existing dents and dings were detected.

Sludge lancing was performed on the secondary side of the tubesheet in all three steam generators during RFO 15. Upon completion of the sludge lancing, a video inspection was performed in this region to identify any foreign objects. No foreign objects detrimental to the steam generator tubing were identified. Minor debris was removed from the steam generators as part of the inspection. The eddy current inspection showed that this debris had not produced any indications of tube wear, nor were there any loose part indications. At the end of the inspection, no foreign objects were known to be present in the steam generators.

As a result of receiving NRC approval to extend the interval between tube inspections from a maximum of 24 to 40 calendar months, no steam generator tube inspections were performed during RFO 16 in 2004. During cycle 16, no primary-to-secondary leakage was detected.

On September 10, 2004, Farley Units 1 and 2 revised the steam generator portion of their technical specifications making them performance-based and consistent with Improved Standard Technical Specifications Change Traveler TSTF-449 (see ADAMS Accession No. ML042570427). As a result of this revision, no steam generator tube inspections were planned for RFO 17 in the fall of 2005.

3.2.11 Ginna

Tables 3-31, 3-32, and 3-33 summarize the information discussed below for Ginna. Table 3-31 provides the number of full-length bobbin inspections and the number of tubes plugged and deplugged during each outage for each of the two steam generators. Table 3-32 lists the reasons why the tubes were plugged. Table 3-33 lists the plugged tubes.

Ginna has two recirculating steam generators designed and fabricated by BWI. The steam generators were put into service in 1996 during RFO 25. The licensee numbers its tube supports as depicted in Figure 2-24.

Before the tubing of the steam generator, one tubesheet hole in steam generator B (row 75, column 91) had several scratches. These scratches, which were on the hot-leg side of the tubesheet, were buffed to remove the raised sharp edges of the scratches, and this location was then tubed.

During hydraulic expansion of the tubes in steam generator A, 28 tubes were expanded beyond the secondary face of the tubesheet as a result of problems with the expansion mandrel. Of these 28 tubes, 25 have a slight bulge above the secondary face of the tubesheet. These tubes were left in service following an assessment which included finite element analysis to determine the residual stresses and accelerated stress-corrosion cracking testing to determine the potential for cracking to occur at these locations.

Approximately 16,000 MBMs were recorded in the preservice inspections. These indications are artifacts of the tube fabrication process and represent repairs of the tube outside diameter surface by light polishing or grinding to remove slight surface imperfections.

Prior to placing the steam generators into service, one tube in each steam generator was plugged. These tubes were plugged with welded plugs.

During RFO 26 in 1998, 100 percent of the tubes in both of the steam generators were inspected full length with a bobbin coil. In addition, a rotating probe equipped with a +Point™ coil was used to inspect bobbin coil indications classified as requiring additional diagnostic testing; the hot-leg expansion transition region of 10 percent of the tubes; the U-bend region of 20 percent of the row 1 and 2 tubes; and peripheral tubes that were identified as potentially being in close proximity. In addition, a visual inspection was performed on one of the plugs. No degradation was observed, and no tubes were plugged as a result of the eddy current inspections.

Both Ginna steam generators received a comprehensive secondary-side internals inspection during RFO 26. The purpose of the inspection was, in part, to verify that no unknown degradation mechanisms were occurring in the steam generators and to determine whether the U-bend tube proximity issue was affecting these steam generators (since the condition had been observed in other BWI steam generators). These inspections confirmed the presence of tube-to-tube contact in five tubes in each of the steam generators. This condition was identified through eddy current testing and confirmed on accessible locations using specially made feeler gauges to determine the proximity of the tubes. No degradation of steam generator internals was detected during the RFO 26 visual inspections, although several tubes were identified in which the U-bends were in close proximity, as discussed above. All of the tubes identified by eddy current testing to be in close proximity and subsequently inspected with feeler gauges were less than 0.100 inch apart.

Before and after water lancing, FOSAR was performed at the top of the tubesheet in both steam generators. Two foreign objects were found in steam generator B during the RFO 26 FOSAR. A 4.25-inch long by 3/8-inch hex-head bolt and matching nut were found near the center of the no-tube lane and were removed. No tube damage was identified (by eddy current testing). The bolt and nut were not native to the steam generator.

During RFO 27 in 1999, more than 50 percent of the tubes in both of the steam generators were inspected full length with a bobbin coil. In addition, a rotating probe equipped with a +Point™ coil was used to examine bobbin coil indications classified as requiring additional diagnostic testing, the hot-leg expansion transition region of 20 percent of the tubes, and peripheral tubes identified as potentially being in close proximity. In addition, a visual inspection was performed on one of the plugs in each of the steam generators. Approximately 20 percent of all hot-leg MBMs with a bobbin voltage greater than 5 volts were inspected with a rotating probe. No degradation was observed and no tubes were plugged during RFO 27. Approximately 37 dents and dings were recorded during this outage. Dents and dings were reported from the bobbin coil data if they exceeded 2 volts.

In steam generator A, 34 locations were identified as being in close proximity with a low frequency bobbin coil technique. The rotating probe inspections of these locations confirmed that 13 of these locations were in close proximity. In steam generator B, 24 locations were identified as being in close proximity with a low frequency bobbin coil technique. The rotating probe inspection of these locations confirmed that 12 of these locations were in close proximity.

During RFO 28 in 2000, no steam generator tube inspections were performed.

During RFO 29 in 2002, approximately 50 percent of the tubes in both of the steam generators were inspected full length with a bobbin coil. In addition, a rotating probe equipped with a +Point™ coil was used to inspect bobbin coil indications classified as requiring additional diagnostic testing; the hot-leg expansion transition region (from 3 inches above the top of the tubesheet to 2 inches below the top of the tubesheet) of 20 percent of the tubes; the U-bend region of 20 percent of the row 1 and 2 tubes (24 tubes); 20 percent of all hot-leg dents and dings which had voltage amplitudes greater than 5 volts (as determined from the bobbin coil probe); 20 percent of hot-leg MBMs with voltage amplitudes greater than 5 volts (as determined from the bobbin coil probe); and in the portion of tubes identified as being in close proximity (13 tubes in steam generator A and 12 tubes in steam generator B). The hot-leg expansion transition inspections included all overexpanded tubes. In addition, a visual inspection was performed on all repaired welded plugs. No degradation was observed and no tubes were plugged as a result of the eddy current inspections.

In addition to the probes discussed above, the X-probe was used in steam generator A to inspect 34 tubes full length and 133 tubes from the hot-leg tube end through the first hot-leg tube support (01H).

In steam generator A, 13 tubes had been previously identified as being in close proximity with the rotating probe. These 13 tubes were confirmed as being in close proximity with a rotating probe in 2002 (although the bobbin coil only identified 10 of the original 13 tubes as being in close proximity). No wear was observed on these tubes. In addition, there was no evidence of change in the location of the close proximity or in the extent of closeness when compared to historical data. 45 tubes were identified with dings (free span dents) or dents (dents at supports). Of the 34 reported dings, 15 were new indications and 20 of the dings are in the U-bend region (i.e., between 08H and 08C). Ten of the dents were new indications and all of the dents are in the U-bend region. None of the dents or dings exhibited any growth when compared to historical data.

In steam generator B, 12 tubes had been previously identified as being in close proximity with the rotating probe. These 12 tubes were confirmed as being in close proximity with a rotating probe in 2002 (although the bobbin coil only identified 9 of the original 12 tubes as being in close proximity). No wear was observed on these tubes. In addition, there was no evidence of change in the location of the close proximity or in the extent of closeness when compared to historical data. 59 tubes were identified with dings (free span dents) or dents (dents at supports). Of the 54 reported dings, 31 were new indications and 16 of the dings are in the U-bend region (i.e., between 08H and 08C). 11 of the dents were new indications and 17 of the dents are in the U-bend region and one is associated with the second hot-leg tube support (02H). None of the dents or dings exhibited any growth when compared to historical data.

During RFO 30 in 2003, no steam generator tube inspections were performed.

3.2.12 Harris

Tables 3-34, 3-35, and 3-36 summarize the information discussed in this section for Harris. Table 3-34 provides the number of full-length bobbin inspections and the number of tubes plugged and deplugged during each outage for each of the three steam generators. Table 3-35 lists the reasons why the tubes were plugged. Table 3-36 lists the plugged tubes.

Harris has three recirculating steam generators designed and fabricated by Westinghouse. The steam generators were put into service in 2001 during RFO 10. The licensee numbers its tube supports as depicted in Figure 2-26.

In 1999, during the fabrication of the steam generators, several dents and dings were detected in the tubes. The majority of the dents are at the upper tube support plate in the outer periphery of the tube bundle; however, some dents and dings are at the lower tube support plates. The dents and dings were attributed to the post-weld heat treatment of the channel head to tubesheet weld. In the steam generator, the shell barrel holds the tube support plates in place, the tube support plates are anchored by stayrods attached to the tubesheet, and the tubes are firmly attached to the tubesheet. It was determined that as the shell barrel elongated during the post-weld heat treatment, the tube support plates deflected slightly at the outer edges while the center of the plate did not deflect as much because the stayrods did not experience as much thermal growth (since they were farther from the heat source). The Harris and South Texas Unit 1 steam generators were the last steam generators manufactured at the Westinghouse Pensacola facility.

Before the steam generators went into service, the tubes received a preservice inspection. The preservice inspection included a full-length bobbin coil inspection of 100 percent of the tubes in each of the steam generators. In addition, a rotating probe equipped with a +Point™ coil was used to examine the hot-leg expansion transition region (from 2 inches above to 2 inches below the top of the tubesheet) of 100 percent of the tubes; 100 percent of all dents with voltage amplitudes of 2 volts or greater (as determined from the bobbin coil probe); the U-bend region of 100 percent of the row 1 tubes; and a sample of benign MBMs and other benign indications. In addition, ultrasonic testing was performed on all potential manufacturing lap indications.

Before the steam generators went into service, two tubes were plugged. One of the tubes was plugged as a result of a defect introduced into the tube following a manual weld repair.

No primary-to-secondary leakage was observed during cycle 11.

During RFO 11 in 2003, 100 percent of the tubes in each of the steam generators were inspected full length with a bobbin coil. In addition, a rotating probe equipped with a +Point™ coil was used to inspect a sample of benign indications at 17 locations. No anomalous dent/ding signals were observed during the inspections (as had been observed at Palo Verde 2 in 2004). Furthermore, no degradation was observed at the time of these inspection and no tubes were plugged. (In May 2004, a subsequent review of the RFO 11 eddy current data revealed one tube with a non-quantifiable indication. This indication was estimated to be 37 percent through-wall from the bobbin data. As discussed below, this indication was attributed to wear from a loose part. A re-review of the RFO 11 eddy current data in May 2004 did not identify any other missed indications.)

A number of benign signals (e.g., MBMs, dings and dents, laps) were detected in 2003. The majority of the benign signals were reported from the bobbin probe and were less than 1 volt (most had amplitudes less than 0.25 volts). These signals have little potential to mask or distort a flaw signal, and none were determined to have through-wall or near through-wall penetration. All of these indications were traceable to the preservice inspection; however, some of these signals exhibited a slight phase rotation when compared to the preservice inspection results. Phase rotation after the first cycle of unit operation (i.e., after the tubes go through a heat cycle) has been observed at other units with replacement steam generators with similar tubing, and the phase rotation is not attributed to tube degradation. The phase rotation can occur as a result of a slight change in the conductivity of the tube (i.e., "relaxing" of the metallurgical condition of the tube), such as can occur from a heat cycle. After the first cycle, these signals tend to stay the same and do not change over time. That is, there is little change in the differential signal amplitude, and sometimes there is a slight change in the absolute signal amplitude, without a phase change. The phase rotation is not observed for all the benign indications, perhaps because the benign indications come from a variety of sources (e.g., a small physical imperfection in the tube during manufacture, a small metallurgical imperfection such as an alloy anomaly or permeability change, a small nick introduced into the tube during installation into the steam generator, or a small nick in the tube that has been buffed or burnished). The heat cycle may alter the conductivity near a metallurgical anomaly, whereas it may not affect a small mechanical imperfection to the degree that it is noticeable in the eddy current data.

In addition to the above, some of the indications detected during RFO 11 had a larger offset in the absolute channel than had been observed during the preservice inspection. For these signals, there was no differential signal change, and the absolute signals showed no phase change between frequencies, again indicating a slight conductivity variation from the nominal tube.

During RFO 11, analysts reviewed the eddy current data from the first 17 innermost rows of tubes to determine if an offset similar to that observed at Seabrook was present. At Seabrook, several tubes were found to have cracklike indications associated with this offset (refer to NRC Information Notice 2002-21, "Axial Outside Diameter Cracking Affecting Thermally Treated Alloy 600 Steam Generator Tubing," dated June 25, 2002, and its supplement dated April 1, 2003, for additional details). The offset was attributed to changes in the residual stress levels in the tube as a result of nonoptimal tube processing. During the RFO 11 review, no signals were identified that indicated a similar manufacturing condition existed at Harris (i.e., similar to that observed at Seabrook).

No eddy current signals indicative of loose parts were detected during the RFO 11 inspection. In addition, a visual inspection performed in all three steam generators in the tubesheet periphery and blowdown lane did not identify any foreign objects with a size or mass such that they would cause steam generator tube degradation; however, two items, of no significant weight or size, were retrieved. These two items had magnetic properties. The steam generators were sludge lanced during RFO 11.

On April 21, 2004, a small (0.42 gpd) primary-to-secondary leak was observed in steam generator C. Over the next two weeks, the leak rate fluctuated primarily between 5 to 10 gpd. On May 6, 2004, the unit tripped due to the failure of a rod control card, and the unit elected to investigate the source of this leak during the shutdown. As part of this investigation, a secondary-side pressure test was performed in which the secondary-side pressure was increased to 60 pounds per square inch gauge (psig) while leakage was monitored on the primary side of the tubesheet. This pressure test revealed that the tube in row 3, column 120 (R3C120) was leaking. Leakage was evident at a pressure between 12 and 20 psig. Subsequent to the pressure test, degradation was identified in three tubes (R1C120, R2C121, and R3C120).

Following the pressure test, a FOSAR was performed in steam generator C. A metallic piece was visually identified above the tubesheet on the cold-leg side of the steam generator and adjacent to a flow-blocking plate. This object contacted several tubes and was removed during the FOSAR. The object was approximately 2 1/4 inches long, magnetic, and irregularly shaped with sharp edges. No other loose parts were found in steam generator C. This loose part caused the degradation discussed above. A review of the videotapes from the RFO 10 FOSAR determined that the part was present during the RFO 10 outage.

Bobbin coil and rotating probe eddy current testing was performed on 15 tubes in the area near the loose part. As discussed above, this testing identified indications in three steam generator tubes at the location of the loose part. The indications were above the top of the cold-leg tubesheet (TSC). Additional details are summarized below.

Row	Column	Location	Bobbin Coil Indication	+Point™	Comments
3	120	TSC + 0.2"	No	73%	Leaking tube
1	120	TSC + 0.7"	Yes	80%	
2	121	TSC + 0.5"	Yes	45%	

The distance in the location column above is the distance to the center of the indication. The bobbin coil technique was not able to detect the through-wall damage in the tube in row 3, column 120, but did detect wear that occurred further away from the top of the TSC in the other two tubes. The bobbin coil also did not show evidence of a loose part at this location. The part was removed before the rotating coil inspections.

The 2003 (RFO 11) eddy current data for the tubes near the loose part were reviewed. This review identified no degradation or loose parts evident in the 2003 bobbin data for the tubes in row 3, column 120, and row 2, column 121. However, the data did show a bobbin coil indication measuring approximately 37 percent through-wall for the tube in row 1, column 120, but both the primary (manual) and secondary (computer) analyst missed this indication. The 37-percent through-wall indication in the 2003 bobbin data was estimated using phase angle analysis. Phase angle analysis of the 2004 bobbin coil data estimated that this indication had grown to 66 percent through-wall (the 2004 rotating probe depth estimate was 80 percent through-wall). Investigation into why the computerized data screening (CDS) system used for secondary analysis had not noted this indication revealed a setup error in the CDS settings. When the inspection parameters were entered into the CDS system in 2003, a ½-inch gap (from ½ inch above the tubesheet to 1 inch above the tubesheet) was inadvertently created which resulted in the software not analyzing this portion of the tube. This CDS input error caused the computerized tube analysis to skip the portion of the tube containing the 37-percent through-wall bobbin indication for the tube in row 1, column 120, during the 2003 analysis. After analysts discovered this situation, they entered the correct CDS settings and reanalyzed the portion of the tube skipped during the 2003 (RFO 11) eddy current data evaluation. This reanalysis identified no indications other than that discussed above.

Because of the 2004 eddy current test results, in situ pressure testing was performed on the defects in tubes R1C120 and R3C120. The tube in R1C120 did not leak at a pressure equal to three times the normal operating differential pressure. The tube in R3C120 leaked at a rate of 50 gpd at a pressure equal to the steam line break differential pressure but did not burst at a pressure equal to three times the normal operating differential pressure. The temperature-corrected leak rate for this tube was approximately 23 gpd at steam line break pressure. These in situ pressure test results demonstrated that the tubes with loose part damage retained adequate structural integrity and provided data supporting the leakage integrity of the steam generator. The three tubes affected by the part were plugged and stabilized.

Based on vendor analysis of preoutage data collected from the loose part monitors from all three steam generators, a FOSAR was also performed at the top of the tubesheet in steam generator A during the 2004 outage. A small part about the size of a washer (less than a quarter of a pound) was detected during this inspection and was removed from the steam generator. This inspection was performed because a review of the tapes from the loose part monitoring system indicated the possible presence of a loose part in steam generator A.

In addition to the investigations into the loose parts, a full-length bobbin inspection of approximately 20 percent of the tubes in steam generator C was performed during the 2004 midcycle outage. A rotating probe was used to inspect selected bobbin indications. The 20-percent sample included all peripheral tubes and selected columns of interior tubes. Other than the three indications discussed above, no degradation was detected.

During RFO 12 in 2004, no steam generator tube inspections were performed.

3.2.13 Indian Point 3

Tables 3-37, 3-38, and 3-39 summarize the information discussed in this section for Indian Point 3. Table 3-37 provides the number of full-length bobbin inspections and the number of tubes plugged and deplugged during each outage for each of the four steam generators. Table 3-38 lists the reasons why the tubes were plugged. Table 3-39 lists the plugged tubes.

Indian Point 3 has four recirculating steam generators designed and fabricated by Westinghouse. The model 44F steam generators were put into service in 1989 during RFO 6. The licensee numbers its tube supports as depicted in Figure 2-28.

Before the steam generators went into service, a preservice inspection of the tubes was performed. This inspection included a full-length bobbin coil inspection of 100 percent of the tubes in each of the steam generators. The inspection found only indications related to manufacturing such as MBMs.

Before the steam generators went into service, two tubes were plugged with welded plugs. These tubes had been damaged when a temporary support fell and dented them.

During RFO 7 in 1990, approximately 20 percent of the tubes in each of the steam generators were inspected full length with a bobbin coil. As a result of these inspections, no inservice imperfections were detected and no tubes were plugged.

During RFO 7, a foreign object was found partially lodged in the tube end of the tube in row 1, column 34, in steam generator D. The object was removed and determined to be a fuel assembly alignment pin from the reactor upper internals. The object had made numerous indentations on the channel head surfaces. All 3214 open tube ends, the tubesheet, the tube-to-tubesheet welds, the divider plate, and the cladding were inspected. Since some tube ends had minor deformation, a structural and thermal hydraulic evaluation was performed which indicated that the tube ends were acceptable, as is. More extensive tube-end gauging was planned for the next RFO to determine whether the tube ends should be rerolled to their original nominal diameter to ensure that a full-diameter eddy current probe and tube plug could pass through the tube end. An evaluation of the channel head condition determined that the structural integrity of the indented components was not degraded and that no repairs were required.

During RFO 8 in 1992, approximately 17 percent of the tubes in each of the steam generators were inspected full length with a bobbin coil. In addition to these full-length inspections, approximately 3 percent of the tubes in each steam generator were inspected with a bobbin coil from the hot-leg tube end through the uppermost tube support plate on the cold-leg side of the steam generator. An additional 3 percent of the tubes in each steam generator were inspected with a bobbin coil from the hot-leg tube end through the uppermost tube support plate on the hot-leg side of the steam generator. These inspections detected no inservice imperfections, and thus no tubes were plugged.

During RFO 8, a visual inspection was performed on the hot-leg channel head of steam generator D (a foreign object had been found in this steam generator during RFO 7). The videotape of this inspection and a similar inspection performed during RFO 7 were compared side-by-side. The inspection and comparative results showed no change in the condition of the channel head.

As a result of an extended shutdown from February 1993 to July 1995, and a subsequent forced outage from September 1995 until April 1996, the next steam generator tube inspections were delayed, with NRC approval, until RFO 9 in 1997.

During RFO 9 in 1997, approximately 61 percent of the tubes in steam generator C and approximately 64 percent of the tubes in steam generator D were inspected full length with a bobbin coil. In addition, a rotating probe equipped with a +Point™ coil was used to inspect the hot-leg expansion transition region of approximately 20 percent of the tubes; the U-bend region of 20 percent of the row 1 and 2 tubes; and all dents at tube support plates (approximately 33 dents). These rotating probe inspections were performed in steam generators C and D. The bobbin inspections included all tubes with dents recorded in prior inspections, all tubes surrounding loose parts, and all tubes that did not have full-length inspections during prior outages. As a result, 100 percent of the tubes in steam generators C and D have now been inspected with a bobbin probe since replacement of the steam generators. A visual inspection of the two welded plugs installed in steam generator D revealed no anomalous conditions.

In addition to the eddy current inspections, the tubesheets were cleaned, a FOSAR was conducted, and an in-bundle inspection was performed in each of the four steam generators. Inspections were also performed on the upper tube support plate, steam drum, feedring, wrapper support, and upper girth weld. Of particular note, these inspections found that the J-tube joints indicated possible initiation of erosion-corrosion. The level of erosion-corrosion was not significant; however, reinspection of these J-tubes was planned for a later outage (RFO 14). Foreign objects were removed from the four steam generators. Objects that could not be removed were evaluated to ensure that continued operation without removal of the objects was appropriate. In addition, the side of one of the welds of a wedge (i.e., located where the wrapper and tube support plate meet) had separated from the shell by approximately 0.5 inch, but the wedge was fully captured by the top weld and the tube support plate.

As a result of the RFO 9 inspections, no new dents were identified and no tubes were plugged. However, several dents, dings, MBMs, free span differential signals, and possible loose part signals were identified. One free span differential signal, located at the tangent point approximately 2.8 inches above the uppermost (sixth) hot-leg tube support plate, exhibited some change since the preservice inspection. This indication, in the tube located in row 8 column 21, was inspected with a rotating probe which resulted in the identification of a small ding at this location. During the eddy current inspection, there were 10 tubes identified in steam generator D as possibly having a loose part next to the tube. Four of these indications were caused by a piece of wire that was later removed. A visual inspection of the remaining 6 tubes revealed small amounts of sludge/scale and no foreign objects. One tube (R21C10) was identified with a dent at the transition area at the top of the tubesheet. This tube was inspected with a rotating probe and there was no detectable degradation at this location. None of the MBMs showed a change since the preservice inspection.

During RFO 10 in 1999, 100 percent of the tubes in steam generators A and B were inspected full length with a bobbin coil, with the exception of the U-bend region of the tubes in rows 1 and 2. In addition, a rotating probe equipped with a +Point™ coil was used to inspect the hot-leg expansion transition region (from 3 inches above to 3 inches below the top of the secondary face of the tubesheet) of approximately 40 percent of the tubes; the U-bend region of 40 percent of the row 1 and 2 tubes; all hot-leg dents and dings in the straight section of the tubing with bobbin voltages greater than 5 volts; all bobbin indications (i.e., I-codes); and PLP indications.

As a result of the RFO 10 inspections, no new dents were identified and no tubes were plugged. Three adjacent tubes in steam generator B had small volumetric wear indications. These indications were sized with a rotating probe and left in service. The deepest indication was 23 percent through-wall. During RFO 10, the indications were attributed to wear from a loose part which was no longer at these locations; however, further analysis during RFO 12 revealed that two of the indications extended slightly below the top of the tubesheet. As a result, it was concluded during RFO 12 that these indications were anomalies from the manufacturing process that were too small to be found with the bobbin probe used during the preservice inspection. The voltages of these indications did not change from RFO 10 to RFO 12 indicating that no growth had occurred, but the calculated depths changed because of the use of a more conservative sizing standard. All three tubes were plugged during RFO 12.

During RFO 10, the tubesheets were cleaned, FOSAR was performed, and an in-bundle inservice inspection was performed. In addition, the upper tube support plate, the steam drum, and the feedring were inspected. Several loose parts were removed from the steam generators during RFO 10.

Also during RFO 10, one tube in steam generator B was classified as having a "trackable anomaly." This anomaly was not considered to be flawlike, and it was dispositioned based on a review of the prior inspection data. Another tube in steam generator B had a permeability variation in the straight section of the tubing on the cold-leg side of the steam generator between the fifth and sixth tube support plates. Since no other degradation was found in the freespan region of the tube, the tube was left in service.

During RFO 11 in 2001, no steam generator tube inspections were performed; however, sludge lancing was performed.

During RFO 12 in 2003, approximately 25 percent of the tubes in each of the four steam generators were inspected full length with a bobbin coil with the exception of the U-bend region of the tubes in rows 1 and 2. In addition, a rotating probe equipped with a +Point™ coil was used to examine the hot-leg expansion transition region (from 3 inches above to 3 inches below the top of the secondary face of the tubesheet) of approximately 20 percent of the tubes in steam generators A and B; the hot-leg expansion transition region (from 3 inches above to 3 inches below the top of the secondary face of the tubesheet) of approximately 30 percent of the tubes in steam generators C and D; the U-bend region of 60 percent of the row 1 and 2 tubes in steam generators A and B; the U-bend region of 100 percent of the row 1 and 2 tubes in steam generators C and D; the cold-leg expansion transition region of 100 percent of the annulus and tube lane peripheral tubes in each of the four steam generators (approximately 270 tubes per steam generator); all hot-leg dents and dings in the straight section of the tubing with bobbin voltages greater than 5 volts (seven tubes); and all bobbin indications (i.e., I-codes) and PLP indications. In addition, two cold-leg expansion transitions, which were identified during the preservice inspection as having transitions that were higher than expected, and three volumetric indications identified in steam generator C during RFO 10, were inspected with a rotating probe.

As a result of these inspections, 12 tubes were plugged with Westinghouse Alloy 690 mechanical plugs. Of these tubes, eight were plugged for wear attributed to contact with sludge-lancing equipment used during RFO 11, one was plugged because of a permeability variation, and three were plugged for volumetric indications located slightly below or at the top of the tubesheet on the hot-leg side of the steam generator. The latter tubes (i.e., the tubes with volumetric indications below or at the top of the tubesheet) were in the periphery of the tube bundle. These indications were initially discovered during RFO 10 and were attributed to wear from loose parts, but closer examination in 2003 revealed that two of the indications extended slightly below the top of the tubesheet. The voltage of the indications had not changed since the RFO 10 inspection. The indications are currently attributed to anomalies from the manufacturing process that were too small to be found with the bobbin probe.

The cause of the wear attributed to contact with sludge lance equipment was determined based on the location of the indication. Visual inspection of two of the wear scars confirmed that the scars faced the tube lane and were about 1.1 inches long by 0.2 inches wide. All of these tubes were in row 1. The damage to these tubes occurred during RFO 11, the first time this particular sludge-lancing equipment was used at Indian Point 3. The deepest indication was 26 percent through-wall.

During the RFO 12 inspections, three tubes in row 3 in steam generator D were identified as having restrictions. These tubes were inspected during the preservice inspection with a 0.740-inch bobbin probe. During the 2003 outage, the U-bends were inspected with a 0.680-inch rotating probe and no degradation was detected. It was assumed that the tubes were restricted as a result of a slightly higher ovality of the tubing in the U-bend region when compared to other tubes. The bobbin probe used in 2003 was of a different design than that used during the preservice inspection. The bend radii of these tubes is 4.656 inches.

During RFO 12, 65 dent and ding indications were identified. Of these, 32 could be traced to indications reported in the preservice inspection. Of the remaining 33 dents and dings, 32 were near or below the reporting threshold of 3 volts used during the preservice inspection for dents and dings, and one ding indication was 6.02 volts (located approximately 4 inches above the uppermost tube support). Although the preservice inspection data were not readily available for reanalysis to determine if this 6-volt ding was present during the preservice inspection, the ding was present during RFO 9 in 1997. The voltage of this ding in 1999 was slightly below that reported in 2003, but it was within the variability of the inspection technique.

During RFO 12, one tube was classified as having a "trackable anomaly." This anomaly was not considered to be flawlike but is being tracked and added to the sample population for the next inspection of that steam generator. The freespan bobbin indication identified in one tube in 1997 in steam generator 4 (row 8, column 21) was not inspected during RFO 12. Four possible loose part indications were identified during the eddy current inspections. Visual inspections at two of these locations did not reveal any loose parts. The other two locations were near the top of the tubesheet and inaccessible to visual inspection, but the indications there were assumed to be from sludge deposits on the tubes.

3.2.14 Kewaunee

Tables 3-40, 3-41, and 3-42 summarize the information discussed in this section for Kewaunee. Table 3-40 provides the number of full-length bobbin inspections and the number of tubes plugged and deplugged during each outage for each of the two steam generators. Table 3-41 lists the reasons why the tubes were plugged. Table 3-42 lists the plugged tubes.

Kewaunee has two recirculating steam generators which were designed by Westinghouse and fabricated at Ansaldo. The steam generators were put into service in 2001 during RFO 24. The licensee numbers its tube supports as depicted in Figure 2-30.

During the preservice inspection in 2001, 100 percent of the tubes in each of the two steam generators were inspected full length with a bobbin coil. In addition, a rotating probe equipped with a +Point™ coil was used to inspect all bobbin indications. A total of 45 indications was identified, 27 in steam generator A and 18 in steam generator B. These indications were attributed to the manufacturing process and included 2 bulge indications, 9 bending machine geometry indications, 28 dings, 1 freespan differential signal, and 5 MBMs. A bulge is defined as a location in the tube where the diameter is greater than nominal. The two bulges are within the tubesheet (one on the hot-leg and one on the cold-leg) and were not present in the tube-mill eddy current data. A bending machine geometry indication is defined as a dentlike indication at the apex of a U-bend. The bending machine geometry indications were present in the tube-mill eddy current data and were in rows 5 (two tubes), 7 (one tube), 9 (two tubes), 11 (one tube), 16 (two tubes), and 22 (one tube). A ding is defined as a location where the diameter is less than nominal. The reporting threshold for dings was 2 volts, and the largest ding was 5.71 volts. The dings are concentrated at upper tube support plate elevations. A freespan differential signal is defined as an indication in the freespan reported on a differential channel. The one freespan differential indication was present in the tube-mill data. An MBM is defined as a material discontinuity at a location which has been repaired in the tube mill by buffing. Before the steam generators went into service, no tubes were plugged.

During cycle 25, no primary-to-secondary leakage was detected.

During RFO 25 in 2003, 100 percent of the tubes in both steam generators were inspected full length with a bobbin coil. In addition, a rotating probe equipped with a +Point™ coil was used to examine the hot-leg expansion transition region (from 2 inches above to 2 inches below the top of the secondary face of the tubesheet) of approximately 20 percent of the tubes with a bias toward peripheral tubes, the U-bend region of 20 percent of the row 1 tubes (i.e., 20 tubes), all bobbin I-code indications, all dings and dents greater than or equal to 5 volts (as measured from the bobbin coil), and all possible loose part indications.

As a result of the RFO 25 inspections, no degradation due to wear or corrosion was identified and no tubes were plugged. The eddy current inspections did, however, result in identifying several possible loose part indications. Tests of all of these PLP indications with a rotating probe found no wear associated with any of the indications. Visual exams performed at the location of the PLP indications did not find any loose parts in steam generator A, which suggests that the signals could be the result of local sludge deposits. However, visual inspection did confirm three objects in steam generator B. Of these three objects, one small machine chip was removed, and the other two parts were left in the steam generator. For the two loose parts left in the steam generator, an analysis was performed which resulted in a conclusion that the identified loose parts are not capable of causing significant damage to the tubes.

In addition to the above, two freespan bobbin indications were identified that exhibited a change since the preservice inspection. The indications were inspected with a rotating probe and no degradation was detected. These tubes also exhibited an offset in the U-bend region which was not present during the preservice inspection. Eight dents (i.e., new signals not present in the preservice inspection) were reported during RFO 25. These dents, located in peripheral tubes above the top of the tubesheet in steam generator A, ranged in voltage from 2.3 to 9.7 volts. Five of the eight dents were greater than 5 volts in magnitude. The dents were thought to be the result of the impact of one or more loose parts at these locations. Since FOSAR identified no foreign objects at the reported dent location, it was assumed that the objects had been either removed from the steam generator, moved to a lower flow field, or simply disintegrated and became part of the sludge pile.

As a result of receiving NRC approval to extend the interval between tube inspections from a maximum of 24 to 40 calendar months, no steam generator tube inspections were performed during RFO 26 in 2004.

During a unit shutdown/cooldown in February 2005, approximately 1000 gallons of service water, which is drawn from Lake Michigan, entered the steam generators at Kewaunee. Additional information regarding this event appears in Preliminary Notification Report, PNO-III-05-003, "Shutdown in Excess of 72 Hours Due to Auxiliary Feedwater System Declared Inoperable Due to High Energy Line Break Concerns (ML050540652)."

At the time of the introduction of the service (Lake Michigan) water, the steam generator pressure was 62 psig in steam generator A and 56 psig in steam generator B. The saturation temperatures associated with these pressures are 309 °F and 304 °F, respectively.

The typical chemical composition of Lake Michigan water is 5 parts per million (ppm) sodium, 38 ppm calcium, 12 ppm magnesium, 144 ppm bicarbonate, 10 ppm chloride, and 22 ppm sulfate. The total dissolved solids is 233 and the pH is 8.5.

The steam generator pH never dropped below 9.5 during this excursion. Steam generator B was considered to have the most off-normal water chemistry. It had 397 parts per billion (ppb) sodium, 2.03 ppm calcium, 699 ppb magnesium, 618 ppb chloride, and 1.47 ppm sulfate. The pH was 9.85. The Electric Power Research Institute water chemistry guidelines were being followed.

After the introduction of the service water from Lake Michigan, the near- and long-term implications of the water chemistry excursion were evaluated, particularly the potential for the water chemistry to result in cracking or pitting of the steam generator tubes. With respect to cracking in the near term, no cracking is anticipated given the Alloy 690 tube material, the low temperature at the time of the service water introduction, and the absence of any appreciable heat flux at the time. With respect to pitting in the near term, no pitting is anticipated since the secondary water was not highly oxidizing or acidic nor did it contain any copper. In the long term, there is a possibility that the service water impurities (e.g., chlorides) could accumulate in deposits including sludge collars on the tubes. Given the recent replacement of the steam generators, the low iron transport into the steam generators, and the minimal heat flux following the service water introduction, the impurities from the steam generators were flushed to avoid any long-term impact from the chemical excursion.

The next steam generator tube inspections at Kewaunee are planned for April 2006.

3.2.15 McGuire 1

Tables 3-43, 3-44, and 3-45 summarize the information discussed in this section for McGuire 1. Table 3-43 provides the number of full-length bobbin inspections and the number of tubes plugged and deplugged during each outage for each of the four steam generators. Table 3-44 lists the reasons why the tubes were plugged. Table 3-45 lists the plugged tubes.

McGuire 1 has four recirculating steam generators designed and fabricated by BWI. The steam generators were put into service in 1997 during RFO 11. The licensee numbers its tube supports as depicted in Figure 2-32.

During the preservice inspection, 100 percent of the tubes in each of the steam generators were inspected full length with a bobbin coil. Before the steam generators went into service, 10 tubes were plugged.

During RFO 12 in 1998, approximately 95 percent of the tubes in each of the four steam generators were inspected full length with a bobbin coil. In addition, partial-length inspections were conducted on approximately 5 percent of the tubes in each of the four steam generators. As a result of these inspections, two tubes were plugged because of atypical wear.

During RFO 13 in 1999, approximately 30 percent of the tubes in all four steam generators were inspected full length with a bobbin coil. The tubes inspected included a 20-percent random sample (every fifth column), tubes around previously plugged tubes, tubes with previous indications, all periphery tubes (two tubes deep), and all tubes in rows 1 to 5. The periphery tubes were inspected because of a loose part found in a Catawba Unit 1 steam generator during RFO 10. In addition to the bobbin coil inspections, a rotating probe equipped with a +Point™ coil was used to examine all bobbin indications and all overexpanded and nonexpanded tubes identified during the preservice inspection, with the exception of one tube in steam generator B.

As a result of these inspections, no tubes were plugged; however, 11 indications of wear were detected. Of these 11 wear indications, 8 were located at fan bars and 3 were at lattice grid tube supports. No pitlike wear indications were identified during RFO 13. During the first inservice inspection of the steam generator tubes at McGuire 1 and 2, pit-like wear indications were identified. These pit-like indications were associated with anomalies on the fan bar.

Various secondary-side components (e.g., feedring, shroud, tubesheet, and various support structures) were inspected with a remote video camera during RFO 13. All components inspected showed no sign of erosion, corrosion, or scale buildup. The secondary side of the tubes at the tubesheet appeared very clean with a slight dusting of sludge in the areas inspected.

During RFO 14 in 2001, 100 percent of the tubes in steam generators B and C were inspected full length with a bobbin coil. In addition, a rotating probe was used to examine bobbin indications (19 locations in steam generator B and 32 in steam generator C). The rotating probe inspections included certain inlet and outlet locations, U-bend locations, and plugs. As a result of these inspections, no tubes were plugged; however, 24 wear indications in 23 tubes were detected (11 wear indications (10 tubes) in steam generator B and 13 wear indications in steam generator C).

During RFO 15 in 2002, no steam generator tube inspections were performed.

During RFO 16 in 2004, 100 percent of the tubes in steam generators A and D and approximately 55 percent of the tubes in steam generators B and C were inspected full length with a bobbin coil. In addition, a rotating probe equipped with a +Point™ coil was used to inspect the hot-leg expansion transition region (from 2 inches above to 8 inches below the top of the secondary face of the tubesheet) of approximately 20 percent of the tubes in all four steam generators. A rotating probe equipped with a +Point™ coil was also used to inspect U-bend special interest locations.

As a result of these inspections, one tube was plugged because of a 54 percent through-wall, single volumetric indication located slightly above the tubesheet on the cold-leg side of the steam generator. A visual inspection of the periphery after sludge lancing did not identify an object in this area; however, this inspection did not specifically target this location (sludge lancing was finished prior to the completion of the eddy current testing on the primary side). Approximately 100 indications of wear at the fan bars were reported during the inspection. These indications are considered typical fan bar wear (refer to Section 2.3.2 which describes the various types of fan bar wear).

No wear degradation resulting from the potential for some tubes to be in close proximity has been observed (refer to Section 2.3.2).

3.2.16 McGuire 2

Tables 3-46, 3-47, and 3-48 summarize the information discussed in this section for McGuire 2. Table 3-46 provides the number of full-length bobbin inspections and the number of tubes plugged and deplugged during each outage for each of the four steam generators. Table 3-47 lists the reasons why the tubes were plugged. Table 3-48 lists the plugged tubes.

McGuire 2 has four recirculating steam generators designed and fabricated by BWI and put into service in 1997 during RFO 11. The licensee numbers its tube supports as depicted in Figure 2-32.

During the preservice inspection, 100 percent of the tubes in each of the steam generators were inspected full length with a bobbin coil. Before the steam generators went into service, two tubes were plugged.

During RFO 12 in 1999, 100 percent of the tubes in each of the steam generators were inspected full length with a bobbin coil. As a result of these inspections, nine tubes were plugged. Most of these tubes had indications near the fan bars (presumably fan bar wear). The maximum depth of these indications was 39 percent through-wall. One of the plugged tubes had a volumetric indication near the third hot-leg tube support.

During RFO 13 in 2000, approximately 26 percent of the tubes in steam generators A and D and less than 1 percent of the tubes in steam generators B and C were inspected full length with a bobbin coil. In addition, partial inspections were conducted for approximately 2 percent of the tubes in steam generator A, 24 percent in steam generator B, 11 percent in steam generator C, and 2 percent in steam generator D. As a result of these inspections, no tubes were plugged.

During RFO 14 in 2002, 100 percent of the tubes in steam generators B and C were inspected full length with a bobbin coil. In addition, a rotating probe was used at various special interest locations (31 locations in steam generator B and 80 locations in steam generator C). The rotating probe inspections included certain inlet, outlet, and U-bend locations. As a result of these inspections, no tubes were plugged; however, 79 fan bar wear indications in 79 tubes were detected (24 wear indications in steam generator B and 55 wear indications in steam generator C). In addition, 2 possible loose part indications in 2 tubes were detected in steam generator B, and 2 possible loose part indications in 2 tubes were detected in steam generator C. There was no wear at the location of the possible loose parts.

During RFO 15 in 2003, no steam generator tube inspections were performed.

3.2.17 Millstone 2

Tables 3-49, 3-50, and 3-51 summarize the information discussed in this section for Millstone 2. Table 3-49 provides the number of full-length bobbin inspections and the number of tubes plugged and deplugged during each outage for each of the two steam generators. Table 3-50 lists the reasons why the tubes were plugged. Table 3-51 lists the plugged tubes.

Millstone 2 has two recirculating steam generators designed and fabricated by BWI and put into service in 1993 during RFO 11. The licensee numbers its tube supports as depicted in Figure 2-34.

During fabrication of the steam generators, one tube in steam generator A was plugged as a result of damage caused by a drill bit that broke while drilling the hot-leg tubesheet. This location was plugged on the hot-leg side of the steam generator, and the associated cold-leg tube hole was never drilled.

Before the steam generators went into service, a preservice inspection of the steam generator tubes was performed. This preservice inspection included a full-length bobbin coil inspection of 100 percent of the tubes in each of the steam generators while they were in a horizontal position before their installation. Immediately following their installation, 21 percent of the tubes in each steam generator were inspected primarily to determine if any tube damage had occurred during the installation process. The results of this inspection showed that the condition of the tubes had not changed since the preinstallation inspection.

During RFO 12 in 1994, approximately 28 percent of the tubes in each of the steam generators were inspected full length with a bobbin coil. As a result of these inspections, no tubes were plugged; however, one tube was reported to have degradation. This tube was located in row 140, column 79, a peripheral tube. The indication was approximately 22 percent through-wall and was located on the portion of the tube within the first hot-leg lattice grid support (i.e., 01H). A rotating probe inspection of this location did not result in the identification of any degradation, which confirmed that this indication was shallow. The tube was left in service. No other

-123-

rotating probe inspections were performed during this outage. As discussed later, this indication no longer exhibited flawlike characteristics during the 1997 tube inspections.

The tubesheet area was inspected both before and after sludge lancing during RFO 12. The inspection before sludge lancing consisted of a 360 degree inspection of the annulus region, views down the blowdown lane from both the hot and cold-leg, inspections down into the blowdown tube holes, and an inspection of the shroud lug-to-shell interface. The purpose of these inspections was to determine the effectiveness of the sludge-lancing process and to search for and retrieve any foreign objects. The annulus region had only a small amount of sludge accumulation, and machine marks were visible on the tubesheet. The blowdown lane and blowdown holes showed little to no sludge buildup. The inner tube bundle area at the top of the tubesheet was inaccessible for visual inspection; however, the tubes that could be inspected were found to be very clean at the top of the tubesheet. The mass of sludge removed during lancing indicated very low levels of sludge accumulation on the tubesheet. A foreign object, thought to be a piece of weld wire, was retrieved from the tubesheet in steam generator B. No damage was associated with this part.

During RFO 12, 3 bulge indications, 13 dents, 281 dings, and 66 MBMs were identified. Various steam generator internal components of both steam generators were visually examined including the steam outlet nozzle; primary and secondary steam separators; primary and secondary decks, supports, and seals; feedwater assembly (as accessible); U-bend support structure; top lattice support (as accessible); and the top of the tubesheet. No degradation was detecting during these inspections.

As a result of an extended shutdown following RFO 12, the licensee asked for NRC approval to delay the next steam generator tube inspections until October 1997. This request was subsequently withdrawn, since the licensee decided to perform the tube inspections during the current midcycle outage.

During a midcycle outage in June 1997, approximately 75 percent of the tubes in steam generator A and 30 percent in steam generator B were inspected full length with a bobbin coil. In addition, a rotating probe was used to inspect 30 locations in steam generator A and 31 locations in steam generator B. The locations inspected were at various elevations along the length of the tube. As a result of these inspections no indications of quantifiable wall loss were identified and no tubes were plugged.

During the midcycle outage, an eddy current signal was identified in the tube in row 140, column 79. During the RFO 12 inspections, this tube had exhibited a shallow flaw signal; however, a review of the midcycle inspection data indicated that this signal was no longer flawlike.

During RFO 13 in 2000, 100 percent of the tubes in steam generator B were inspected full length with a bobbin coil. In addition, a rotating probe equipped with a +Point™ coil was used to examine approximately 87 locations in steam generator B. The rotating probe inspections were performed at locations of special interest identified during the bobbin coil inspection (e.g., hot-leg tubesheet, bulges, dents, and dings). These inspections detected no degradation, and no tubes were plugged.

During RFO 14 in 2002, 100 percent of the tubes in steam generator A were inspected full length with a bobbin coil. In addition, a rotating probe equipped with a +Point™ coil to examine approximately 57 locations in steam generator A. These rotating probe inspections were performed at 37 locations of special interest identified during the bobbin coil inspection (e.g., hot-leg tubesheet, bulges, MBMs, dents, and dings) and 20 locations for PLPs. As a result of these inspections, two tubes were identified with indications of wear at fan bar locations. The wear scars measured 9 percent through-wall. The affected tubes were in row 40, column 155, and row 140, column 93. No tubes were plugged during this outage.

During the outage, 2 bulge indications (at tube support 07H), 9 dents, 13 dings, and 77 nonquantifiable indications (previously referred to as MBMs) were identified. None of the nonquantifiable indications exhibited any change when compared to the preservice data. In addition to these indications, one tube was identified with a partial tube expansion (i.e., the tube was not fully expanded into the tubesheet region).

In response to a loose part monitoring system alarm in February 2002, the typical visual inspection of the secondary side of the steam generator was augmented. The inspection included an extensive visual inspection of the top of the tubesheet, tubesheet annulus, blowdown holes, and inner bundle. The visual inspection identified only one small object lodged between several tubes (row 23, column 102; row 24, column 103; row 24, column 102; and row 24, column 101). A review of the eddy current data from this outage confirmed that a marginal PLP signal was present in three tubes. A review of eddy current data from prior cycles indicated that the signal was present in 1994 and 1997. No tube damage was associated with the loose part signal. The part was not retrieved, but it was assessed from a tube integrity standpoint. Since there was no physical evidence of a part in the lower portion of the secondary side of the steam generator that would have caused the loose part monitoring system alarm, the upper steam drum was inspected. This inspection included a check of each secondary separator and a complete visual inspection of the accessible areas above the primary deck. All components were intact. The upper tube bundle was then inspected. No abnormalities were identified in the U-bend region. Specific attention was paid to the eighth fan bar intersection for the tube in row 140, column 93 (this was one of the tubes with a wear scar). the inspection found no evidence of unusual support conditions that could have explained the loose part monitoring system alarm. Subsequent testing of the loose part monitoring system before the unit returned to full power indicated that a cable malfunction had caused the alarm.

During RFO 15 in 2003, 100 percent of the tubes in steam generator B were inspected full length with a bobbin coil. In addition, a rotating probe equipped with a +Point™ coil was used to inspect approximately 214 tubes (227 locations) in steam generator B. These rotating probe inspections were performed at 87 locations of special interest identified during the bobbin coil inspection (e.g., hot-leg tubesheet, bulges, MBMs, dents, dings, and PLPs) and 140 locations where past inspections had shown a sludge pile to exist. As a result of these eddy current inspections, no tubes were plugged.

During the eddy current inspections, two tubes were identified with indications of wear at fan bar locations. These tubes were in row 37, column 120, and row 99, column 80. The deepest wear scar at the fan bars measured 11 percent through-wall.

Five tubes were identified with wear indications or single volumetric indications attributed to loose parts. These wear indications were located from seven inches below the first lattice grid support to the support. The deepest wear scar measured 24 percent through-wall. A visual inspection indicated that the loose parts were flexitallic gaskets, and all parts at these locations were removed from the steam generator during the FOSAR. The visual inspections identified irretrievable flexitallic gaskets in four other locations. This gasket material was located on the tubesheet. In addition to the five tubes with wear associated with loose parts, several other tubes were identified that had eddy current indications of PLPs but with no indication of wear. These indications were located slightly above the tubesheet on the hot-leg side of the steam generator. All these PLP indications were identified and evaluated during RFO 13. These eddy current indications are attributed to flexitallic gaskets, weld rods, weld slag, and spacers.

During RFO 15, 22 dents, dings, and bulges were identified. Most of the dents and dings were traceable to the preservice inspection. Rotating probe inspections at dent and ding locations has not resulted in the identification of any tube degradation. Three tubes were identified with bulges located at either lattice grid support 07H or near the top of the tubesheet on the hot-leg side of the steam generator.

3.2.18 North Anna 1

Tables 3-52, 3-53, and 3-54 summarize the information discussed in this section for North Anna 1. Table 3-52 provides the number of full-length bobbin inspections and the number of tubes plugged and deplugged during each outage for each of the three steam generators. Table 3-53 lists the reasons why the tubes were plugged. Table 3-54 lists the plugged tubes.

North Anna 1 has three recirculating steam generators designed and fabricated by Westinghouse. The model 54F steam generators were put into service in 1993. The licensee numbers its tube supports as depicted in Figure 2-36.

Before the steam generators went into service, no tubes were plugged.

During RFO 10 in 1994, approximately 50 percent of the tubes in steam generators A and C were inspected full length with a bobbin coil. No tubes were plugged as a result of these inspections, and no indications were recorded (i.e., no degradation was identified).

During RFO 11 in 1996, approximately 50 percent of the tubes in steam generator B were inspected full length with a bobbin coil. In addition, a rotating probe was used to inspect the hot-leg expansion transition region of approximately 9 percent of the tubes in steam generator B. No tubes were plugged as a result of these inspections, and no indications were recorded.

During RFO 12 in 1997, approximately 50 percent of the tubes in steam generator A were inspected full length with a bobbin coil. In addition, a rotating probe was used to inspect the hot-leg expansion transition region of approximately 9 percent of the tubes in steam generator A. No tubes were plugged as a result of these inspections, and no indications were recorded.

During RFO 12, it was determined that the eddy current data for two tubes in steam generator A did not match signatures reported during the preservice inspection. Investigations determined that these two tubes had not received a preservice inspection, and the data for other tubes were misreported as being for these tubes. The affected tubes were located in row 21, column 23, and row 22, column 23.

During RFO 13 in 1998, 50 percent of the tubes in steam generator C were inspected full length with a bobbin coil. In addition, a rotating probe was used to inspect the hot-leg expansion transition region of approximately 9 percent of the tubes in steam generator C and the U-bend region of 10 tubes. One tube was plugged during this outage as a result of a small volumetric indication classified as a "pit." The indication was inspected with a rotating probe but was not depth sized because the signal was too poorly formed on the sizing channel of the bobbin data.

No primary-to-secondary leakage was observed between the 1998 and 2000 refueling outages.

During RFO 14 in 2000, 50 percent of the tubes in steam generator B were inspected full length with a bobbin coil. In addition, a rotating probe was used to inspect the hot-leg expansion transition region (from 3 inches above to 3 inches below the top of the secondary face of the tubesheet) of approximately 20 percent of the tubes in steam generator B and the U-bend region of 100 percent of the row 1 tubes (98 tubes) in steam generator B. No tubes were plugged as a result of these inspections, and no degradation was identified.

During the rotating probe inspections at the top of the tubesheet, one single axial anomaly and one MBM were observed. The axial anomaly in the tube at row 3, column 58, appeared to be produced by a ding in the transition area. The manufacturing mark in the tube at row 10, column 52, was located above the transition and could be traced back to the preservice bobbin data.

During the rotating probe inspection in the U-bends of the row 1 tubes, six circumferential anomalies were observed. Further examination of these indications with a rotating probe equipped with a +Point™ coil showed no signals indicative of cracks so the tubes were left in service.

Secondary-side inspections were performed during this outage. The uppermost tube support (i.e., the seventh), including wedge blocks, backup bars, and support structures, was inspected, and no issues were identified. A light oxide deposit covered all tube support plate surfaces examined with no appreciable loose sludge. Tube surfaces showed minimal oxide buildup and no evidence of scale. No blockage or sludge buildup was noted within the tube support plate broached holes.

No primary-to-secondary leakage was observed during the operating cycle between RFO 14 and 15.

During RFO 15 in 2001, approximately 60 percent of the tubes in steam generator A were inspected full length with a bobbin coil. In addition, a rotating probe equipped with a +Point™ coil was used to inspect the hot-leg expansion transition region (from 3 inches above to 3 inches below the top of the secondary face of the tubesheet) of approximately 20 percent of the tubes in steam generator A, the U-bend region of 100 percent of the row 1 tubes (98 tubes) in steam generator A, and selected bobbin indications in steam generator A. No tubes were plugged as a result of these inspections, and no degradation was identified.

No wear at the AVBs was observed during the 2001 inspection. A total of 78 tubes were reported with dent signals exceeding 2 volts (approximately 21 tubes have dents exceeding 5 volts). Nineteen tubes were reported to have MBMs and one tube was reported with a non-quantifiable indication that has existed since the preservice inspection.

During RFO 16 in 2003, no steam generator tube inspections were performed.

No primary-to-secondary leakage was observed during the operating cycle between RFO 16 and 17.

During RFO 17 in 2004, 100 percent of the tubes in steam generator C were inspected full length with a bobbin coil, with the exception of the U-bend region of the tubes in row 1. In addition, a rotating probe equipped with a +Point™ coil was used to inspect the hot-leg expansion transition region (from 3 inches above to 3 inches below the top of the secondary face of the tubesheet) of approximately 20 percent of the tubes in steam generator C, the U-bend region of 100 percent of the row 1 tubes in steam generator C, 27 percent of the dents greater than 2 volts in steam generator C, and selected bobbin indications in steam generator C. The dents selected for examination (approximately 31 locations) consisted of dent or bulge signals not meeting a voltage and/or phase change criteria (3 locations) and a sample of locations with no significant voltage or phase change.

As a result of these inspections, one tube was plugged because of wear associated with a loose part. This was the only degradation mechanism identified during the outage. This tube contained an indication measuring 43 percent through-wall and located approximately 4 inches above the seventh cold-leg tube support plate. There was also an indication of a PLP at this location and on a neighboring tube, which visual examinations confirmed as a loose part. The part was a steel wire measuring 14.2 inches long and 3/32 inch in diameter. The part, presumed to be an original fabrication or construction installation remnant was removed. A bobbin inspection of the tubes in two columns surrounding the exit path of the part confirmed that no damage had occurred during the removal process. A reanalysis of the 1998 inspection data revealed a signal that could be consistent with an object adjacent to the tube; however, there was no tube wear at this location. In addition to this loose part, two other objects were identified and removed from the steam generator.

No indications of AVB wear were detected, and none have been observed in any of the three steam generators since they were put into service. In addition, no AVB wear has been observed at North Anna 2.

3.2.19 North Anna 2

Tables 3-55, 3-56, and 3-57 summarize the information discussed in this section for North Anna 2. Table 3-55 provides the number of full-length bobbin inspections and the number of tubes plugged and deplugged during each outage for each of the three steam generators. Table 3-56 lists the reasons why the tubes were plugged. Table 3-57 lists the plugged tubes.

North Anna 2 has three recirculating steam generators designed and fabricated by Westinghouse. The model 54F steam generators were put into service in 1995. The licensee numbers its tube supports as depicted in Figure 2-36.

Before the steam generators went into service, no tubes were plugged.

During RFO 1 in 1996 (designated as RFO 1 to indicate the first RFO after replacement), 50 percent of the tubes in steam generators B and C were inspected full length with a bobbin coil. In addition, a rotating probe was used to inspect the hot-leg expansion transition region of approximately 4.5 percent of the tubes in steam generators B and C. No tubes were plugged as a result of these inspections, and no indications were recorded.

During RFO 2 in 1998, approximately 50 percent of the tubes in steam generator A were inspected full length with a bobbin coil. In addition, a rotating probe was used to inspect the expansion transition region of approximately 9 percent of the tubes in steam generator A. No tubes were plugged as a result of these inspections.

During RFO 3 in 1999, 50 percent of the tubes in steam generator B were inspected full length with a bobbin coil. In addition, a rotating probe was used to inspect the hot-leg expansion transition region (from 3 inches above to 3 inches below the top of the secondary face of the tubesheet) of approximately 20 percent of the tubes in steam generator B and the U-bend region of approximately 20 percent of the row 1 tubes (20 tubes) in steam generator B. No tubes were plugged as a result of these inspections, and no degradation was identified.

No primary-to-secondary leakage was observed during the operating cycle between RFO 3 and 4.

During RFO 4 in 2001, approximately 60 percent of the tubes in steam generator C were inspected full length with a bobbin coil. In addition, a rotating probe equipped with a +Point™ coil was used to inspect the hot-leg expansion transition region (from 3 inches above to 3 inches below the top of the secondary face of the tubesheet) of approximately 20 percent of the tubes in steam generator C, the U-bend region of 100 percent of the row 1 tubes in steam generator C, and selected bobbin indications in steam generator C. One tube was plugged as a result of these inspections. This tube had a 30 percent through-wall volumetric indication at the fifth cold-leg tube support. The bobbin voltage associated with this indication was 1.61 volts. The indication, which was near the upper edge of the support and coincided with one of the lands of the support plate, appeared to be mechanical in origin. The indication likely initiated during the first cycle of operation and might have resulted from wear against a small manufacturing burr or some other small discrete particle located at the edge of one of the quatrefoil lands.

No wear at the AVBs was observed during the 2001 inspection. Seven tubes were reported with dent signals, and nine tubes were reported to have MBMs.

No primary-to-secondary leakage was observed between the 2001 and 2002 RFOs.

During RFO 5 in 2002, approximately 60 percent of the tubes in steam generator A were inspected full length with a bobbin coil with the exception of the U-bend region of the tubes in row 1. These bobbin coil inspections included 50 percent of the tubes previously examined during the preservice inspection and an additional 10 percent of the tubes inspected in 1998. The inspections focused on the peripheral and tube lane areas and other random locations. In addition, a rotating probe equipped with a +Point™ coil was used to inspect the hot-leg expansion transition region (from 3 inches above to 3 inches below the top of the secondary face of the tubesheet) of approximately 20 percent of the tubes in steam generator A, the U-bend region of 100 percent of the row 1 tubes in steam generator A, and all dents greater than 2 volts in steam generator A (four locations). The dents selected for examination consisted of dent or bulge signals not meeting a voltage and/or phase change criteria and a sample of locations with no significant voltage or phase change.

One tube was plugged as a result of these inspections. This tube was plugged due to a permeability signal which extended the full length of the area inspected with the rotating probe. The permeability signal was not present in the bobbin coil data, because the bobbin coil is magnetically biased and the rotating probe is not. The use of a magnetically biased rotating probe reduced the permeability signal but did not eliminate it. Since the permeability signal may prohibit unambiguous interpretation of the +Point™ data at the expansion transition area (i.e., it could reduce the ability of the eddy current probe to detect actual flaws that could develop in the tube), the tube was plugged.

A visual inspection was performed of the steam drum area, the interface between the feedwater distribution ring and the J nozzles, and the seventh tube support plate during RFO 5. In addition to these visual inspections, ultrasonic thickness measurements were performed on selected feedwater distribution ring components. These inspections identified no conditions that would potentially compromise tube integrity.

The 100-kHz bobbin coil data exhibited very minimal signal distortion during the outage indicating that there is no appreciable amount of sludge at, or within, the steam generator support structures. Visual inspections of the seventh tube support plate confirmed this.

Sludge lancing and top of tubesheet video inspections were not conducted during this outage; however, data from the +Point™ coil were reviewed for evidence of localized sludge deposits. Most locations examined showed no evidence of sludge accumulation, although some sludge was present in the tubes in the baffle plate "cut-out" region and in the region bounded by columns 39 through 54 and extending from rows 4 through 12.

Four tubes with dent signals were identified during the bobbin coil inspection. The dents ranged in magnitude from 2.0 volts to 2.8 volts. A rotating probe equipped with a +Point™ coil was used to further examine the dent locations to obtain an appropriate baseline for future inspections. No degradation was identified at these locations, so the four tubes with dent signals remained in service.

Seventeen tubes were identified with MBMs. A comparison of these indications with the 1995 preservice inspection results revealed no change from the preservice inspection. No wear was observed at the AVBs.

During RFO 6 in 2004, no steam generator tube inspections were performed.

3.2.20 Oconee 1

Tables 3-58, 3-59, and 3-60 summarize the information discussed in this section for Oconee 1. Table 3-58 provides the number of full-length bobbin inspections and the number of tubes plugged and deplugged during each outage for each of the two steam generators. Table 3-59 lists the reasons why the tubes were plugged. Table 3-60 lists the plugged tubes.

Oconee 1 has two once-through steam generators designed and fabricated by BWI. The once-through steam generators went into service in 2004 during RFO 21. The licensee numbers its tube supports as depicted in Figure 2-38.

During the preservice inspection, 100 percent of the tubes in each of the two steam generators were inspected full length with a bobbin coil and X-probe (a 2x14 pancake array probe) simultaneously. The X-probe technique was used to characterize reportable and special interest indications discovered with the bobbin coil inspection technique. One tube in each steam generator was removed from service by plugging because of damage that occurred during manufacturing.

During the preservice inspection, profilometry was performed on each tubesheet hole. This analysis consisted of evaluating the tube expansion profile to determine if the tube was expanded, where the closure gap location was with relation to the top of the tubesheet, the diameter of the expansion, and any abnormal signals within the tubesheet expansion. In steam generators A and B, all tubes were expanded, and no tube holes were overexpanded (i.e., no indications within the tubesheet had a deviation of 0.005 inches or greater from the nominal tube expansion diameter of each tube hole).

As a result of the preservice inspection, 1265 MBMs in 1044 tubes were identified in steam generator A and 920 MBMs in 835 tubes were identified in steam generator B. These indications may have been caused by tube conditioning at the tube manufacturer to remove localized tubing imperfections on the outside diameter of the tube or to remove imperfections caused by the installation of the tubes into the tube bundle. No dents, dings, or permeability variations were observed in steam generator A. One ding and one permeability variation were reported in steam generator B.

During RFO 22 in 2005, the first inservice inspection of the steam generator tubing following replacement of the original steam generators was performed. During RFO 22, 100 percent of the tubes in each of the two steam generators were inspected full length with a bobbin coil.

These inspections found that approximately 11.5 percent of the tubes in steam generator A and 9.6 percent of the tubes in steam generator B had indications of wear at the tube support plate elevations. Most of the indications were located between the ninth (009) and eleventh (011) tube supports. In addition, most of the indications were shallow (less than 20 percent through-wall), and all of the tubes had adequate structural and leakage integrity. Some of the tubes had multiple indications at the same tube support elevation, and some tubes had indications at multiple support plate elevations. Most of the indications are in the periphery of the tube bundle; however, indications are spread throughout the interior portion of the tube bundle. The largest indications are located approximately five tubes in from the periphery. The maximum depth reported for the wear indications was 42 percent through-wall.

The repair criteria for these wear indications were assessed during the outage. As part of this assessment, various analysis methodologies were investigated. The maximum size a flaw could be and still have adequate structural integrity was estimated to be 80.6 percent through-wall (based on a tapered wear scar of 1.0 inch). The repair criterion implemented during the outage was 28 percent through-wall. This criterion was developed assuming the maximum wear rate observed (i.e., 42 percent through-wall per cycle). This criterion resulted in the plugging of 30 tubes in steam generator A and 18 tubes in steam generator B. All plugged tubes were stabilized for the full length of the tube.

A root cause investigation was commenced during the outage to aid in determining the extent of condition and the factors that led to the large number of tubes affected by wear. At the time of the writing of this report, the root cause investigation was ongoing.

3.2.21 Oconee 2

Tables 3-61, 3-62, and 3-63 summarize the information discussed in this section for Oconee 2. Table 3-61 provides the number of full-length bobbin inspections and the number of tubes plugged and deplugged during each outage for each of the two steam generators. Table 3-62 lists the reasons why the tubes were plugged. Table 3-63 lists the plugged tubes.

Oconee 2 has two once-through steam generators designed and fabricated by BWI which were put into service in 2004 during RFO 20. The licensee numbers its tube supports as depicted in Figure 2-38.

During the preservice inspection, 100 percent of the tubes in each of the two steam generators were inspected full length with a bobbin coil and X-probe (a 2x14 pancake array probe) simultaneously. The X-probe technique was analyzed only for special interest areas and a sample of the MBM indications. Four tubes in steam generator A and one tube in steam generator B were plugged because of misdrilling of the tubesheet or because of damage that occurred during manufacturing.

During the preservice inspection, profilometry was performed on each tubesheet hole. This analysis consisted of evaluating the tube expansion profile to determine if the tube was expanded, where the closure gap location was with relation to the top of the tubesheet, the diameter of the expansion, and any abnormal signals within the tubesheet expansion. In steam generators A and B, all tubes were expanded and no tube holes were overexpanded (i.e., no indications within the tubesheet had a deviation of 0.005 inches or greater from the nominal tube expansion diameter of each tube hole).

The preservice inspection identified 1166 MBMs in 1048 tubes in steam generator A and 580 MBMs in 475 tubes in steam generator B. These indications may be the result of tube conditioning at the tube manufacturer to remove localized tubing imperfections on the outside diameter of the tube or to remove imperfections caused by the installation of the tubes into the tube bundle. No dents, dings, or permeability variations were observed in steam generator A or B.

During the preservice inspection, four tubes were found to have a deposit-like signal in steam generator A. These signals resulted from melted plugs. After removal of the deposits, the four tubes were reinspected and left in service.

3.2.22 Oconee 3

Tables 3-64, 3-65, and 3-66 summarize the information discussed in this section for Oconee 3. Table 3-64 provides the number of full-length bobbin inspections and the number of tubes plugged and deplugged during each outage for each of the two steam generators. Table 3-65 lists the reasons why the tubes were plugged. Table 3-66 lists the plugged tubes.

Oconee 3 has two once-through steam generators designed and fabricated by BWI which were put into service in 2004 during RFO 21. The licensee numbers its tube supports as depicted in Figure 2-38.

During the preservice inspection, 100 percent of the tubes in each of the two steam generators were inspected full length with a bobbin coil and X-probe (a 2x14 pancake array probe) simultaneously. The X-probe technique was analyzed only for special interest areas and a sample of the MBM indications greater than 1 volt. Before the steam generators went into service, no tubes were plugged.

During the preservice inspection, profilometry was performed on each tubesheet hole. This analysis consisted of evaluating the tube expansion profile to determine if the tube was expanded, where the closure gap location was with relation to the top of the tubesheet, the diameter of the expansion, and any abnormal signals within the tubesheet expansion. In steam generators A and B, all tubes were expanded, and no tube holes were overexpanded (i.e., no indications within the tubesheet had a deviation of 0.005 inches or greater from the nominal tube expansion diameter of each tube hole).

As a result of the preservice inspection, 383 MBMs in 357 tubes were identified in steam generator A and 990 MBMs in 880 tubes were identified in steam generator B. These indications may be caused by tube conditioning at the tube manufacturer to remove localized tubing imperfections on the outside diameter of the tube or to remove imperfections caused by the installation of the tubes into the tube bundle.

3.2.23 Palo Verde 2

Tables 3-67, 3-68, and 3-69 summarize the information discussed in this section for Palo Verde 2. Table 3-67 provides the number of full-length bobbin inspections and the number of tubes plugged and deplugged during each outage for each of the three steam generators. Table 3-68 lists the reasons why the tubes were plugged. Table 3-69 lists the plugged tubes.

Palo Verde 2 has two recirculating steam generators designed by Combustion Engineering and fabricated by Ansaldo. The steam generators were put into service in 2003 during RFO 11. The licensee numbers its tube supports as depicted in Figure 2-40.

One tube in steam generator B was plugged at the factory with a welded plug.

During the preservice inspection, 100 percent of the tubes in both steam generators were inspected full length with a bobbin coil (except for the U-bend region of the tubes in rows 1 through 3). In addition, a rotating probe was used to inspect the hot-leg and cold-leg expansion transition region (from 2 inches above to 3 inches below the top of the secondary face of the tubesheet) of 100 percent of the tubes and the U-bend region of 100 percent of the row 1, 2, and 3 tubes. In addition, a rotating probe was used to inspect selected locations including dents, MBMs, and bulges. The dent inspections focused on dents located in the hot-leg region and larger dents.

As a result of these inspections, several tubes were found to have been (1) over expanded above the top of the tubesheet, (2) not hard rolled, or (3) not hydraulically expanded. These conditions were determined to be acceptable. Several tubes were also identified with PLP indications, and some of these loose parts were removed from the steam generator. An engineering evaluation was performed to address tube integrity concerns related to the PLPs that were not removed.

During the preservice inspection, a total of 22 tubes were plugged. Of these 22 tubes, 16 tubes were plugged to support installation of the robotic fixture for inspection and repair of the steam generator tubes (referred to as "rail plugs"), 5 tubes were plugged because of the detection of a groove on the outside diameter surface of the tube, and 1 tube was plugged because of a large dent at a lower hot-leg support (03H).

When the unit started up following the replacement of the steam generators in December 2003, a small primary-to-secondary leak measuring approximately 0.6 gpd was observed. Over the following 2 months, the leak rate varied between 0.4 and 0.7 gpd until February 19, 2004, when the leak rate increased from approximately 0.7 to 11 gpd in a 38-minute timeframe. Although the leak rate did not exceed the technical specification limit, the unit was shut down to identify the source of the leak.

While the unit was shut down, the secondary side of the steam generator was pressurized to 600 pounds per square inch (psi) to assist in the identification of the leaking tube or tubes.

During this pressure test, leakage was easily observed coming from a peripheral tube. This tube was subsequently inspected with both a bobbin and a rotating probe. These inspections did not reveal evidence of inservice degradation but did confirm the presence of a dent near a vertical support in the middle of the horizontal run of the tube which had been detected in the preservice inspection. This dent signal was considered anomalous because it differed from a typical dent signal in that it exhibited some flawlike characteristics (i.e., it had a vertical component). A comparison of the preservice bobbin and rotating probe inspection data to the data obtained during the outage revealed no significant differences in the dent signal. Although the dent signal was anomalous, there was no distinct indication of material volume loss.

Since the eddy current inspections of the affected tube did not provide conclusive evidence of a through-wall flaw, additional testing was performed. This testing included primary- and secondary-side visual inspections and an in situ pressure test. The visual inspections confirmed the presence of a dent which did not appear to be the result of the fabrication of the support structure since the dent was not located directly next to a support strap and did not appear to be the result of impact or leverage. During an in situ pressure test of the entire tube, leakage of 0.08 gallons per minute (gpm) was observed at the differential pressure associated with postulated accident conditions (e.g., a main steam line break), and the tube did not burst at three times the differential pressure associated with normal operating conditions. These tests confirmed that the tube had adequate structural integrity. In addition, the leakage from this tube was well below the allowable leakage under postulated accident conditions. Following the in situ pressure test, the leaking tube was plugged and stabilized.

Because of the findings regarding the leaking tube, the rotating probe data for all dent signals obtained during the preservice inspection were reviewed to ascertain whether similar anomalous dent signals existed. In addition, rotating probe inspections were performed at dents whose voltages exceeded a specific voltage (e.g., 2 to 5 volts) if these dents had not been inspected with a rotating probe during the preservice inspection. These efforts identified one additional tube with an anomalous signal, but this was not conclusively similar to the other indication with respect to the vertical presentation of the eddy current signal. This tube was plugged during the preservice inspection because the dent obstructed the passage of the normal-sized bobbin probe, and there was concern for the future inspectability of this location.

Additional efforts were made to determine the root cause of the leak in the tube. These efforts included reviewing steam generator manufacturing records and developing mockup specimens to simulate the anomalous eddy current signal in the leaking tube.

During the examination of the manufacturing records of the steam generator, reviewers determined that one tube was scrapped during the fabrication of the replacement steam generators because it had been damaged (or pierced) by a packing screw. Screws were used in the packing crate in which the tubes were shipped from the tubing manufacturer to the steam generator fabrication facility. The affected portion of this tube was returned to the tubing manufacturer and corrective actions were taken; however, at the time of the discovery of this damaged tube, all of the tubes in one of the Unit 2 steam generators had been installed, and the other steam generator was in the process of being fabricated.

To simulate the anomalous dent signal in the leaking tube, a series of dents was fabricated in a mockup facility. The simulation included impact dents from a nail, a screw, and a drill bit. A wood screw, similar to that used in the tube manufacturer's crate, was driven through a piece of wood and into the sample tube. Eddy current testing performed on these specimens revealed that the damage caused by the wood screw yielded a similar anomalous signal to that found in the leaking tube.

The formal root cause evaluation confirmed that the tube packing crate used wood spacers and cross brace materials that were assembled using common screws as the tubes were loaded into the crate. The design of this packing material placed the screws in proximity to specific locations on some tubes, and the location, shape, and size of the deformation in the leaking tube are consistent with damage that would occur if a screw penetrated completely through the packing material and came in contact with the tube.

As a result of the findings, many corrective actions were taken, including inspecting selected tubes, plugging and stabilizing the leaking tube, adding additional quality control inspectors at the steam generator fabrication facility (since this facility is fabricating replacement steam generators for Unit 1), modifying the receipt inspections performed (including procedural changes) on the tubes at the fabrication facility, evaluating and modifying the packing procedure/design, identifying the tubes that were shipped in package locations where packing screw damage was possible, and initiating additional mockup testing to improve the capability to identify and characterize volumetric flaws located within a dent (e.g., puncture-type defects).

After concluding there was reasonable assurance of tube integrity, the unit was returned to service. The primary-to-secondary leak rate following startup was near the detection threshold (i.e., less than 0.1 gpd). In addition, following unit startup, six additional tubes that had been pierced by a packing crate screw were found at the fabrication facility during the unpacking of tubes for the Palo Verde Unit 1 replacement steam generators. These tubes were not installed in any of the steam generators being fabricated.

During the 2004 midcycle inspection, approximately 95 tubes were inspected with a bobbin coil. These 95 tubes included tubes adjacent to the leaking tube (R156C143), tubes in the cold-leg corner of the steam generator (a region in the original steam generator that was susceptible to wear), and tubes near the stay cylinder (a region in the original steam generator that was susceptible to wear near the batwings). In addition to the bobbin probe inspections, a rotating probe equipped with a +Point™ coil was used to inspect various dents. In steam generator A, all dents greater than 2 volts that had not been inspected with a rotating probe during the preservice inspection were inspected during the midcycle. In addition, eight dents in steam generator B were inspected with a rotating probe during the midcycle inspection. All dents greater than 5 volts in steam generator B were inspected during the preservice inspection or during the midcycle inspection.

As a result of these inspections, only one tube was plugged (i.e., the leaking tube). The inspections did detect a loose part which was subsequently removed from the steam generator. The part appeared to be a thin shim of ferritic material which crumbled upon touch.

3.2.24 Point Beach 2

Tables 3-70, 3-71, and 3-72 summarize the information discussed in this section for Point Beach 2. Table 3-70 provides the number of full-length bobbin inspections and the number of tubes plugged and deplugged during each outage for each of the two steam generators. Table 3-71 lists the reasons why the tubes were plugged. Table 3-72 lists the plugged tubes.

Point Beach 2 has two recirculating steam generators designed and fabricated by Westinghouse. The model Delta 47 steam generators were put into service in 1997 during RFO 22. The licensee numbers its tube supports as depicted in Figure 2-42.

During the preservice inspection, 100 percent of the tubes (except for the U-bend region of the row 1 and 2 tubes) in each of the steam generators were inspected full length with a bobbin coil. In addition, a rotating probe equipped with a +Point™ coil was used to inspect the U-bend region of 100 percent of the row 1 and 2 tubes (i.e., 105 tubes); at least one tube from each heat (34 tubes in steam generator A and 24 tubes in steam generator B) at the expansion transition, at the flow distribution baffle, and at each of the support plates; and MBMs. Before the steam generators went into service, no tubes were plugged.

During RFO 23 in 1999, 100 percent of the tubes (except for the U-bend region of the row 1 and 2 tubes) in each of the steam generators were inspected full length with a bobbin coil. In addition, a rotating probe equipped with a +Point™ coil was used to inspect the hot-leg expansion transition region (from 3 inches above to 2 inches below the top of the secondary face of the tubesheet) of approximately 20% of the tubes, and the U-bend region of 100 percent of the row 1 and 2 tubes (i.e., 105 tubes). As a result of these inspections, two tubes were plugged and stabilized as a result of wear associated with a loose part. The wear occurred at the top of the tubesheet on the hot-leg side of the steam generator. Additional rotating probe inspections performed in the adjacent tubes confirmed that no other tubes had experienced wear as a result of this loose part. The loose part was removed during sludge lancing, and subsequent visual inspections verified that the loose part was no longer present. Several small foreign objects were removed from the sludge-lance filters. These objects were believed to have been introduced during the steam generator replacement outage. Six dents were recorded during the RFO 23 inspections.

During RFO 24 in 2000, 100 percent of the tubes in each of the steam generators were inspected full length with a bobbin coil. In addition, a rotating probe equipped with a +Point™ coil was used to inspect the hot-leg expansion transition region (from tube end to more than 2 inches above the top of the tubesheet) of approximately 40 percent of the tubes; the U-bend region of 20 percent of the row 1 tubes (11 tubes per steam generator); all dents with bobbin voltages greater than 5 volts; permeability variation signals; and deposit indications. Twenty dents were recorded during the RFO 24 inspections. As a result of these inspections, two tubes were plugged because of excessive noise associated with the eddy current data. The noise in the data for these two tubes was present at the same level during the preservice inspection.

In addition to the steam generator tube inspections during RFO 24, the steam generator swirl vane and moisture separators were inspected. No degradation was noted during these inspections.

During RFO 25 in 2002, no steam generator tube inspections were performed.

During RFO 26 in 2003, approximately 50 percent of the tubes in each of the steam generators were inspected full length with a bobbin coil. In addition, a rotating probe equipped with a +Point™ coil was used to inspect the hot-leg expansion transition region (from 2 inches above to 2 inches below the top of the secondary face of the tubesheet) of approximately 25 percent of the tubes; the U-bend region of 25 percent of the row 1 tubes; the U-bend region of 15 percent of the row 2 tubes; and an additional 379 tubes at the hot- and cold-leg expansion transitions (i.e., these 379 tubes were in addition to the 25 percent sample) in each steam generator. The 379 tubes are peripheral tubes and tubes associated with the "no-tube" lane (i.e., the open area between the row 1 tubes).

As a result of these inspections, no degradation was detected and no tubes were plugged. However, three freespan dents/dings were reported in steam generator A and one in steam generator B. None of these dents/dings was greater than 5 volts, and none had significantly changed since the first inservice inspection.

In addition to the steam generator tube inspections during RFO 26, visual inspections were performed on the top support plate and the U-bends of the low-row tubes in steam generator A. The area was free from soft and hard sludge with a very thin deposit film indicative of good thermal performance. In addition, the trifoil lobes of the tube support plate were open and free of deposits.

No significant foreign objects were found during visual inspections of the secondary side; however, they did detect some very small wires on the tubesheet.

3.2.25 Prairie Island 1

Tables 3-73, 3-74, and 3-75 summarize the information discussed in this section for Prairie Island 1. Table 3-73 provides the number of full-length bobbin inspections and the number of tubes plugged and deplugged during each outage for each of the two steam generators. Table 3-74 lists the reasons why the tubes were plugged. Table 3-75 lists the plugged tubes.

Prairie Island 1 has two recirculating steam generators designed and fabricated by Framatome in France. The model 56/19 steam generators were put into service in 2004 during RFO 23.

During the preservice inspection, 100 percent of the tubes in each of the steam generators were inspected full length with a bobbin coil. In addition, a rotating probe equipped with a +Point™ coil was used to inspect the hot-leg and cold-leg expansion transition region (from 3 inches above to 3 inches below the top of the secondary face of the tubesheet) of 100 percent of the tubes, the U-bend region of 100 percent of the tubes in rows 1 through 9, and approximately 850 special interest locations.

During the preservice inspection in steam generator A, 286 dings, 14 bulges, 7 MBMs, 1 over-expansion, 24 pilger drift signals, and 1 percent through-wall indication (measuring 15 percent) were detected with a bobbin coil. A bulge is a location where the localized tube diameter is greater than nominal. An overexpansion indicates that the expansion transition is beyond the top of the tubesheet. As a result of the rotating probe inspections, five geometric distortions (surface blemish on the inside diameter of the tube), four permeability variations (non-nominal magnetic permeability), and one single volumetric indication were detected. Three of the reported dings had voltages in excess of 1.0 volt.

During the preservice inspection in steam generator B, 473 dings, 39 bulges, 8 MBMs, and 18 pilger drift signals were detected with a bobbin coil. As a result of the rotating probe inspections, one geometric distortion was detected. Four of the reported dings had voltages in excess of 1.0 volt.

The criteria for reporting indications included, (1) any signal measuring 7 percent wall-loss or greater, (2) any ding signal measuring 1.0 volt or greater in amplitude, (3) any MBM measuring 0.5 volts or greater, or (4) other signals measuring 2.0 volts or greater in amplitude (except for pilger drift signals). Based on these criteria, seven dings in excess of 1.0 volt, one overexpansion, two bulges, and one 15-percent through-wall indication were reported.

3.2.26 Sequoyah 1

Tables 3-76, 3-77, and 3-78 summarize the information discussed in this section for Sequoyah 1. Table 3-76 provides the number of full-length bobbin inspections and the number of tubes plugged and deplugged during each outage for each of the four steam generators. Table 3-77 lists the reasons why the tubes were plugged. Table 3-78 lists the plugged tubes.

Sequoyah 1 has four recirculating steam generators designed by Combustion Engineering and fabricated by Doosan in Korea. The model 57AG steam generators were put into service in 2003 during RFO 12.

During the fabrication process, 100 percent of the tubes received a volumetric inspection designed to detect indications greater than or equal to 0.003 inches in depth. Tubes were rejected if they had one or more flaws with a depth greater than 0.004 inches. In addition, the signal-to-noise ratio was measured for the entire tube length for all of the tubes before bending. A minimum signal-to-noise ratio of 30:1 in any straight fixed 1/6-meter length was required for acceptance. This criterion was established to ensure the detection of extremely small flaws during inservice inspections.

During the preservice inspection, 100 percent of the tubes in each of the steam generators were inspected full length with a bobbin coil. In addition, a rotating probe equipped with a +Point™ coil was used to inspect the hot-leg expansion transition region (from 3 inches above to 3 inches below the top of the secondary face of the tubesheet) of 100 percent of the tubes, the U-bend region of 100 percent of the tubes in rows 1, 2, and 3, and approximately 240 special interest locations. The special interest locations included approximately 50 tube support locations. The tube support locations were inspected to provide eddy current data at various locations near internal bundle features to give future analysts a baseline signal to evaluate the new type of support structure. The special interest locations also included bobbin signals that either could not be characterized or were dings or MBMs. A rotating Ghent probe (a transmit/receive probe) was also used to inspect a few permeability variations.

These inspections identified approximately 350 dings, ranging from 0.2 volts to 42.9 volts. All dings greater than or equal to 2 volts were examined with a rotating probe equipped with a +Point™ coil as part of the inspection of the special interest locations (approximately 130 dings). The inspections found 28 dings in steam generator A, 42 in steam generator B, 166 in steam generator C, and 107 in steam generator D.

Profilometry was performed on 100 percent of the tubes from the tube end through the first tube support. The profilometry was done during fabrication in both the hot and cold-legs of the steam generator using a specialized bobbin probe with profiling software. No bulges were identified; however, several tubes were identified as requiring re-expansion (i.e., the tube was not fully expanded as a result of original application of the hydraulic expansion).

Before the steam generators went into service, 20 tubes were plugged in the four steam generators (4 in steam generator A, 6 in steam generator B, 5 in steam generator C, and 5 in steam generator D). One tube was plugged as a result of a geometry/lift-off signal in the U-bend region (i.e., between the second and third vertical support). The remaining 19 tubes were plugged and stabilized because of a condition discovered during fabrication in which the upper bundle support structure, called lock bars, had cracked on certain peripheral tubes. Portions of some of these lock bars had to be cut out. A flow analysis of the final support structure determined that specific tubes would need plugging and stabilization prior to operation. Of the 19 tubes that were plugged and stabilized, 1 had a 22-percent through-wall indication (i.e., a mar on the tube) as a direct result of the lock bar cutting operation.

Since the replacement of the steam generators, there has been no primary-to-secondary leakage as of May 2004.

During RFO 13 in 2004, 100 percent of the tubes in each of the steam generators were inspected full length with a bobbin coil. In addition, a rotating probe equipped with a +Point™ coil was used to inspect approximately 180 special interest locations. As a result of these inspections, 11 tubes were plugged for wear at the AVBs.

3.2.27 South Texas Project 1

Tables 3-79, 3-80, and 3-81 summarize the information discussed in this section for South Texas Project 1. Table 3-79 provides the number of full-length bobbin inspections and the number of tubes plugged and deplugged during each outage for each of the four steam generators. Table 3-80 lists the reasons why the tubes were plugged. Table 3-81 lists the plugged tubes.

South Texas Project 1 has four recirculating steam generators designed and fabricated by Westinghouse. The model D94 steam generators were put into service in 2000 during RFO 9. The licensee numbers its tube supports as depicted in Figure 2-44.

Before the steam generators went into service, 108 tubes were plugged. These tubes were plugged because of manufacturing phenomena such as laps which resulted in an eddy current signal that could represent a reduction in the tube wall thickness. A lap can be described as a fold in the metal caused by an imperfection in the tube surface as it is formed. Laps are not cracklike, and they lie parallel to the tube surface.

During RFO 10 in 2001, 100 percent of the tubes in each of the steam generators were inspected full length with a bobbin coil. In addition, a rotating probe equipped with a +Point™ coil was used to inspect all bobbin I-codes. The MBMs (greater than or equal to 1 volt) and dents/dings (greater than 0.75 volts) were compared to the preservice inspection signals, and if any changes were detected, they were assigned I-codes. Approximately 80 rotating probe inspections were performed at dents (5 locations), dings (9 locations), MBMs (32 locations), other nonquantifiable indications (28 locations), and PLPs (1 location).

As a result of these inspections, no tubes were plugged. In addition, no wear was detected in any of the four steam generators. However, many dents and dings were reported in the steam generators (51 dings and 765 dents). One ding in steam generator B measured 58.55 volts. A rotating probe inspection of this location revealed no degradation. In addition, one tube had a bulge, four tubes had permeability variations (five indications), and one tube had a PLP indication. The PLP was located below the sixth hot-leg tube support. The tube was deep in the tube bundle and could not be visually investigated. The affected tube and surrounding tubes were inspected with a rotating probe and no degradation or additional signals indicative of PLPs were detected.

During RFO 10, sludge lancing was performed on the secondary-side tubesheet region of all four steam generators. Inspections before sludge lancing identified several small (less than 1.5inches long) pieces of spiral-wound metal gasket banding in steam generator A. No tube wear had occurred, and the lancing process removed the material.

A visual inspection of secondary-side internals was performed during RFO 10 in one steam generator. The objectives were to verify that the upper steam generator internal welds and parts had not cracked or eroded during the first cycle of operation and to obtain data on deposits. No problems were identified during this inspection.

During RFO 10, the eddy current probes and guide tube became contaminated with cobalt from the tube's inside surface. The cobalt was suspected to have come from the unusually high particulate corrosion product release from the reactor core during shutdown.

As a result of receiving NRC approval to extend the interval between tube inspections from a maximum of 24 to 44 calendar months, no steam generator tube inspections were performed during RFO 11 in 2003. However, a FOSAR was performed in steam generator D during RFO 11.

On November 24, 2004, South Texas Projects 1 and 2 revised the steam generator portion of their technical specifications making them performance-based, consistent with Improved Standard Technical Specifications Change Traveler TSTF-449 (see ADAMS Accession Nos. ML043290311 and ML043370370).

During RFO 12 in 2005, sludge lancing and FOSAR were scheduled for all four steam generators; however, no steam generator tube inspections were planned. The FOSAR included a video inspection of all four steam generators before and after sludge lancing.

In steam generators A, B, and C, several foreign objects (between 9 and 11 small objects per steam generator) were identified. These objects were stainless steel flexitallic spiral wound gasket material. Several larger objects (approximately 2.5 to 3.5 inches long) were also identified in peripheral locations of the tube bundle. Video examination of the tubes in the vicinity of these objects revealed no wear indications. All of the identified foreign objects in steam generators A, B, and C were removed except for one piece which was too fragile to be removed and crumbled when grasped by a robotic arm.

The specific source of the gasket material identified and removed from steam generators A, B, and C was not identified. The introduction of gasket material is not believed to be an ongoing, age-related process; rather, damage to the gasket is suspected to have occurred during the installation process with degradation occurring soon after the unit was returned to service. As a result of these findings, plans were made to replace any noncontained gaskets with a different style gasket that is not susceptible to this type of degradation.

In steam generator D, several hundred small wire fragments were identified. The wire fragments were attributed to a feedwater heater tube stabilizing cable. Portions of the stabilizing cable, which had been inserted into a feedwater heater tube during a previous outage, were found wrapped around a regulating valve cage and were removed. The stabilizing cable was damaged when the valve was manually closed during RFO 11. As a result of the damage to the stabilizing cable by the valve closure, a piece of the cable had been severed and swept downstream in many small fragments. The missing piece of cable was approximately 13 inches in length and 7/16 of an inch in diameter.

The wire fragments from the stabilizing cable entered only steam generator D. Fragments of the severed stabilizing cable were recovered from the steam generator D feedring spray cans and sludge collector. Video inspections revealed that fragments of wire were able to pass through the feedring and enter steam generator D. Sludge lancing was performed seven times to remove the bulk of the wire fragments from within the tube bundle. After sludge lancing, another video inspection of the tube bundle was performed. Wire fragments were still present in the steam generator after sludge lancing, and some of the fragments could not be retrieved.

An analysis showed that wire fragments less than or equal to 3 inches in length would not challenge tube integrity if left in the tube bundle. Although one wire fragment measuring approximately 4 inches in length could not be removed, it was determined that this larger piece of wire would not cause tube damage because it was not located near the bundle periphery (which experiences the highest flow velocity and is therefore the location most susceptible to wear caused by loose parts). The analysis employed conservative values for a range of fluid velocities and object sizes.

The video inspection performed after sludge lancing revealed one indication of tube wear near the top of the tubesheet on the hot-leg side of the bundle in steam generator D. Because of this indication of wear, 20 percent of the tubes in steam generator D were inspected full length with a bobbin coil. In addition, 791 tubes in the center of the bundle (60-tube buffer zone) and 435 tubes in the periphery of the bundle (plus an 87-tube buffer zone) were inspected full length with a bobbin coil. In addition to the bobbin coil inspections, a rotating probe equipped with a +Point™ coil was used to inspect 100 percent of tubes where the gap velocities exceeded 8.5 feet per second (fps) which includes at least the two tubes around the outer periphery of the tube bundle, 20 percent of tubes where the gap velocities are between 7 and 8.5 fps, and 702 special interest locations. As a result of these inspections, one tube was identified with damage due to wear caused by a wire fragment. This tube was plugged, and the associated wire fragment was removed.

An inspection was also performed to assess the condition of the feedwater heaters which were the original source of the wire fragments. All plugged tubes in the feedwater heaters were inspected, and no other plugs were found missing or loose.

3.2.28 South Texas Project 2

Tables 3-82, 3-83, and 3-84 summarize the information discussed in this section for South Texas Project 2. Table 3-82 provides the number of full-length bobbin inspections and the number of tubes plugged and deplugged during each outage for each of the four steam generators. Table 3-83 lists the reasons why the tubes were plugged. Table 3-84 lists the plugged tubes.

South Texas Project 2 has four recirculating steam generators designed by Westinghouse and fabricated by ENSA in Spain. The model D94 steam generators were put into service in 2002 during RFO 9. The licensee numbers its tube supports as depicted in Figure 2-44.

During the preservice inspection, 100 percent of the tubes in each of the steam generators were inspected full length with a bobbin coil. In addition, a rotating probe equipped with a +Point™ coil was used to inspect all bobbin indications (i.e, I-codes), all freespan signals greater then 3 volts on the bobbin coil 150-kHz absolute channel that were not verified to be less than 5 percent through-wall by the manufacturer, all bobbin coil support suppression mix channel eddy current ding or dent indications greater than 3 volts and a sampling of others to cumulatively equal 20 percent but not more than 200 tubes in each steam generator, all crevices at the top of the tubesheet with a measured depth in excess of 0.25 inches, all MBMs and dings and dents in the U-bend region of rows 1 and 2, all MBMs and dings and dents exceeding 3 volts, and a total of 200 hot-leg top-of-tubesheet locations. The hot-leg top-of-tubesheet locations included the 20 largest bulge indications which were inspected from 1 inch above to 4 inches below the secondary face, any overexpansions located more than 0.005 inches above the top of the tubesheet which were inspected from 1 inch above the secondary face of the tubesheet to the tube end, and locations selected from below the baffle plate cut-out region and on the periphery of the bundle which were inspected from 1 inch above to 4 inches below the top of the tubesheet, and approximately 200 other hot-leg top-of-tubesheet locations per steam generator.

As a result of these inspections, MBMs, dings, dents, bulges, non-quantifiable signals, and permeability variations were reported. However, in the tubesheet region there were no over expansions, no unexpanded tubes, no expansion skip rolls, no bulges, and no crevice depths exceeding 0.25-inch identified.

Before the steam generators went into service, six tubes were plugged with Alloy 690 welded plugs.

During RFO 10 in 2004, 100 percent of the tubes in each of the steam generators were inspected full length with a bobbin coil. In addition, a rotating probe equipped with a +Point™ coil was used to inspect the hot-leg expansion transition region (from tube end to more than 3 inches above the top of the tubesheet) of 3 percent of the tubes; the U-bend region of 20 percent of the row 1 tubes (16 tubes); 20 percent of dings less than or equal to 5 volts (as measured from the bobbin coil); and 100 percent of dings greater than 5 volts (as measured from the bobbin coil) in each steam generator. A total of 231 dings were inspected.

As a result of these inspections, no tubes were plugged and no tubes had wear at the AVBs. In addition, there were no PLP indications reported (although several loose parts were visually identified as discussed below). However, many (594) dents and dings were reported in the steam generators (most of which had been present since the preservice inspection). In addition, two tubes had a bulge, one tube had a permeability variation, and several tubes (approximately 150) had MBMs.

During RFO 10, an upper steam drum visual inspection of the main feedwater and auxiliary feedwater spray cans were performed, along with the support structure for the main feedwater header. No anomalies were found as a result of these inspections. Sludge lancing and a

FOSAR was also performed in all four steam generators. The sludge lancing also included a center stay rod lancing process to enhance the removal of deposits directly adjacent to the stay rod in-bundle areas. No foreign objects were identified in steam generators C and D; however, loose parts were retrieved from steam generators A and B. No tube wear and no PLP indications were associated with these foreign objects. Of the five foreign objects found, three were in the hot- or cold-leg annulus region and two were in-bundle (row 126–127/column 86–87; row 92, column 78). All known foreign objects were removed from the steam generators.

On November 24, 2004, South Texas Projects 1 and 2 revised the steam generator portion of their technical specifications, making them performance-based to be consistent with Improved Standard Technical Specifications Change Traveler TSTF-449 (see ADAMS Accession Nos. ML043290311 and ML043370370).

3.2.29 St. Lucie 1

Tables 3-85, 3-86, and 3-87 summarize the information discussed in this section for St. Lucie 1. Table 3-85 provides the number of full-length bobbin inspections and the number of tubes plugged and deplugged during each outage for each of the three steam generators. Table 3-86 lists the reasons why the tubes were plugged. Table 3-87 lists the plugged tubes.

St. Lucie 1 has two steam generators designed and fabricated by Babcock and Wilcox International. The steam generators were put into service in 1998 during RFO 15.

Before the steam generators went into service, no tubes were plugged.

During RFO 16 in 1999, approximately 52 percent of the tubes in each of the steam generators were inspected full length with a bobbin coil. As a result of these inspections, 11 tubes were plugged. Of these 11 tubes, 10 were plugged because of wear at the fan bars, and 1 was plugged because of a volumetric indication at the hot-leg collector bar. This latter indication is attributed to a manufacturing or installation anomaly. The maximum depth reported for the fan bar wear indications was 34 percent through-wall.

During RFO 17 in 2001, approximately 56 percent of the tubes in steam generator A and 53 percent of the tubes in steam generator B were inspected full length with a bobbin coil. In addition, a rotating probe equipped with a +Point™ coil was used to inspect the U-bend region of approximately 30 percent of the row 1 and 2 tubes and all previously reported dents (one in each steam generator). As a result of these inspections, no tubes were plugged.

During RFO 18 in 2002, no steam generator tube inspections were performed.

During RFO 19 in 2004, approximately 57 percent of the tubes in steam generator A and 54 percent of the tubes in steam generator B were inspected full length with a bobbin coil (with the exception of the tubes in row 3 where the bobbin was used to inspect the straight sections of tube and a rotating probe equipped with a +Point™ coil was used to inspect the U-bend region). The bobbin inspections were performed in all active tubes in odd-numbered columns (except for the row 3 tubes) and all active peripheral tubes. In addition, a rotating probe equipped with a +Point™ coil was used to inspect the hot-leg expansion transition region (from 3 inches above to 2 inches below the top of the secondary face of the tubesheet) of approximately 50 percent of the tubes; the U-bend region of approximately 50 percent of the row 3 tubes; 50 percent of the dings greater than 3 volts in the straight section of tubing on the hot-leg side of the steam generator; and other special interest locations. As a result of these inspections, three steam generator tubes were plugged because of wear in the U-bend region. The depth of degradation was less than 39 percent through-wall. An additional 21 wear indications were detected at U-bend support structures. These indications measured less than 20 percent through-wall and were left in service.

In addition to these eddy current examinations, a visual examination of all installed tube plugs and a FOSAR of the tubesheet annulus and blowdown lane was performed during RFO 19.

3.2.30 Summer

Tables 3-88, 3-89, and 3-90 summarize the information discussed in this section for Summer. Table 3-88 provides the number of full-length bobbin inspections and the number of tubes plugged and deplugged during each outage for each of the three steam generators. Table 3-89 lists the reasons why the tubes were plugged. Table 3-90 lists the plugged tubes.

Summer has three model Delta 75 steam generators designed and fabricated by Westinghouse. The steam generators were put into service in 1994 during RFO 8. The licensee numbers its tube supports as depicted in Figure 2-46.

Before the steam generators went into service, three tubes were plugged. These tubes were plugged because the tubes were not fully expanded into the tubesheet.

During RFO 9 in 1996, approximately 22 percent of the tubes in steam generator A and approximately 16 percent of the tubes in steam generator B were inspected full length with a bobbin coil. As a result of these inspections, no tubes were plugged. No rotating probe inspections were performed during this outage.

A visual inspection performed on the secondary side of all three steam generators included the moisture separator area, the upper tube bundle, and support plates. No abnormalities were noted.

During RFO 10 in 1997, approximately 30 percent of the tubes in steam generator C were inspected full length with a bobbin coil. In addition, a rotating probe was used to examine 100 locations to investigate bobbin coil indications. These inspections were performed in a 6-inch area at the hot-leg expansion transition region. As a result of these inspections, no tubes were plugged.

A total of 161 imperfections were recorded during the inspection, including 11 dings, 138 dents, and 12 MBMs. The dings and dents were minor and not indicative of service-induced anomalies. All MBMs were attributed to the manufacturing process and showed no appreciable change since the preservice inspection.

During RFO 11 in 1999, approximately 40 percent of the tubes in steam generators A and B were inspected full length with a bobbin coil. In addition, a rotating probe was used to examine 100 locations to investigate bobbin coil indications. As a result of these inspections, no tubes were plugged.

A review of the eddy current inspection data identified a PLP near the tubes located in row 89, column 62, and row 88, column 63. The loose part signal was near the top of the tubesheet. During RFO 11, the part was evaluated and left in place; however, during RFO 12, a small piece of wire, 0.5 inches long, was removed from this location. There was no wear on any of the adjacent tubes.

During RFO 12 in 2000, 100 percent of the tubes in each of the three steam generators were inspected full length with a bobbin coil. In addition, a rotating probe was used to inspect the hot-leg expansion transition region (from 3 inches above to 3 inches below the top of the secondary face of the tubesheet) of approximately 5 percent of the tubes in steam generator B; the U-bend region of approximately 20 percent of the row 1 tubes in steam generator C (14 tubes); and approximately 65 other locations. As a result of these inspections, five tubes were plugged because the tubes were not expanded in the tubesheet area.

In steam generator A, three wear indications were identified in two tubes. These indications were at the AVBs and have not changed significantly since the preservice inspection. The maximum depth reported for these indications is 9 percent through-wall.

During RFO 12, FOSAR and sludge lancing were performed on all three steam generators. As a result of the FOSAR and eddy current examinations, only one loose part was identified (as discussed above). This part was removed from the steam generator. In addition, visual inspections were performed in the upper steam drum, the ninth tube support and U-bend, the middle and lower steam drum, the flow distribution plate, tubesheet, and lower tube bundle regions.

As a result of identifying a crack in the reactor coolant system hot-leg piping during RFO 12, which extended the RFO by approximately 4 months, and the improved steam generator design, the next steam generator tube inspections were delayed, with NRC approval, until 58 months after the RFO 12 inspections (i.e., until RFO 15 in spring 2005).

As of July 2003, there has been no primary-to-secondary leakage at Summer since the steam generators were replaced.

Table 3-1: ANO 2 Full-Length Bobbin Exams

Outage	Completion Date	Cumul. EFPY	SG A Insp.	SG A Plug	SG A DePl	SG B Insp.	SG B Plug	SG B DePl	Total Plug	Total DePl	Cumul. Plugged	Percent Plugged	Notes
Pre-op				0			1		1	0	1	0.00	
RFO 15	04/24/02		10637	0		10636	0		0	0	1	0.00	
RFO 16	10/14/03							0	0	0	1	0.00	

Totals: 0   0   0   1   0   1   0

Plant Data
Model: D109
T-hot (approximate): 609 F
Tubes per steam generator: 10637
Number of steam generators: 2

Notes

Acronyms
Pre-op = prior to operation
Cumul. = cumulative
Insp. = number of tubes inspected
Plug = number of tubes plugged
DePl = number of tubes deplugged
RFO = refueling outage

-144-

Table 3-2: ANO 2 Causes of Tube Plugging

Cause of Tube Plugging/Outage		Year			Totals
		Pre-Op	2002 RFO 15	2003 RFO 16	
Wear	AVB				0
	Tube Support				
Loose Part Wear	Confirmed, Periphery				0
	Confirmed, Interior				
	Not Confirmed, Periphery				
	Not Confirmed, Interior				
Obstruction Restriction	From PSI, No Progression				0
	Service-Induced				
Manufacturing/ Maintenance	Pre-Operation	1			1
	Other				
Inspection Issues	Probe Lodged				0
	Data Quality				
	Dent/Geometry				
	Permeability				
	Not Inspected				
Other	Top of Tubesheet				0
	Freespan				
	TSP				
	Other/Not Reported				
SCC	ID				0
	OD				
TOTALS		1	0		1

Notes:

Table 3-3: ANO 2: Tubes Plugged

STEAM GENERATOR A				
Tube	Location	RFO #	Characterization	Stabilized[1]

STEAM GENERATOR B				
Tube	Location	RFO #	Characterization	Stabilized[1]
23-8		Pre-op	Failure of mandrel during expansion process	

[1] An empty cell indicates that it was not reported whether the tube was stabilized or not.

Table 3-4: Braidwood 1 Full-Length Bobbin Exams

Outage	Completion Date	Cumul. EFPY	SG A Insp.	SG A Plug	SG A DePl	SG B Insp.	SG B Plug	SG B DePl	SG C Insp.	SG C Plug	SG C DePl	SG D Insp.	SG D Plug	SG D DePl	Total Plug	Total DePl	Cumul. Plugged	Percent Plugged	Notes
Pre-op				1			2			0			0		3	0	3	0.01	
RFO 8	03/28/00	1.29	6632	1		6631	0		6633	0		6633	0		1	0	4	0.02	
RFO 9	10/12/01														0	0	4	0.02	
RFO 10	04/24/03		6631	8		3582	10		3582	3		3582	0		21	0	25	0.09	
RFO 11	10/18/04	5.64		0		6621	2			2			1		5	0	30	0.11	1
Totals:				10	0		14	0		5	0		1	0	30	0			

Plant Data
Model: BWI 7720
T-hot (approximate): 618 F
Tubes per steam generator: 6633
Number of steam generators: 4

Acronyms
Pre-op = prior to operation
Cumul. = cumulative
Insp. = number of tubes inspected
Plug = number of tubes plugged
DePl = number of tubes deplugged
RFO = refueling outage

Notes
1. Approximately 22 percent of the hot-leg portion of the tubes in steam generators A, C, and D was inspected with bobbin coil. Approximately 16 percent of the tubes in these steam generators straight portion of the cold-leg side of the tubes inspected with a bobbin probe.

Table 3-5: Braidwood 1 Causes of Tube Plugging

Cause of Tube Plugging/Outage		Pre-Op	2000 RFO 8	2001 RFO 9	2003 RFO 10	2004 RFO 11		Totals	Totals
Wear	AVB		1					1	1
	Tube Support							0	
Loose Part Wear	Confirmed, Periphery				21			21	26
	Confirmed, Interior							0	
	Not Confirmed, Periphery					2		2	
	Not Confirmed, Interior					3		3	
Obstruction Restriction	From PSI, No Progression							0	0
	Service-Induced							0	
Manufacturing/ Maintenance	Pre-Operation	3						3	3
	Other							0	
Inspection Issues	Probe Lodged							0	0
	Data Quality							0	
	Dent/Geometry							0	
	Permeability							0	
	Not Inspected							0	
Other	Top of Tubesheet							0	0
	Freespan							0	
	TSP							0	
	Other/Not Reported							0	
SCC	ID							0	0
	OD							0	
TOTALS		3	1		21	5		30	30
Notes:				1					

Notes

1. Six tubes plugged due to foreign objects. (No wear associated w/ these loose parts.)

Table 3-6: Braidwood 1: Tubes Plugged

STEAM GENERATOR A				
Tube	Location	RFO #	Characterization	Stabilized[1]
42-135	TSC+0.48	10	25% TW, Confirmed Loose Part	N
44-135	TSC+0.37	10	48% TW, Confirmed Loose Part	Y
44-137	TSC+0.15	10	5% TW, Confirmed Loose Part	Y
45-134	TSC+0.28	10	15% TW, Confirmed Loose Part	N
45-136	TSC+0.09	10	20% TW, Confirmed Loose Part	Y
46-135	TSC+0.23	10	12% TW, Confirmed Loose Part	Y
46-137	TSC+0.28	10	14% TW, Confirmed Loose Part	Y
47-136		10	Confirmed Loose Part	Y
87-54	FB5+1.24	8	<10% Fan Bar Wear	
104-107		Pre-op	Plugged prior to operation	

STEAM GENERATOR B				
Tube	Location	RFO #	Characterization	Stabilized[1]
1-116	6C+0.40	11	42% TW, Non-Confirmed Loose Part	N
48-89	6H-1.42	11	27% TW, Non-Confirmed Loose Part	N
51-12	TSH+1.17	10	7% TW, Confirmed Loose Part	N
52-11	TSH+1.39	10	11% TW, Confirmed Loose Part	N
52-13	TSH+0.38	10	4% TW, Confirmed Loose Part	N
53-12	TSH+1.50	10	12% TW, Confirmed Loose Part	N
54-11	TSH+0.92	10	13% TW, Confirmed Loose Part	N
73-104		Pre-op	Plugged prior to operation	
86-79		Pre-op	Plugged prior to operation	
90-29	TSH+0.26	10	7% TW, Confirmed Loose Part	Y
91-28		10	Confirmed Loose Part	Y
91-30		10	Confirmed Loose Part	Y
92-29	TSH+0.75	10	17% TW, Confirmed Loose Part	Y
93-30	TSH+0.49	10	7% TW, Confirmed Loose Part	Y

Table 3-6: Braidwood 1: Tubes Plugged (cont'd)

			STEAM GENERATOR C		
Tube	Location	RFO #	Characterization		Stabilized[1]
2-49	TSH+0.07	10	Confirmed Loose Part		Y
3-48	TSH+0.21	10	Confirmed Loose Part		Y
4-49	TSH+0.07	10	Confirmed Loose Part		Y
66-13	1H-0.72	11	17%TW, Non-Confirmed Loose Part		Y
69-50	4H-1.71	11	9% TW, Non-Confirmed Loose Part		N

			STEAM GENERATOR D		
Tube	Location	RFO #	Characterization		Stabilized[1]
70-41	2H-1.57	11	25%TW, Non-Confirmed Loose Part		N

[1] An empty cell indicates that it was not reported whether the tube was stabilized or not.

Table 3-7: Byron 1 Full-Length Bobbin Exams

Outage	Completion Date	Cumul. EFPY	SG A			SG B			SG C			SG D			Total Plug	Total DePl	Cumul. Plugged	Percent Plugged	Notes
			Insp.	Plug	DePl	Insp.	Plug	DePl	Insp.	Plug	DePl	Insp.	Plug	DePl					
Pre-op					0			0		1				0	1	0	1	0.00	
RFO 9	04/09/99		6633		0	6633	0	0	6632	0	0	6633	0	0	0	0	1	0.00	
RFO 10	10/01/00		6633	0											0	0	1	0.00	
RFO 11	03/20/02	3.79	3582	0	0	3582	0	0	3582	0	0	3582	0	0	0	0	1	0.00	
RFO 12	10/14/03														0	0	1	0.00	
Totals:				0	0		0	0		1	0		0	0	1	0			

Plant Data
Model: BWI 7720
T-hot (approximate):
Tubes per steam generator: 6633
Number of steam generators: 4

Acronyms
Pre-op = prior to operation
Cumul. = cumulative
Insp. = number of tubes inspected
Plug = number of tubes plugged
DePl = number of tubes deplugged
RFO = refueling outage

-151-

Table 3-8: Byron 1 Causes of Tube Plugging

Cause of Tube Plugging/Outage		Pre-Op	1999 RFO 9	2000 RFO 10	2002 RFO 11	2003 RFO 12			Totals	Totals
Wear	AVB									0
	Tube Support									0
Loose Part Wear	Confirmed, Periphery									
	Confirmed, Interior									0
	Not Confirmed, Periphery									
	Not Confirmed, Interior									
Obstruction Restriction	From PSI, No Progression									0
	Service-Induced									
Manufacturing/ Maintenance	Pre-Operation	1							1	1
	Other									
Inspection Issues	Probe Lodged									0
	Data Quality									
	Dent/Geometry									
	Permeability									
	Not Inspected									
Other	Top of Tubesheet									0
	Freespan									
	TSP									
	Other/Not Reported									
SCC	ID									0
	OD									0
TOTALS		1	0	0	0				1	1

Notes:

Table 3-9: Byron 1: Tubes Plugged

STEAM GENERATOR A				
Tube	Location	RFO #	Characterization	Stabilized[1]

STEAM GENERATOR B				
Tube	Location	RFO #	Characterization	Stabilized[1]

STEAM GENERATOR C				
Tube	Location	RFO #	Characterization	Stabilized[1]
34-129		Pre-op	Plugged prior to operation	

STEAM GENERATOR D				
Tube	Location	RFO #	Characterization	Stabilized[1]

[1]An empty cell indicates that it was not reported whether the tube was stabilized or not.

Table 3-10: Calvert Cliffs 1 Full-Length Bobbin Exams

Outage	Completion Date	Cumul. EFPY	SG A			SG B			Total Plug	Total DePl	Cumul. Plugged	Percent Plugged	Notes
			Insp.	Plug	DePl	Insp.	Plug	DePl					
Pre-op											0	0.00	
RFO 16	05/08/04	1.76	8471	0	0	8471	0	0	0	0	0	0.00	

Totals: 0   0   0   0   0   0

Plant Data
Model: BWI
T-hot (approximate):
Tubes per steam generator: 8471
Number of steam generators: 2

Acronyms
Pre-op = prior to operation
Cumul. = cumulative
Insp. = number of tubes inspected
Plug = number of tubes plugged
DePl = number of tubes deplugged
RFO = refueling outage

Table 3-11: Calvert Cliffs 1 Causes of Tube Plugging

Cause of Tube Plugging/Outage		Year Pre-Op	2004 RFO 16	Totals
Wear	AVB			0
	Tube Support			
Loose Part Wear	Confirmed, Periphery			0
	Confirmed, Interior			
	Not Confirmed, Periphery			
	Not Confirmed, Interior			
Obstruction Restriction	From PSI, No Progression			0
	Service-Induced			0
Manufacturing/ Maintenance	Pre-Operation			
	Other			0
Inspection Issues	Probe Lodged			
	Data Quality			
	Dent/Geometry			
	Permeability			
	Not Inspected			
Other	Top of Tubesheet			
	Freespan			0
	TSP			
	Other/Not Reported			
SCC	ID			0
	OD			
TOTALS		0	0	0

Notes:

-155-

Table 3-12: Calvert Cliffs 1: Tubes Plugged

STEAM GENERATOR A				
Tube	Location	RFO #	Characterization	Stabilized[1]

STEAM GENERATOR B				
Tube	Location	RFO #	Characterization	Stabilized[1]

[1]An empty cell indicates that it was not reported whether the tube was stabilized or not.

Table 3-13: Calvert Cliffs 2 Full-Length Bobbin Exams

Outage	Completion Date	Cumul. EFPY	SG A Insp.	SG A Plug	SG A DePl	SG B Insp.	SG B Plug	SG B DePl	Total Plug	Total DePl	Cumul. Plugged	Percent Plugged	Notes
Pre-op				3			0		3	0	3	0.02	

Totals: 3   0   0   0   3   0

Plant Data
Model: BWI
T-hot (approximate):
Tubes per steam generator: 8471
Number of steam generators: 2

Acronyms
Pre-op = prior to operation
Cumul. = cumulative
Insp. = number of tubes inspected
Plug = number of tubes plugged
DePl = number of tubes deplugged
RFO = refueling outage

Table 3-14: Calvert Cliffs 2 Causes of Tube Plugging

Cause of Tube Plugging/Outage		Year		Totals	Totals
		Pre-Op			
Wear	AVB			0	0
	Tube Support			0	
Loose Part Wear	Confirmed, Periphery			0	0
	Confirmed, Interior			0	
	Not Confirmed, Periphery			0	
	Not Confirmed, Interior			0	
Obstruction Restriction	From PSI, No Progression			0	0
	Service-Induced			0	
Manufacturing/ Maintenance	Pre-Operation	3		3	3
	Other			0	
Inspection Issues	Probe Lodged			0	0
	Data Quality			0	
	Dent/Geometry			0	
	Permeability			0	
	Not Inspected			0	
Other	Top of Tubesheet			0	0
	Freespan			0	
	TSP			0	
	Other/Not Reported			0	
SCC	ID			0	0
	OD			0	
TOTALS		3		3	3

Notes:

Table 3-15: Calvert Cliffs 2: Tubes Plugged

STEAM GENERATOR A				
Tube	Location	RFO #	Characterization	Stabilized[1]
***		Pre-op	3 tubes were plugged prior to operation (specific tubes not identified). 2 of the 3 tubes were potentially in contact.	

STEAM GENERATOR B				
Tube	Location	RFO #	Characterization	Stabilized[1]

[1]An empty cell indicates that it was not reported whether the tube was stabilized or not.

Table 3-16: Catawba 1 Full-Length Bobbin Exams

Outage	Completion Date	Cumul. EFPY	SG A Insp.	SG A Plug	SG A DePl	SG B Insp.	SG B Plug	SG B DePl	SG C Insp.	SG C Plug	SG C DePl	SG D Insp.	SG D Plug	SG D DePl	Total Plug	Total DePl	Cumul. Plugged	Percent Plugged	Notes
Pre-op				8	0		0	0		7	0		4	0	19	0	19	0.07	
RFO 10	12/01/97		6625	0		6633	0		6626	0		6629	0		0	0	19	0.07	
RFO 11	05/23/99		1455	0		1382	0		1365	0		1381	0		0	0	19	0.07	1
RFO 12	11/20/00					5316	0		5317	0					0	0	19	0.07	
RFO 13	05/17/02																19	0.07	
RFO 14	12/18/03		6625	0			0			7		6629	0		7	0	26	0.10	2
Totals:				8	0		0	0		14	0		4	0	26	0			

Plant Data
Model: BWI CFR80
T-hot (approximate):
Tubes per steam generator: 6633
Number of steam generators: 4

Notes
1. During RFO 11, 487, 109, and 480 tubes in SG A, C, and D, respectively, were partially inspected.
2. During RFO 14, 100 percent of the tubes in steam generators B and C were inspected from both tube ends through the second lattice grid.

Table 3-17: Catawba 1 Causes of Tube Plugging

Cause of Tube Plugging/Outage		Pre-Op	1997 RFO 10	1999 RFO 11	2000 RFO 12	2002 RFO 13	2003 RFO 14	Totals
Wear	AVB							0
	Tube Support							0
Loose Part Wear	Confirmed, Periphery							0
	Confirmed, Interior						7	7
	Not Confirmed, Periphery							0
	Not Confirmed, Interior							0
Obstruction Restriction	From PSI, No Progression							0
	Service-Induced							0
Manufacturing/ Maintenance	Pre-Operation	19						19
	Other							0
Inspection Issues	Probe Lodged							0
	Data Quality							0
	Dent/Geometry							0
	Permeability							0
	Not Inspected							0
Other	Top of Tubesheet							0
	Freespan							0
	TSP							0
	Other/Not Reported							0
SCC	ID							0
	OD							0
	TOTALS	19	0	0	0	0	7	26
	Notes:						1	26

Notes

1. Loose part on hot leg side. Could not be removed.

Table 3-18: Catawba 1: Tubes Plugged

			STEAM GENERATOR A		
Tube	Location	RFO #	Characterization	Stabilized[1]
•••		Pre-op	7 additional tubes were plugged prior to operation (specific tubes not identified)	
4-125		Pre-op	Expanded above the tubesheet on the hot-leg side	Y

			STEAM GENERATOR B		
Tube	Location	RFO #	Characterization	Stabilized[1]

			STEAM GENERATOR C		
Tube	Location	RFO #	Characterization	Stabilized[1]
•••		Pre-op	7 tubes were plugged prior to operation (specific tubes not identified)	
78-23	TSH + 0.84	14	Confirmed loose part that could not be removed	Y
79-24	TSH + 0.55	14	Confirmed loose part that could not be removed	Y
80-23	TSH + 1.72	14	Confirmed loose part that could not be removed	Y
81-24	TSH + 1.14	14	Confirmed loose part that could not be removed	Y
82-23	TSH + 0.4	14	Confirmed loose part that could not be removed	Y
83-24	TSH + 0.10	14	Confirmed loose part that could not be removed	Y
84-23	TSH + 0.36	14	Confirmed loose part that could not be removed	Y

			STEAM GENERATOR D		
Tube	Location	RFO #	Characterization	Stabilized[1]
•••			3 additional tubes were plugged during fabrication (specific tubes not identified)	
111-64		Pre-op	Scratches on cold-leg tube end from a stuck expansion mandrel	

[1]An empty cell indicates that it was not reported whether the tube was stabilized or not.

Table 3-19: Cook 1 Full-Length Bobbin Exams

Outage	Completion Date	Cumul. EFPY	SG A Insp.	SG A Plug	SG A DePl	SG B Insp.	SG B Plug	SG B DePl	SG C Insp.	SG C Plug	SG C DePl	SG D Insp.	SG D Plug	SG D DePl	Total Plug	Total DePl	Cumul. Plugged	Percent Plugged	Notes
Pre-Op					0		0			0			0		0	0	0	0.00	
RFO 18	06/08/02		3496	2		3496	0		3496	1		3496	1		4	0	4	0.03	
RFO 19	11/25/03											701	0		0	0	4	0.03	
Totals:				2	0		0	0		1	0		1	0	4	0			

Plant Data
Model: BWI
T-hot (approximate):
Tubes per steam generator: 3496
Number of steam generators: 4

Acronyms
Pre-op = prior to operation
Cumul. = cumulative
Insp. = number of tubes inspected
Plug = number of tubes plugged
DePl = number of tubes deplugged
RFO = refueling outage

Table 3-20: Cook 1 Causes of Tube Plugging

Cause of Tube Plugging/Outage		Pre-Op	2002 RFO18	2003 RFO19	Totals	Totals
Wear	AVB				0	0
	Tube Support				0	
Loose Part Wear	Confirmed, Periphery				0	0
	Confirmed, Interior				0	
	Not Confirmed, Periphery				0	
	Not Confirmed, Interior				0	
Obstruction Restriction	From PSI, No Progression				0	0
	Service-Induced				0	
Manufacturing/ Maintenance	Pre-Operation				0	0
	Other				0	
Inspection Issues	Probe Lodged				0	0
	Data Quality				0	
	Dent/Geometry				0	
	Permeability				0	
	Not Inspected				0	
Other	Top of Tubesheet				0	4
	Freespan		4		4	
	TSP				0	
	Other/Not Reported				0	
SCC	ID				0	0
	OD				0	
TOTALS		0	4	0	4	4

Notes:

Notes
1. Four tubes plugged since signal response changed since pre-operational inspection.

-164-

Table 3-21: Cook 1: Tubes Plugged

STEAM GENERATOR A				
Tube	Location	RFO #	Characterization	Stabilized[1]
13-85	TSC +14.97	18	Change in MBM signal response since preservice	
16-40	6C + 34.48 5C + 37.06	18	Change in MBM signal response since preservice	

STEAM GENERATOR B				
Tube	Location	RFO #	Characterization	Stabilized[1]

STEAM GENERATOR C				
Tube	Location	RFO #	Characterization	Stabilized[1]
19-61	6C + 19.45	18	Change in MBM signal response since preservice	

STEAM GENERATOR D				
Tube	Location	RFO #	Characterization	Stabilized[1]
69-45	1C+12.06	18	Change in MBM signal response since preservice	

[1]An empty cell indicates that it was not reported whether the tube was stabilized or not.

Table 3-22: Cook 2 Full-Length Bobbin Exams

Outage	Completion Date	Cumul. EFPY	SG A Insp.	SG A Plug	SG A DePl	SG B Insp.	SG B Plug	SG B DePl	SG C Insp.	SG C Plug	SG C DePl	SG D Insp.	SG D Plug	SG D DePl	Total Plug	Total DePl	Cumul. Plugged	Percent Plugged	Notes
Pre-Op					0		1			0			0		1	0	1	0.01	
RFO 7	1990					67	0		68	0					0	0	1	0.01	1
RFO 8	03/28/92		71	0								69	0		0	0	1	0.01	2
RFO 9	10/01/94					67	3		67	6					9	0	10	0.07	3
RFO 10	1996																		
RFO 11	11/05/97	5.8	1798	1		1796	0		1796	0		1798	4		5	0	15	0.10	
RFO 12	02/25/02		1796	0		1796	0		1796	0		1796	0		0	0	15	0.10	
RFO 13	06/19/03																		
RFO 14	11/08/04		900	0		900	1		900	0		900	0		1	0	16	0.11	
Totals:				1	0		5	0		6	0		4	0	16	0			

Plant Data
Model: 54F
T-hot (approximate): 607 F
Tubes per steam generator: 3592
Number of steam generators: 4

Acronyms
Pre-op = prior to operation
Cumul. = cumulative
Insp. = number of tubes inspected
Plug = number of tubes plugged
DePl = number of tubes deplugged
RFO = refueling outage

Notes
1. During RFO 7, 169 tubes in steam generator B and 167 tubes in steam generator C were inspected from the hot-leg tube end and through the top tube support on the cold-leg side.
2. During RFO 8, 164 tubes in steam generator A and 166 tubes in steam generator D were inspected from the hot-leg tube end and through the top tube support on the cold-leg side.
   In addition, four of these tubes in steam generator A and two of these tubes in steam generator D were inspected from the hot-leg tube end to support 6C and one of these tubes in both steam generators A and D was inspected to support 5C.
3. During RFO 9, 168 tubes in steam generators B and C were inspected from the hot-leg tube end and through the top tube support on the cold-leg side. In addition, 108 tubes in steam generators B and C were inspected from the hot-leg and cold-leg tube ends to 3 inches above the flow distribution baffle.

Table 3-23: Cook 2 Causes of Tube Plugging

Cause of Tube Plugging/Outage		Pre-Op	1990 RFO 7	1992 RFO 8	1994 RFO 9	1996 RFO 10	1997 RFO 11	2002 RFO 12	2003 RFO 13	2004 RFO 14	Totals	Totals
Wear	AVB										0	0
	Tube Support										0	
Loose Part Wear	Confirmed, Periphery						1			1	2	6
	Confirmed, Interior										0	
	Not Confirmed, Periphery						4				4	
	Not Confirmed, Interior										0	
Obstruction Restriction	From PSI, No Progression										0	0
	Service-Induced										0	
Manufacturing/ Maintenance	Pre-Operation	1									1	10
	Other				9						9	
Inspection Issues	Probe Lodged										0	0
	Data Quality										0	
	Dent/Geometry										0	
	Permeability										0	
	Not Inspected										0	
Other	Top of Tubesheet										0	0
	Freespan										0	
	TSP										0	
	Other/Not Reported										0	
SCC	ID										0	0
	OD										0	
TOTALS		1	0	0	9		5	0		1	16	16

Notes

1. Nine tubes plugged due to interaction with pressure pulse cleaning equipment.

Table 3-24: Cook 2: Tubes Plugged

STEAM GENERATOR A				
Tube	Location	RFO #	Characterization	Stabilized[1]
1-70	1C - 0.16	11	28% through-wall indication due to burr or small foreign object	

STEAM GENERATOR B				
Tube	Location	RFO #	Characterization	Stabilized[1]
1-92		9	Mechanical damage from pressure pulse cleaning equipment.	
1-93		9	Mechanical damage from pressure pulse cleaning equipment.	
1-94		9	Mechanical damage from pressure pulse cleaning equipment.	
8-3		Pre-op	Plugged prior to operation	
46-37	FBH - 0.36	14	33% through-wall indication from a confirmed loose part	

STEAM GENERATOR C				
Tube	Location	RFO #	Characterization	Stabilized[1]
1-5		9	Mechanical damage from pressure pulse cleaning equipment.	
1-6		9	Mechanical damage from pressure pulse cleaning equipment.	
1-7		9	Mechanical damage from pressure pulse cleaning equipment.	
1-92		9	Mechanical damage from pressure pulse cleaning equipment.	
1-93		9	Mechanical damage from pressure pulse cleaning equipment.	
1-94		9	Mechanical damage from pressure pulse cleaning equipment.	

STEAM GENERATOR D				
Tube	Location	RFO #	Characterization	Stabilized[1]
33-15	TSH + 0.49	11	Suspect loose part wear.	
33-16	TSH + 0.05	11	Suspect loose part wear.	
33-17	TSH + 0.84	11	Suspect loose part wear.	
38-72	TSH + 0.07	11	Confirmed loose part wear.	

[1] An empty cell indicates that it was not reported whether the tube was stabilized or not.

Table 3-25: Farley 1 Full-Length Bobbin Exams

Outage	Completion Date	Cumul. EFPY	SG A Insp.	Plug	DePl	SG B Insp.	Plug	DePl	SG C Insp.	Plug	DePl	Total Plug	Total DePl	Cumul. Plugged	Percent Plugged	Notes
Pre-op					0			0		0	0	0	0	0	0.00	
RFO 17	10/18/01	1.33	3592	0	0	3592	0	0	3592	0	0	0	0	0	0.00	
RFO 18	04/30/03												0	0	0.00	
RFO 19	11/15/04												0	0	0.00	
Totals:				0	0		0	0		0	0	0	0			

Plant Data
Model: 54F
T-hot (approximate): 607 F
Tubes per steam generator: 3592
Number of steam generators: 3

Acronyms
Pre-op = prior to operation
Cumul. = cumulative
Insp. = number of tubes inspected
Plug = number of tubes plugged
DePl = number of tubes deplugged
RFO = refueling outage

Table 3-26: Farley 1 Causes of Tube Plugging

Cause of Tube Plugging/Outage		Pre-Op	2001 RFO17	2003 RFO 18	2004 RFO 19			Totals
Wear	AVB							0
	Tube Support							
Loose Part Wear	Confirmed, Periphery							0
	Confirmed, Interior							
	Not Confirmed, Periphery							
	Not Confirmed, Interior							
Obstruction Restriction	From PSI, No Progression							0
	Service-Induced							
Manufacturing/ Maintenance	Pre-Operation							0
	Other							
Inspection Issues	Probe Lodged							0
	Data Quality							
	Dent/Geometry							
	Permeability							
	Not Inspected							
Other	Top of Tubesheet							0
	Freespan							
	TSP							
	Other/Not Reported							
SCC	ID							0
	OD							
TOTALS		0	0					0

Notes:

-170-

Table 3-27: Farley 1: Tubes Plugged

STEAM GENERATOR A				
Tube	Location	RFO #	Characterization	Stabilized[1]

STEAM GENERATOR B				
Tube	Location	RFO #	Characterization	Stabilized[1]

STEAM GENERATOR C				
Tube	Location	RFO #	Characterization	Stabilized[1]

[1] An empty cell indicates that it was not reported whether the tube was stabilized or not.

Table 3-28: Farley 2 Full-Length Bobbin Exams

Outage	Completion Date	Cumul. EFPY	SG A			SG B			SG C			Total Plug	Total DePl	Cumul. Plugged	Percent Plugged	Notes
			Insp.	Plug	DePl	Insp.	Plug	DePl	Insp.	Plug	DePl					
Pre-Op					0			0			0	0	0	0	0.00	
RFO 15	09/27/02	1.33	3592		0	3592		0	3592		0	0	0	0	0.00	
RFO 16	04/12/04													0	0.00	
Totals:			0	0	0	0	0	0	0	0	0	0	0			

Plant Data
Model: 54F
T-hot (approximate): 607 F
Tubes per steam generator: 3592
Number of steam generators: 3

Acronyms
Pre-op = prior to operation
Cumul. = cumulative
Insp. = number of tubes inspected
Plug = number of tubes plugged
DePl = number of tubes deplugged
RFO = refueling outage

-172-

Table 3-29: Farley 2 Causes of Tube Plugging

Cause of Tube Plugging/Outage		Year Pre-Op	2002 RFO15	2004 RFO16		Totals
Wear	AVB					0
	Tube Support					
Loose Part Wear	Confirmed, Periphery					0
	Confirmed, Interior					
	Not Confirmed, Periphery					
	Not Confirmed, Interior					
Obstruction	From PSI, No Progression					0
Restriction	Service-Induced					0
Manufacturing/ Maintenance	Pre-Operation					0
	Other					
Inspection Issues	Probe Lodged					
	Data Quality					
	Dent/Geometry					
	Permeability					
	Not Inspected					
Other	Top of Tubesheet					0
	Freespan					
	TSP					
	Other/Not Reported					
SCC	ID					0
	OD					
TOTALS		0	0	0		0

Notes:

-173-

Table 3-30: Farley 2: Tubes Plugged

STEAM GENERATOR A				
Tube	Location	RFO #	Characterization	Stabilized[1]

STEAM GENERATOR B				
Tube	Location	RFO #	Characterization	Stabilized[1]

STEAM GENERATOR C				
Tube	Location	RFO #	Characterization	Stabilized[1]

[1]An empty cell indicates that it was not reported whether the tube was stabilized or not.

Table 3-31: Ginna Full-Length Bobbin Exams

Outage	Completion Date	Cumul. EFPY	SG A Insp.	Plug	DePl	SG B Insp.	Plug	DePl	Total Plug	Total DePl	Cumul. Plugged	Percent Plugged	Notes
Pre-op				1			1		2	0	2	0.02	
RFO 26	01/06/98		4764	0		4764	0		0	0	2	0.02	
RFO 27	04/21/99		2383	0		2383	0		0	0	2	0.02	
RFO 28	10/17/00								0	0	2	0.02	
RFO 29	04/19/02	5.0	2487	0		2486	0		0	0	2	0.02	
RFO 30	10/14/03			0			0		0	0	2	0.02	
Totals:				1	0		1	0	2	0			

Plant Data
Model: BWI
T-hot (approximate): 590 F
Tubes per steam generator: 4765
Number of steam generators: 2

Acronyms
Pre-op = prior to operation
Cumul. = cumulative
Insp. = number of tubes inspected
Plug = number of tubes plugged
DePl = number of tubes deplugged
RFO = refueling outage

-175-

Table 3-32: Ginna Causes of Tube Plugging

Cause of Tube Plugging/Outage		Pre-Op	1998 RFO 26	1999 RFO 27	2000 RFO 28	2002 RFO 29	2003 RFO 30	Totals	Totals
Wear	AVB							0	0
	Tube Support							0	
Loose Part Wear	Confirmed, Periphery							0	0
	Confirmed, Interior							0	
	Not Confirmed, Periphery							0	
	Not Confirmed, Interior							0	
Obstruction Restriction	From PSI, No Progression							0	0
	Service-Induced							0	
Manufacturing/ Maintenance	Pre-Operation	2						2	2
	Other							0	
Inspection Issues	Probe Lodged							0	0
	Data Quality							0	
	Dent/Geometry							0	
	Permeability							0	
	Not Inspected							0	
Other	Top of Tubesheet							0	0
	Freespan							0	
	TSP							0	
	Other/Not Reported							0	
SCC	ID							0	0
	OD							0	
TOTALS		2	0	0		0		2	2

Notes:

Table 3-33: Ginna: Tubes Plugged

STEAM GENERATOR A				
Tube	Location	RFO #	Characterization	Stabilized[1]
52-14		Pre-op	Wall Loss greater than 15% between the first and second lattice grids on hot-leg side	

STEAM GENERATOR B				
Tube	Location	RFO #	Characterization	Stabilized[1]
67-17		Pre-op	Undercut below the secondary face of the tubesheet that reduced a tubesheet ligament below the minimum allowable per the ASME Code. This hole was tubed normally except that a shorter expansion was performed.	

[1]An empty cell indicates that it was not reported whether the tube was stabilized or not.

Table 3-34: Harris Full-Length Bobbin Exams

Outage	Completion Date	Cumul. EFPY	SG A Insp.	SG A Plug	SG A DePl	SG B Insp.	SG B Plug	SG B DePl	SG C Insp.	SG C Plug	SG C DePl	Total Plug	Total DePl	Cumul. Plugged	Percent Plugged	Notes
Pre-op				1			1					2	0	2	0.01	
RFO 11	05/07/03		6306	0		6306	0		6307	0		0	0	2	0.01	
Mid-Cycle	05/14/04								1268	3	3	3	0	5	0.03	1
RFO 12	11/07/04											0	0	5	0.03	
Totals:				1	0		1	0		3	0	5	0			

Plant Data
Model: D75
T-hot (approximate):
Tubes per steam generator: 6307
Number of steam generators: 3

Acronyms
Pre-op = prior to operation
Cumul. = cumulative
Insp. = number of tubes inspected
Plug = number of tubes plugged
DePl = number of tubes deplugged
RFO = refueling outage

Notes
1. Following a reactor trip due to an unrelated issue, steam generator tube inspections were performed to investigate the source of a small primary-to-secondary leak.

Table 3-35: Harris Causes of Tube Plugging

Cause of Tube Plugging/Outage		Pre-Op	2003 RFO 11	2004 Mid	2004 RFO 12	Totals
Wear	AVB					0
	Tube Support					
Loose Part Wear	Confirmed, Periphery			3		3
	Confirmed, Interior					
	Not Confirmed, Periphery					
	Not Confirmed, Interior					
Obstruction Restriction	From PSI, No Progression					0
	Service-Induced					
Manufacturing/ Maintenance	Pre-Operation	2				2
	Other					
Inspection Issues	Probe Lodged					0
	Data Quality					
	Dent/Geometry					
	Permeability					
	Not Inspected					
Other	Top of Tubesheet					0
	Freespan					
	TSP					
	Other/Not Reported					
SCC	ID					0
	OD					
TOTALS		2	0	3		5
Notes:				1		

Notes
1. Following a reactor trip due to an unrelated issue, steam generator tube inspections were performed to investigate the source of a small primary-to-secondary leak.

Table 3-36: Harris: Tubes Plugged

STEAM GENERATOR A				
Tube	Location	RFO #	Characterization	Stabilized[1]
106-85		Pre-op	Plugged prior to operation	

STEAM GENERATOR B				
Tube	Location	RFO #	Characterization	Stabilized[1]
114-73		Pre-op	Plugged prior to operation	

STEAM GENERATOR C				
Tube	Location	RFO #	Characterization	Stabilized[1]
1-120	TSC+0.75"	Mid Cycle (2004)	80% through-wall. Confirmed loose part.	Y
2-121	TSC+0.5"	Mid Cycle (2004)	45% through-wall. Confirmed loose part.	Y
3-120	TSC+0.2"	Mid Cycle (2004)	73% through-wall. Confirmed loose part. Leaking tube.	Y

[1]An empty cell indicates that it was not reported whether the tube was stabilized or not.

Table 3-37: Indian Point 3 Full-Length Bobbin Exams

Outage	Completion Date	Cumul. EFPY	SG A Insp.	SG A Plug	SG A DePl	SG B Insp.	SG B Plug	SG B DePl	SG C Insp.	SG C Plug	SG C DePl	SG D Insp.	SG D Plug	SG D DePl	Total Plug	Total DePl	Cumul. Plugged	Percent Plugged	Notes
Pre-op					0			0			0		2	2	2	0	2	0.02	
RFO 7	09/01/90		688	0	0	644	0	0	717	0	0	644	0	0	0	0	2	0.02	
RFO 8	05/01/92		552	0	0	554	0	0	555	0	0	551	0	0	0	0	2	0.02	
RFO 9	05/01/97					3214	0		1966	0		2065	0		0	0	2	0.02	
RFO 10	10/02/99		3214	0	0										0	0	2	0.02	
RFO 11	05/24/01														0	0	2	0.02	
RFO 12	04/12/03	8.8	756	1		756	6		756	3		756	2		12	0	14	0.11	
Totals:				1	0		6	0		3	0		4	0	14	0			

Plant Data
Model: 44F
T-hot (approximate): 600 F
Tubes per steam generator: 3214
Number of steam generators: 4

Acronyms
Pre-op = prior to operation
Cumul. = cumulative
Insp. = number of tubes inspected
Plug = number of tubes plugged
DePl = number of tubes deplugged
RFO = refueling outage

Notes

Table 3-38: Indian Point 3 Causes of Tube Plugging

Cause of Tube Plugging/Outage		Pre-Op	1990 RFO 7	1992 RFO 8	1997 RFO 9	1999 RFO 10	2001 RFO 11	2003 RFO 12	Totals	Totals
Wear	AVB								0	0
	Tube Support								0	
Loose Part Wear	Confirmed, Periphery								0	0
	Confirmed, Interior								0	
	Not Confirmed, Periphery								0	
	Not Confirmed, Interior								0	
Obstruction Restriction	From PSI, No Progression								0	0
	Service-Induced								0	
Manufacturing/ Maintenance	Pre-Operation	2							2	13
	Other							11	11	
Inspection Issues	Probe Lodged								0	1
	Data Quality								0	
	Dent/Geometry								0	
	Permeability							1	1	
	Not Inspected								0	
Other	Top of Tubesheet								0	0
	Freespan								0	
	TSP								0	
	Other/Not Reported								0	
SCC	ID								0	0
	OD								0	
TOTALS		2	0	0	0	0	0	12	14	14
Notes:								1,2		

Notes
1. Eight tubes plugged due to wear associated with sludge-lancing equipment.
2. Three tubes plugged due to volumetric indications at the top of the tubesheet on the hot-leg side. The indications are attributed to manufacturing anomalies.

Table 3-39: Indian Point 3: Tubes Plugged

STEAM GENERATOR A				
Tube	Location	RFO #	Characterization	Stabilized[1]
28-29	TSH-4.96 to 5.04"	12	Permeability	

STEAM GENERATOR B				
Tube	Location	RFO #	Characterization	Stabilized[1]
1-9	TSC+16.01"	12	8% through-wall wear indication from sludge lancing equipment	
1-66	TSC+18.16"	12	13% through-wall wear indication from sludge lancing equipment	
1-85	TSH+16.70"	12	11% through-wall wear indication from sludge lancing equipment	
40-29	TSH+0.00"	12	32% through-wall volumetric manufacturing anomaly	
41-28	TSH+0.15"	12	34% through-wall volumetric manufacturing anomaly	
41-29	TSH+0.05"	12	24% through-wall volumetric manufacturing anomaly	

STEAM GENERATOR C				
Tube	Location	RFO #	Characterization	Stabilized[1]
1-8	TSC+16.51"	12	9% through-wall wear indication from sludge lancing equipment	
1-27	TSC+17.86" TSH+18.04"	12	12% and 16% through-wall wear indications from sludge lancing equipment	
1-66	TSH+15.62"	12	26% through-wall wear indication from sludge lancing equipment	

STEAM GENERATOR D				
Tube	Location	RFO #	Characterization	Stabilized[1]
1-8	TSH+16.69"	12	10% through-wall wear indication from sludge lancing equipment	
1-84	TSC+16.92"	12	11% through-wall wear indication from sludge lancing equipment	
44-57		Pre-op	Damage from temporary support	
45-52		Pre-op	Damage from temporary support	

[1]An empty cell indicates that it was not reported whether the tube was stabilized or not.

Table 3-40: Kewaunee Full-Length Bobbin Exams

Outage	Completion Date	Cumul. EFPY	SG A Insp.	SG A Plug	SG A DePI	SG B Insp.	SG B Plug	SG B DePI	Total Plug	Total DePI	Cumul. Plugged	Percent Plugged	Notes
Pre-op				0			0		0	0	0	0.00	
RFO 25	04/19/03		3592	0		3592	0		0	0	0	0.00	
RFO 26	12/03/04			0			0		0	0	0	0.00	

Totals: 0 0 0 0 0 0

Plant Data
Model: 54F
T-hot (approximate): 592 F
Tubes per steam generator: 3592
Number of steam generators: 2

Acronyms
Pre-op = prior to operation
Cumul. = cumulative
Insp. = number of tubes inspected
Plug = number of tubes plugged
DePI = number of tubes deplugged
RFO = refueling outage

Table 3-41: Kewaunee Causes of Tube Plugging

Cause of Tube Plugging/Outage		Pre-Op	2003 RFO 25	2004 RFO 26		Totals
Wear	AVB					0
	Tube Support					
Loose Part Wear	Confirmed, Periphery					0
	Confirmed, Interior					
	Not Confirmed, Periphery					
	Not Confirmed, Interior					
Obstruction Restriction	From PSI, No Progression					0
	Service-Induced					
Manufacturing/ Maintenance	Pre-Operation					0
	Other					
Inspection Issues	Probe Lodged					0
	Data Quality					
	Dent/Geometry					
	Permeability					
	Not Inspected					
Other	Top of Tubesheet					0
	Freespan					
	TSP					
	Other/Not Reported					
SCC	ID					0
	OD					
TOTALS		0	0			0

Notes:

Table 3-42: Kewaunee: Tubes Plugged

STEAM GENERATOR A				
Tube	Location	RFO #	Characterization	Stabilized[1]

STEAM GENERATOR B				
Tube	Location	RFO #	Characterization	Stabilized[1]

[1]An empty cell indicates that it was not reported whether the tube was stabilized or not.

Table 3-43: McGuire 1 Full-Length Bobbin Exams

Outage	Completion Date	Cumul. EFPY	SG A Insp.	SG A Plug	SG A DePl	SG B Insp.	SG B Plug	SG B DePl	SG C Insp.	SG C Plug	SG C DePl	SG D Insp.	SG D Plug	SG D DePl	Total Plug	Total DePl	Cumul. Plugged	Percent Plugged	Notes
Pre-op				1			1			4			4		10	0	10	0.04	
RFO 12	07/31/98		6290	0	0	6334	1	0	6335	1	0	6306	0	0	2	0	12	0.05	1
RFO 13	11/04/99		2017	0	0	2024	0	0	2069	0	0	2017	0	0	0	0	12	0.05	
RFO 14	04/16/01					6631	0	0	6628	0					0	0	12	0.05	
RFO 15	10/07/02														0	0	12	0.05	
RFO 16	04/11/04		6632	1		3649	0		3743	0		6629	0		1	0	13	0.05	
Totals:				2	0		2	0		5	0		4	0	13	0			

Plant Data
Model: BWI CFR80
T-hot (approximate):
Tubes per steam generator: 6633
Number of steam generators: 4

Acronyms
Pre-op = prior to operation
Cumul. = cumulative
Insp. = number of tubes inspected
Plug = number of tubes plugged
DePl = number of tubes deplugged
RFO = refueling outage

Notes
1. During RFO 12, 342, 298, 294, and 323 tubes were partially inspected in steam generators A, B, C, and D, respectively.

-187-

Table 3-44: McGuire 1 Causes of Tube Plugging

Cause of Tube Plugging/Outage		Year Pre-Op	1998 RFO 12	1999 RFO 13	2001 RFO 14	2002 RFO 15	2004 RFO 16	Totals	Totals
Wear	AVB		2					2	2
	Tube Support							0	
Loose Part Wear	Confirmed, Periphery							0	1
	Confirmed, Interior							0	
	Not Confirmed, Periphery						1	1	
	Not Confirmed, Interior							0	
Obstruction Restriction	From PSI, No Progression							0	0
	Service-Induced							0	
Manufacturing/ Maintenance	Pre-Operation	10						10	10
	Other							0	
Inspection Issues	Probe Lodged							0	0
	Data Quality							0	
	Dent/Geometry							0	
	Permeability							0	
	Not Inspected							0	
Other	Top of Tubesheet							0	0
	Freespan							0	
	TSP							0	
	Other/Not Reported							0	
SCC	ID							0	0
	OD							0	
TOTALS		10	2	0	0	0	1	13	13

Notes:

Table 3-45: McGuire 1: Tubes Plugged

STEAM GENERATOR A				
Tube	Location	RFO #	Characterization	Stabilized[1]
1-10		Pre-op	Plugged prior to operation	
96-33	TSC + 3.15	16	54% through-wall indication attributed to foreign object wear	

STEAM GENERATOR B				
Tube	Location	RFO #	Characterization	Stabilized[1]
24-121[2]		Pre-op	Plugged prior to operation	
24-121[2]		12	Atypical wear	

STEAM GENERATOR C				
Tube	Location	RFO #	Characterization	Stabilized[1]
45-122		Pre-op	Outside Diameter Indication	
51-68	FB6 + 2.13	12	Atypical fan bar wear	
78-21		Pre-op	Outside Diameter Indication	
93-98		Pre-op	Plugged prior to operation	
94-99		Pre-op	Plugged prior to operation	

STEAM GENERATOR D				
Tube	Location	RFO #	Characterization	Stabilized[1]
31-90		Pre-op	Outside Diameter Indication	
81-106		Pre-op	Plugged prior to operation	
95-34		Pre-op	Tube not expanded	
106-39		Pre-op	Plugged prior to operation	

[1]An empty cell indicates that it was not reported whether the tube was stabilized or not.

[2]Tube 24-121 was identified as being plugged both prior to operation and during RFO 12. The tube plugged during RFO 12 was plugged due to atypical wear.

Table 3-46: McGuire 2 Full-Length Bobbin Exams

Outage	Completion Date	Cumul. EFPY	SG A Insp.	SG A Plug	SG A DePl	SG B Insp.	SG B Plug	SG B DePl	SG C Insp.	SG C Plug	SG C DePl	SG D Insp.	SG D Plug	SG D DePl	Total Plug	Total DePl	Cumul. Plugged	Percent Plugged	Notes
Pre-op				0			1			0			1		2	0	2	0.01	
RFO 12	04/12/99		6633	3	0	6632	2	0	6633	1	0	6632	3	0	9	0	11	0.04	
RFO 13	10/12/00		1693	0	0	35	0	0	23	0	0	1697	0	0	0	0	11	0.04	1
RFO 14	03/26/02					6630	0	0	6632	0	0				0	0	11	0.04	
RFO 15	10/04/03														0	0	11	0.04	
Totals:				3	0		3	0		1	0		4	0	11	0			

Plant Data
Model: BWI CFR80
T-hot (approximate):
Tubes per steam generator: 6633
Number of steam generators: 4

Acronyms
Pre-op = prior to operation
Cumul. = cumulative
Insp. = number of tubes inspected
Plug = number of tubes plugged
DePl = number of tubes deplugged
RFO = refueling outage

Notes
1. During RFO 13, 146, 1571, 738, and 147 tubes were partially inspected in steam generators A, B, C, and D, respectively.

Table 3-47: McGuire 2 Causes of Tube Plugging

Cause of Tube Plugging/Outage		Pre-Op	1999 RFO 12	2000 RFO 13	2002 RFO 14	2003 RFO 15	Totals
Wear	AVB		8				8
	Tube Support						0
Loose Part Wear	Confirmed, Periphery						
	Confirmed, Interior						
	Not Confirmed, Periphery						
	Not Confirmed, Interior						0
Obstruction Restriction	From PSI, No Progression						
	Service-Induced						0
Manufacturing/ Maintenance	Pre-Operation	2					
	Other						2
Inspection Issues	Probe Lodged						
	Data Quality						
	Dent/Geometry						
	Permeability						
	Not Inspected						0
Other	Top of Tubesheet						
	Freespan						
	TSP		1				1
	Other/Not Reported						
SCC	ID						
	OD						0
TOTALS		2	9	0	0	0	11

Notes:

Table 3-48: McGuire 2: Tubes Plugged

STEAM GENERATOR A				
Tube	Location	RFO #	Characterization	Stabilized[1]
66-61	FB4 + 1.09	12	18% through-wall indication	
68-75	FB4 + 1.04	12	5% through-wall indication	
81-62	FB5 + 1.26	12	39% through-wall indication	

STEAM GENERATOR B				
Tube	Location	RFO #	Characterization	Stabilized[1]
12-117		Pre-op	Plugged prior to operation	
43-100	03H - 1.60	12	Volumetric indication	
108-69	FB3 + 1.85	12	6% through-wall indication	

STEAM GENERATOR C				
Tube	Location	RFO #	Characterization	Stabilized[1]
70-67	FB5 + 1.50	12	13% through-wall indication	

STEAM GENERATOR D				
Tube	Location	RFO #	Characterization	Stabilized[1]
8-23		Pre-op	Plugged prior to operation	
64-61	FB5 - 1.62	12	31% through-wall indication	
81-82	FB5 + 1.38	12	33% through-wall indication	
89-70	FB8 - 1.91	12	8% through-wall indication	

[1] An empty cell indicates that it was not reported whether the tube was stabilized or not.

Table 3-49: Millstone 2 Full-Length Bobbin Exams

Outage	Completion Date	Cumul. EFPY	SG A Insp.	SG A Plug	SG A DePl	SG B Insp.	SG B Plug	SG B DePl	Total Plug	Total DePl	Cumul. Plugged	Percent Plugged	Notes
Pre-op				1				0	1	0	1	0.01	
RFO 12	10/20/94		2511	0		2380	0		0	0	1	0.01	
Mid-Cycle	06/26/97		6408	0		2565	0		0	0	1	0.01	
RFO 13	05/09/00					8523	0		0	0	1	0.01	
RFO 14	03/03/02		8522	0					0	0	1	0.01	
RFO 15	10/25/03					8523	0		0	0	1	0.01	
Totals:				1	0		0	0	1	0			

Plant Data
Model: BWI
T-hot (approximate):
Tubes per steam generator: 8523
Number of steam generators: 2

Notes

Acronyms
Pre-op = prior to operation
Cumul. = cumulative
Insp. = number of tubes inspected
Plug = number of tubes plugged
DePl = number of tubes deplugged
RFO = refueling outage

-193-

Table 3-50: Millstone 2 Causes of Tube Plugging

	Year	Pre-Op	1994 RFO 12	1997 Mid	2000 RFO 13	2002 RFO 14	2003 RFO 15	Totals
Cause of Tube Plugging/Outage								
Wear	AVB							0
	Tube Support							0
Loose Part Wear	Confirmed, Periphery							0
	Confirmed, Interior							
	Not Confirmed, Periphery							
	Not Confirmed, Interior							
Obstruction Restriction	From PSI, No Progression							0
	Service-Induced							
Manufacturing/ Maintenance	Pre-Operation	1						1
	Other							
Inspection Issues	Probe Lodged							0
	Data Quality							
	Dent/Geometry							
	Permeability							
	Not Inspected							
Other	Top of Tubesheet							0
	Freespan							
	TSP							
	Other/Not Reported							
SCC	ID							0
	OD							
TOTALS		1	0	0	0	0	0	0

Totals	Totals
1	1

Notes:

Table 3-51: Millstone 2: Tubes Plugged

STEAM GENERATOR A				
Tube	Location	RFO #	Characterization	Stabilized[1]
57-156	TSH	Pre-op	Broken drill bit in hot-leg Tube was not drilled on the cold-leg.	

STEAM GENERATOR B				
Tube	Location	RFO #	Characterization	Stabilized[1]

[1]An empty cell indicates that it was not reported whether the tube was stabilized or not.

Table 3-52: North Anna 1 Full-Length Bobbin Exams

Outage	Completion Date	Cumul. EFPY	SG A Insp.	SG A Plug	SG A DePl	SG B Insp.	SG B Plug	SG B DePl	SG C Insp.	SG C Plug	SG C DePl	Total Plug	Total DePl	Cumul. Plugged	Percent Plugged	Notes
Pre-op					0			0			0	0	0	0	0.00	
RFO 10	09/30/94		1803	0					1796	0		0	0	0	0.00	
RFO 11	03/01/96	2.7				1798	0					0	0	0	0.00	
RFO 12	06/01/97		1804	0								0	0	0	0.00	
RFO 13	10/01/98	5.1							1796	1		1	0	1	0.01	
RFO 14	03/01/00	6.5				1796	0					0	0	1	0.01	
RFO 15	10/01/01		2156	0								0	0	1	0.01	
RFO 16	04/17/03											0	0	1	0.01	
RFO 17	10/06/04	10.5							3591	1		1	0	2	0.02	
Totals:				0	0		0	0		2	0	2	0			

Plant Data
Model: 54F
T-hot (approximate):
Tubes per steam generator: 3592
Number of steam generators: 3

Acronyms
Pre-op = prior to operation
Cumul. = cumulative
Insp. = number of tubes inspected
Plug = number of tubes plugged
DePl = number of tubes deplugged
RFO = refueling outage

Table 3-53: North Anna 1 Causes of Tube Plugging

Cause of Tube Plugging/Outage		Year Pre-Op	1994 RFO 10	1996 RFO 11	1997 RFO 12	1998 RFO 13	2000 RFO 14	2001 RFO 15	2003 RFO 16	2004 RFO 17	Totals
Wear	AVB										0
	Tube Support										
Loose Part Wear	Confirmed, Periphery									1	1
	Confirmed, Interior										
	Not Confirmed, Periphery										
	Not Confirmed, Interior										
Obstruction Restriction	From PSI, No Progression										0
	Service-Induced										
Manufacturing/ Maintenance	Pre-Operation										0
	Other										
Inspection Issues	Probe Lodged										0
	Data Quality										
	Dent/Geometry										
	Permeability										
	Not Inspected										
Other	Top of Tubesheet										1
	Freespan					1					
	TSP										
	Other/Not Reported										
SCC	ID										0
	OD										
TOTALS		0	0	0	0	1	0	0	0	1	2

Notes:

Notes
1. Classified as pit (small volumetric indication).

Table 3-54: North Anna 1: Tubes Plugged

STEAM GENERATOR A				
Tube	Location	RFO #	Characterization	Stabilized[1]

STEAM GENERATOR B				
Tube	Location	RFO #	Characterization	Stabilized[1]

STEAM GENERATOR C				
Tube	Location	RFO #	Characterization	Stabilized[1]
36-22	TSH+10"	4	Pit (small volumetric indication)	
37-79	AV6+22.36"	8	43% TW, Confirmed Loose Part	

[1]An empty cell indicates that it was not reported whether the tube was stabilized or not.

Table 3-55: North Anna 2 Full-Length Bobbin Exams

Outage	Completion Date	Cumul. EFPY	SG A Insp.	SG A Plug	SG A DePl	SG B Insp.	SG B Plug	SG B DePl	SG C Insp.	SG C Plug	SG C DePl	Total Plug	Total DePl	Cumul. Plugged	Percent Plugged	Notes
Pre-op					0		0	0		0	0	0	0	0	0.00	
RFO 1	10/01/96	1.2				1796	0	0	1796	0	0	0	0	0	0.00	
RFO 2	04/01/98	2.6	1803	0								0	0	0	0.00	
RFO 3	09/01/99	3.8				1796	0					0	0	0	0.00	
RFO 4	04/01/01	5.3							2159	1		1	0	1	0.01	
RFO 5	10/01/02	6.7	2156	1								1	0	2	0.02	
RFO 6	05/30/04											0	0	2	0.02	
Totals:				1	0		0	0		1	0	2	0			

Plant Data
Model: 54F
T-hot (approximate):
Tubes per steam generator: 3592
Number of steam generators: 3

Acronyms
Pre-op = prior to operation
Cumul. = cumulative
Insp. = number of tubes inspected
Plug = number of tubes plugged
DePl = number of tubes deplugged
RFO = refueling outage

Table 3-56: North Anna 2 Causes of Tube Plugging

Cause of Tube Plugging/Outage		Pre-Op	1996 RFO 1	1998 RFO 2	1999 RFO 3	2001 RFO 4	2002 RFO 5	2004 RFO 6	Totals	Totals
Wear	AVB								0	1
	Tube Support					1			1	
Loose Part Wear	Confirmed, Periphery								0	0
	Confirmed, Interior								0	
	Not Confirmed, Periphery								0	
	Not Confirmed, Interior								0	
Obstruction Restriction	From PSI, No Progression								0	0
	Service-Induced								0	
Manufacturing/ Maintenance	Pre-Operation								0	0
	Other								0	
Inspection Issues	Probe Lodged								0	1
	Data Quality								0	
	Dent/Geometry								0	
	Permeability						1		1	
	Not Inspected								0	
Other	Top of Tubesheet								0	0
	Freespan								0	
	TSP								0	
	Other/Not Reported								0	
SCC	ID								0	0
	OD								0	
TOTALS		0	0	0	0	1	1		2	2

Notes:

Table 3-57: North Anna 2: Tubes Plugged

STEAM GENERATOR A				
Tube	Location	RFO #	Characterization	Stabilized[1]
9-45	TSH	5	Permeability	

STEAM GENERATOR B				
Tube	Location	RFO #	Characterization	Stabilized[1]

STEAM GENERATOR C				
Tube	Location	RFO #	Characterization	Stabilized[1]
43-56	5C	4	30% TW volumetric indication at fifth cold-leg tube support. Indication corresponds with land of tube support (i.e., mechanical wear).	

[1]An empty cell indicates that it was not reported whether the tube was stabilized or not.

Table 3-58: Oconee 1 Full-Length Bobbin Exams

Outage	Completion Date	Cumul. EFPY	SG A			SG B			Total Plug	Total DePl	Cumul. Plugged	Percent Plugged	Notes
			Insp.	Plug	DePl	Insp.	Plug	DePl					
Pre-op				1			1		2	0	2	0.01	

Totals: 1   0   1   0   2   0

Plant Data
Model: BWI OTSG
T-hot (approximate): 604 F
Tubes per steam generator: 15631
Number of steam generators: 2

Acronyms
Pre-op = prior to operation
Cumul. = cumulative
Insp. = number of tubes inspected
Plug = number of tubes plugged
DePl = number of tubes deplugged
RFO = refueling outage

-202-

Table 3-59: Oconee 1 Causes of Tube Plugging

Cause of Tube Plugging/Outage		Year Pre-Op	Totals	Totals
Wear	AVB		0	0
	Tube Support		0	
Loose Part Wear	Confirmed, Periphery		0	0
	Confirmed, Interior		0	
	Not Confirmed, Periphery		0	
	Not Confirmed, Interior		0	
Obstruction Restriction	From PSI, No Progression		0	0
	Service-Induced		0	
Manufacturing/ Maintenance	Pre-Operation	2	2	2
	Other		0	
Inspection Issues	Probe Lodged		0	0
	Data Quality		0	
	Dent/Geometry		0	
	Permeability		0	
	Not Inspected		0	
Other	Top of Tubesheet		0	0
	Freespan		0	
	TSP		0	
	Other/Not Reported		0	
SCC	ID		0	0
	OD		0	
TOTALS		2	2	2

Notes:

Table 3-60: Oconee 1: Tubes Plugged

STEAM GENERATOR A				
Tube	Location	RFO #	Characterization	Stabilized[1]
118-73		Pre-op	Plugged prior to operation	

STEAM GENERATOR B				
Tube	Location	RFO #	Characterization	Stabilized[1]
121-105		Pre-op	Plugged prior to operation	

[1] An empty cell indicates that it was not reported whether the tube was stabilized or not.

Table 3-61: Oconee 2 Full-Length Bobbin Exams

Outage	Completion Date	Cumul. EFPY	SG A			SG B			Total Plug	Total DePl	Cumul. Plugged	Percent Plugged	Notes
			Insp.	Plug	DePl	Insp.	Plug	DePl					
Pre-op				4			1		5	0	5	0.02	
Totals:				4	0		1	0	5	0			

Plant Data
Model: BWI OTSG
T-hot (approximate): 604 F
Tubes per steam generator: 15631
Number of steam generators: 2

Acronyms
Pre-op = prior to operation
Cumul. = cumulative
Insp. = number of tubes inspected
Plug = number of tubes plugged
DePl = number of tubes deplugged
RFO = refueling outage

Table 3-62: Oconee 2 Causes of Tube Plugging

Cause of Tube Plugging/Outage		Year		Totals	Totals
		Pre-Op			
Wear	AVB			0	0
	Tube Support			0	
Loose Part Wear	Confirmed, Periphery			0	0
	Confirmed, Interior			0	
	Not Confirmed, Periphery			0	
	Not Confirmed, Interior			0	
Obstruction Restriction	From PSI, No Progression			0	0
	Service-Induced			0	
Manufacturing/ Maintenance	Pre-Operation	5		5	5
	Other			0	
Inspection Issues	Probe Lodged			0	0
	Data Quality			0	
	Dent/Geometry			0	
	Permeability			0	
	Not Inspected			0	
	Top of Tubesheet			0	
Other	Freespan			0	0
	TSP			0	
	Other/Not Reported			0	
SCC	ID			0	0
	OD			0	
TOTALS		5		5	5

Notes:

-206-

Table 3-63: Oconee 2: Tubes Plugged

STEAM GENERATOR A				
Tube	Location	RFO #	Characterization	Stabilized[1]
104-1		Pre-op	Mis-drilled tubesheet	
104-8		Pre-op	Mis-drilled tubesheet	
104-15		Pre-op	Mis-drilled tubesheet	
104-22		Pre-op	Mis-drilled tubesheet	

STEAM GENERATOR B				
Tube	Location	RFO #	Characterization	Stabilized[1]
147-39		Pre-op	Plugged prior to operation	

[1]An empty cell indicates that it was not reported whether the tube was stabilized or not.

Table 3-64: Oconee 3 Full-Length Bobbin Exams

Outage	Completion Date	Cumul. EFPY	SG A			SG B			Total Plug	Total DePl	Cumul. Plugged	Percent Plugged	Notes
			Insp.	Plug	DePl	Insp.	Plug	DePl					
Pre-op				0	0		0	0	0	0	0	0.00	
Totals:				0	0		0	0	0	0			

Plant Data
Model: BWI OTSG
T-hot (approximate): 604 F
Tubes per steam generator: 15631
Number of steam generators: 2

Acronyms
Pre-op = prior to operation
Cumul. = cumulative
Insp. = number of tubes inspected
Plug = number of tubes plugged
DePl = number of tubes deplugged
RFO = refueling outage

Table 3-65: Oconee 3 Causes of Tube Plugging

Cause of Tube Plugging/Outage		Year	Pre-Op		Totals
Wear	AVB				0
	Tube Support				
Loose Part Wear	Confirmed, Periphery				0
	Confirmed, Interior				
	Not Confirmed, Periphery				
	Not Confirmed, Interior				
Obstruction Restriction	From PSI, No Progression				0
	Service-Induced				
Manufacturing/ Maintenance	Pre-Operation				0
	Other				
Inspection Issues	Probe Lodged				0
	Data Quality				
	Dent/Geometry				
	Permeability				
	Not Inspected				
Other	Top of Tubesheet				0
	Freespan				
	TSP				
	Other/Not Reported				
SCC	ID				0
	OD				
TOTALS		0			0

Notes:

Table 3-66: Oconee 3: Tubes Plugged

STEAM GENERATOR A				
Tube	Location	RFO #	Characterization	Stabilized[1]

STEAM GENERATOR B				
Tube	Location	RFO #	Characterization	Stabilized[1]

[1]An empty cell indicates that it was not reported whether the tube was stabilized or not.

Table 3-67: Palo Verde 2 Full-Length Bobbin Exams

Outage	Completion Date	Cumul. EFPY	SG A			SG B			Total Plug	Total DePl	Cumul. Plugged	Percent Plugged	Notes
			Insp.	Plug	DePl	Insp.	Plug	DePl					
Pre-op				10			13		23	0	23	0.09	
Mid-Cycle	03/08/04			1			0		1	0	24	0.10	1
Totals:				11	0		13	0	24	0			

Plant Data
Model:
T-hot (approximate):
Tubes per steam generator: 12580
Number of steam generators: 2

Acronyms
Pre-op = prior to operation
Cumul. = cumulative
Insp. = number of tubes inspected
Plug = number of tubes plugged
DePl = number of tubes deplugged
RFO = refueling outage

Notes
1. Limited bobbin inspections were conducted since purpose of outage was to investigate the source of a leak.

Table 3-68: Palo Verde 2 Causes of Tube Plugging

Cause of Tube Plugging/Outage		Year 2004 Pre-Op	Mid		Totals	Totals
Wear	AVB				0	0
	Tube Support				0	
Loose Part Wear	Confirmed, Periphery				0	0
	Confirmed, Interior				0	
	Not Confirmed, Periphery				0	
	Not Confirmed, Interior				0	
Obstruction Restriction	From PSI, No Progression				0	0
	Service-Induced				0	
Manufacturing/ Maintenance	Pre-Operation	23			23	24
	Other		1		1	
Inspection Issues	Probe Lodged				0	0
	Data Quality				0	
	Dent/Geometry				0	
	Permeability				0	
	Not Inspected				0	
Other	Top of Tubesheet				0	0
	Freespan				0	
	TSP				0	
	Other/Not Reported				0	
SCC	ID				0	0
	OD				0	
TOTALS		23	1		24	24

Notes:

Notes
1. Tube damaged by screw used in packing crate.

-212-

Table 3-69: Palo Verde 2: Tubes Plugged

			STEAM GENERATOR A	
Tube	Location	RFO #	Characterization	Stabilized[1]
17-76		Pre-Op	Preventive (robotic fixture installation)	
17-128		Pre-Op	Preventive (robotic fixture installation)	
23-78		Pre-Op	Preventive (robotic fixture installation)	
23-126		Pre-Op	Preventive (robotic fixture installation)	
38-87		Pre-Op	Preventive (robotic fixture installation)	
38-117		Pre-Op	Preventive (robotic fixture installation)	
44-95		Pre-Op	Preventive (robotic fixture installation)	
44-109		Pre-Op	Preventive (robotic fixture installation)	
50-11		Pre-Op	Groove on tube outside diameter	
96-123		Pre-Op	Groove on tube outside diameter	
156-143	VS3	2004 Mid-cycle	Dent with 100% through-wall defect (screw hole from packaging)	

			STEAM GENERATOR B	
Tube	Location	RFO #	Characterization	Stabilized[1]
17-76		Pre-Op	Preventive (robotic fixture installation)	
17-128		Pre-Op	Preventive (robotic fixture installation)	
23-78		Pre-Op	Preventive (robotic fixture installation)	
23-126		Pre-Op	Preventive (robotic fixture installation)	
31-176		Pre-Op	Groove on tube outside diameter or factory defect	
37-86		Pre-Op	Groove on tube outside diameter or factory defect	
38-87		Pre-Op	Preventive (robotic fixture installation)	
38-117		Pre-Op	Preventive (robotic fixture installation)	
41-200	03H	Pre-Op	52 volt dent which obstructed the passage of a normal sized bobbin probe	Y
44-95		Pre-Op	Preventive (robotic fixture installation)	
44-109		Pre-Op	Preventive (robotic fixture installation)	
98-133		Pre-Op	Groove on tube outside diameter or factory defect	
171-96		Pre-Op	Groove on tube outside diameter or factory defect	

[1] An empty cell indicates that it was not reported whether the tube was stabilized or not.

Table 3-70: Point Beach 2 Full-Length Bobbin Exams

Outage	Completion Date	Cumul. EFPY	SG A			SG B			Total Plug	Total DePl	Cumul. Plugged	Percent Plugged	Notes
			Insp.	Plug	DePl	Insp.	Plug	DePl					
Pre-op				0	0		0	0	0	0	0	0.00	
RFO 23	01/08/99		3499	0	0	3499	2	2	2	0	2	0.03	
RFO 24	11/02/00		3499	0	0	3497	2	2	2	0	4	0.06	
RFO 25	05/13/02								0	0	4	0.06	
RFO 26	10/23/03	5.1	1750	0	0	1759	0	0	0	0	4	0.06	
Totals:				0	0		4	4	4	0			

Plant Data
Model: D47
T-hot (approximate):
Tubes per steam generator: 3499
Number of steam generators: 2

Acronyms
Pre-op = prior to operation
Cumul. = cumulative
Insp. = number of tubes inspected
Plug = number of tubes plugged
DePl = number of tubes deplugged
RFO = refueling outage

Table 3-71: Point Beach 2 Causes of Tube Plugging

Cause of Tube Plugging/Outage		Pre-Op	1999 RFO 23	2000 RFO 24	2002 RFO 25	2003 RFO 26	Totals
Wear	AVB						0
	Tube Support						
Loose Part Wear	Confirmed, Periphery						2
	Confirmed, Interior		2				
	Not Confirmed, Periphery						
	Not Confirmed, Interior						
Obstruction Restriction	From PSI, No Progression						0
	Service-Induced						
Manufacturing/ Maintenance	Pre-Operation						0
	Other						
Inspection Issues	Probe Lodged						2
	Data Quality			2			
	Dent/Geometry						
	Permeability						
	Not Inspected						
Other	Top of Tubesheet						0
	Freespan						
	TSP						
	Other/Not Reported						
SCC	ID						0
	OD						
TOTALS		0	2	2		0	4

Notes:

Table 3-72: Point Beach 2: Tubes Plugged

STEAM GENERATOR A				
Tube	Location	RFO #	Characterization	Stabilized[1]

STEAM GENERATOR B				
Tube	Location	RFO #	Characterization	Stabilized[1]
31-48	TSH	24	Excessive noise in +Point™ data	
55-70	TSH	23	10% through-wall wear from a confirmed loose part	Y
56-71	TSH	23	10% through-wall wear from a confirmed loose part	Y
73-58		24	Excessive noise in bobbin coil data from the hot-leg tube end to just below the fourth tube support	

[1]An empty cell indicates that it was not reported whether the tube was stabilized or not.

Table 3-73: Prairie Island 1 Full-Length Bobbin Exams

Outage	Completion Date	Cumul. EFPY	SG A Insp.	SG A Plug	SG A DePl	SG B Insp.	SG B Plug	SG B DePl	Total Plug	Total DePl	Cumul. Plugged	Percent Plugged	Notes
Pre-op				0	0		0	0	0	0	0	0.00	

Totals: 0 0 0 0 0 0

Plant Data
Model: 56/19
T-hot (approximate):
Tubes per steam generator: 4868
Number of steam generators: 2

Acronyms
Pre-op = prior to operation
Cumul. = cumulative
Insp. = number of tubes inspected
Plug = number of tubes plugged
DePl = number of tubes deplugged
RFO = refueling outage

Table 3-74: Prairie Island 1 Causes of Tube Plugging

Cause of Tube Plugging/Outage		Year	Pre-Op		Totals
Wear	AVB				0
	Tube Support				
Loose Part Wear	Confirmed, Periphery				0
	Confirmed, Interior				
	Not Confirmed, Periphery				
	Not Confirmed, Interior				
Obstruction Restriction	From PSI, No Progression				0
	Service-Induced				
Manufacturing/ Maintenance	Pre-Operation				0
	Other				
Inspection Issues	Probe Lodged				0
	Data Quality				
	Dent/Geometry				
	Permeability				
	Not Inspected				
Other	Top of Tubesheet				0
	Freespan				
	TSP				
	Other/Not Reported				
SCC	ID				0
	OD				
			TOTALS	0	0

Notes:

Table 3-75: Prairie Island 1: Tubes Plugged

STEAM GENERATOR A				
Tube	Location	RFO #	Characterization	Stabilized[1]

STEAM GENERATOR B				
Tube	Location	RFO #	Characterization	Stabilized[1]

[1]An empty cell indicates that it was not reported whether the tube was stabilized or not.

Table 3-76: Sequoyah 1 Full-Length Bobbin Exams

Outage	Completion Date	Cumul. EFPY	SG A			SG B			SG C			SG D			Total Plug	Total DePl	Cumul. Plugged	Percent Plugged	Notes
			Insp.	Plug	DePl	Insp.	Plug	DePl	Insp.	Plug	DePl	Insp.	Plug	DePl					
Pre-op				4			6			5			5		20	0	20	0.10	
RFO 13	11/20/04		4979	10	0	4977	0	0	4978	1	0	4978	0	0	11	0	31	0.16	

Totals: 14 0 6 0 6 0 5 0 31 0

Plant Data
Model: ABB/Doosan
T-hot (approximate):
Tubes per steam generator: 4983
Number of steam generators: 4

Acronyms
Pre-op = prior to operation
Cumul. = cumulative
Insp. = number of tubes inspected
Plug = number of tubes plugged
DePl = number of tubes deplugged
RFO = refueling outage

-220-

Table 3-77: Sequoyah 1 Causes of Tube Plugging

Cause of Tube Plugging/Outage		Year			Totals	Totals
		2004				
		Pre-Op	RFO 13			
Wear	AVB		11		11	11
	Tube Support				0	
Loose Part Wear	Confirmed, Periphery				0	0
	Confirmed, Interior				0	
	Not Confirmed, Periphery				0	
	Not Confirmed, Interior				0	
Obstruction Restriction	From PSI, No Progression				0	0
	Service-Induced				0	
Manufacturing/ Maintenance	Pre-Operation	20			20	20
	Other				0	
Inspection Issues	Probe Lodged				0	0
	Data Quality				0	
	Dent/Geometry				0	
	Permeability				0	
	Not Inspected				0	
Other	Top of Tubesheet				0	0
	Freespan				0	
	TSP				0	
	Other/Not Reported				0	
SCC	ID				0	0
	OD				0	
TOTALS		20	11		31	31

Notes:

Table 3-78: Sequoyah 1: Tubes Plugged

STEAM GENERATOR A				
Tube	Location	RFO #	Characterization	Stabilized[1]
***		Pre-op	3 additional tubes were plugged due to lock bar modification (specific tubes not identified)	Y
59-33		Pre-op	Geometry/lift-off signal between the second and third vertical support	N

STEAM GENERATOR B				
Tube	Location	RFO #	Characterization	Stabilized[1]
***		Pre-op	6 tubes were plugged due to lock bar modification (specific tubes not identified)	Y

STEAM GENERATOR C				
Tube	Location	RFO #	Characterization	Stabilized[1]
***		Pre-op	5 tubes were plugged due to lock bar modification (specific tubes not identified)	Y

STEAM GENERATOR D				
Tube	Location	RFO #	Characterization	Stabilized[1]
***		Pre-op	4 additional tubes were plugged due to lock bar modification (specific tubes not identified)	Y
42-118		Pre-op	Modification of lock bar, 22% through-wall indication	Y

[1]An empty cell indicates that it was not reported whether the tube was stabilized or not.

Table 3-79: South Texas Project 1 Full-Length Bobbin Exams

Outage	Completion Date	Cumul. EFPY	SG A			SG B			SG C			SG D			Total Plug	Total DePl	Cumul. Plugged	Percent Plugged	Notes
			Insp.	Plug	DePl	Insp.	Plug	DePl	Insp.	Plug	DePl	Insp.	Plug	DePl					
Pre-op				33			40			26			9		108	0	108	0.36	
RFO 10	10/23/01		7552		0	7545		0	7559		0	7576		0	0	0	108	0.36	
RFO 11	04/20/03																108	0.36	
Totals:				33	0		40	0		26	0		9	0	108	0			

Plant Data
Model: D94
T-hot (approximate): 620 F
Tubes per steam generator: 7585
Number of steam generators: 4

Acronyms
Pre-op = prior to operation
Cumul. = cumulative
Insp. = number of tubes inspected
Plug = number of tubes plugged
DePl = number of tubes deplugged
RFO = refueling outage

-223-

Table 3-80: South Texas Project 1 Causes of Tube Plugging

Cause of Tube Plugging/Outage		Year			Totals
		Pre-Op	2001 RFO 10	2003 RFO 11	
Wear	AVB				0
	Tube Support				
Loose Part Wear	Confirmed, Periphery				0
	Confirmed, Interior				
	Not Confirmed, Periphery				
	Not Confirmed, Interior				
Obstruction Restriction	From PSI, No Progression				0
	Service-Induced				
Manufacturing/ Maintenance	Pre-Operation	108			108
	Other				
Inspection Issues	Probe Lodged				0
	Data Quality				
	Dent/Geometry				
	Permeability				
	Not Inspected				
Other	Top of Tubesheet				0
	Freespan				
	TSP				
	Other/Not Reported				
SCC	ID				0
	OD				
TOTALS		108	0		108

Notes:

-224-

Table 3-81: South Texas Project 1: Tubes Plugged

STEAM GENERATOR A				
Tube	Location	RFO #	Characterization	Stabilized[1]
A		Pre-op	33 tubes plugged prior to operation due to a manufacturing phenomenon (such as laps)	

STEAM GENERATOR B				
Tube	Location	RFO #	Characterization	Stabilized[1]
A		Pre-op	40 tubes plugged prior to operation due to a manufacturing phenomenon (such as laps)	

STEAM GENERATOR C				
Tube	Location	RFO #	Characterization	Stabilized[1]
A		Pre-op	26 tubes plugged prior to operation due to a manufacturing phenomenon (such as laps)	

STEAM GENERATOR D				
Tube	Location	RFO #	Characterization	Stabilized[1]
A		Pre-op	9 tubes plugged prior to operation due to a manufacturing phenomenon (such as laps)	

[1]An empty cell indicates that it was not reported whether the tube was stabilized or not.

"A" Prior to placing the steam generators in service the following tubes were plugged:

Steam Generator A:4-54, 4-60, 8-6, 10-86, 11-105, 15-89, 18-92, 20-60, 21-89, 25-49, 29-81, 32-122, 34-72 44-144, 47-43, 48-50, 52-60, 54-128, 55-123, 62-76, 74-132, 78-114, 79-135, 81-121, 88-36, 91-39, 92-82, 94-30, 106-112, 110-108, 112-88, 114-46, 118-86.

Steam Generator B: 1-53, 5-27, 15-87, 24-114, 25-109, 26-82, 30-8, 33-37, 38-108, 40-24, 41-7, 42-24, 47-107, 48-146, 49-123, 52-82, 52-140, 60-38, 62-56, 62-106, 68-72, 68-96, 73-27, 73-37, 80-20, 80-32, 87-39, 88-32, 88-110, 90-70, 92-90, 96-90, 96-102, 99-97, 104-62, 110-102, 110-112, 119-79, 121-85, 126-78.

Steam Generator C: 3-59, 3-103, 9-99, 11-27, 11-41, 15-35, 15-123, 18-62, 28-130, 29-139, 31-107, 53-75, 61-15, 64-106, 67-45, 76-70, 78-40, 83-27, 92-130, 93-59, 103-83, 104-102, 105-85, 106-36, 108-40, 120-58.

Steam Generator D: 14-16, 35-67, 43-45, 55-35, 64-128, 66-14, 95-41, 109-63, 115-45.

Table 3-82: South Texas Project 2 Full-Length Bobbin Exams

Outage	Completion Date	Cumul. EFPY	SG A Insp.	Plug	DePl	SG B Insp.	Plug	DePl	SG C Insp.	Plug	DePl	SG D Insp.	Plug	DePl	Total Plug	Total DePl	Cumul. Plugged	Percent Plugged	Notes
Pre-Op																			
RFO 10	04/27/04		7584	1	0	7583	2	0	7582	3	0	7585	0	0	6	0	6	0.02	
																		0.02	

Totals: 1   0   2   0   3   0   0   0   6   0   6

Acronyms
Pre-op = prior to operation
Cumul. = cumulative
Insp. = number of tubes inspected
Plug = number of tubes plugged
DePl = number of tubes deplugged
RFO = refueling outage

Plant Data
Model: D94
T-hot (approximate): 620 F
Tubes per steam generator: 7585
Number of steam generators: 4

Table 3-83: South Texas Project 2 Causes of Tube Plugging

Cause of Tube Plugging/Outage		Year 2004 Pre-Op	Year 2004 RFO 10	Totals
Wear	AVB			0
	Tube Support			0
Loose Part Wear	Confirmed, Periphery			0
	Confirmed, Interior			0
	Not Confirmed, Periphery			0
	Not Confirmed, Interior			0
Obstruction Restriction	From PSI, No Progression			0
	Service-Induced			0
Manufacturing/ Maintenance	Pre-Operation	6		6
	Other			0
Inspection Issues	Probe Lodged			0
	Data Quality			0
	Dent/Geometry			0
	Permeability			0
	Not Inspected			0
Other	Top of Tubesheet			0
	Freespan			0
	TSP			0
	Other/Not Reported			0
SCC	ID			0
	OD			0
TOTALS		6	0	6

Notes:

Table 3-84: South Texas Project 2: Tubes Plugged

STEAM GENERATOR A				
Tube	Location	RFO #	Characterization	Stabilized[1]
127-87		Pre-op	Plugged prior to operation	

STEAM GENERATOR B				
Tube	Location	RFO #	Characterization	Stabilized[1]
51-9		Pre-op	Plugged prior to operation	
73-113		Pre-op	Plugged prior to operation	

STEAM GENERATOR C				
Tube	Location	RFO #	Characterization	Stabilized[1]
70-22		Pre-op	Plugged prior to operation	
75-77		Pre-op	Plugged prior to operation	
76-48		Pre-op	Plugged prior to operation	

STEAM GENERATOR D				
Tube	Location	RFO #	Characterization	Stabilized[1]

[1] An empty cell indicates that it was not reported whether the tube was stabilized or not.

Table 3-85: St. Lucie 1 Full-Length Bobbin Exams

Outage	Completion Date	Cumul. EFPY	SG A Insp.	SG A Plug	SG A DePl	SG B Insp.	SG B Plug	SG B DePl	Total Plug	Total DePl	Cumul. Plugged	Percent Plugged	Notes
Pre-op				0	0		0	0	0	0	0	0.00	
RFO 16	09/26/99		5055	11	0	4665	0	0	11	0	11	0.06	
RFO 17	04/13/01		4764	0	0	4525	0	0	0	0	11	0.06	
RFO 18	10/24/02								0	0	11	0.06	
RFO 19	04/10/04		4834	3		4640	0		3	0	14	0.08	

Totals: 14   0   0   0   14   0

Plant Data
Model:
T-hot (approximate):
Tubes per steam generator: 8523
Number of steam generators: 2

Acronyms
Pre-op = prior to operation
Cumul. = cumulative
Insp. = number of tubes inspected
Plug = number of tubes plugged
DePl = number of tubes deplugged
RFO = refueling outage

Table 3-86: St. Lucie 1 Causes of Tube Plugging

Cause of Tube Plugging/Outage		Pre-Op	1999 RFO 16	2001 RFO 17	2002 RFO 18	2004 RFO 19	Totals	Totals
Wear	AVB		10				10	10
	Tube Support						0	
Loose Part Wear	Confirmed, Periphery						0	0
	Confirmed, Interior						0	
	Not Confirmed, Periphery						0	
	Not Confirmed, Interior						0	
Obstruction Restriction	From PSI, No Progression						0	0
	Service-Induced						0	
Manufacturing/ Maintenance	Pre-Operation						0	1
	Other		1				1	
Inspection Issues	Probe Lodged						0	0
	Data Quality						0	
	Dent/Geometry						0	
	Permeability						0	
	Not Inspected						0	
Other	Top of Tubesheet						0	3
	Freespan						0	
	TSP						0	
	Other/Not Reported					3	3	
SCC	ID						0	0
	OD						0	
TOTALS		0	11	0		3	14	14

Notes:

Notes
1. Cause or location of indications not reported.

Table 3-87: St. Lucie 1: Tubes Plugged

STEAM GENERATOR A				
Tube	Location	RFO #	Characterization	Stabilized[1]
27-100	CBH+1.2 CBH+0.9	16	30% through-wall manufacturing anomaly	
69-92	F4-0.8 F9-2.6	16	21% and 22% through-wall wear	
85-92	F9-2.6	16	22% through-wall wear	
87-92	F9-2.6	16	20% through-wall wear	
89-92	F9-2.2	16	24% through-wall wear	
102-77	F10+0.6	16	30% through-wall wear	
105-92	F5-0.4	16	28% through-wall wear	
106-77	F10+0.7	16	34% through-wall wear	
116-77	F10+0.6	16	21% through-wall wear	
118-77	F10+0.6	16	30% through-wall wear	
122-77	F10+0.6	16	24% through-wall wear	
***		19	3 additional tubes were plugged during RFO 19 (specific tubes not identified).	

STEAM GENERATOR B				
Tube	Location	RFO #	Characterization	Stabilized[1]

[1] An empty cell indicates that it was not reported whether the tube was stabilized or not.

Table 3-88: Summer Full-Length Bobbin Exams

Outage	Completion Date	Cumul. EFPY	SG A Insp.	SG A Plug	SG A DePl	SG B Insp.	SG B Plug	SG B DePl	SG C Insp.	SG C Plug	SG C DePl	Total Plug	Total DePl	Cumul. Plugged	Percent Plugged	Notes
Pre-op				0	0		1			2		3	0	3	0.02	
RFO 9	04/29/96		1393	0	0	1039	0					0	0	3	0.02	1
RFO 10	10/20/97								1892	0		0	0	3	0.02	
RFO 11	04/24/99		2527	0	0	2527	0					0	0	3	0.02	
RFO 12	10/27/00	5.4	6307	3		6306	0		6305	2		5	0	8	0.04	
RFO 13	06/02/02											0	0	8	0.04	
RFO 14	11/23/03											0	0	8	0.04	
Totals:				3	0		1	0		4	0	8	0			

Plant Data
Model: D75
T-hot (approximate):
Tubes per steam generator: 6307
Number of steam generators: 3

Acronyms
Pre-op = prior to operation
Cumul. = cumulative
Insp. = number of tubes inspected
Plug = number of tubes plugged
DePl = number of tubes deplugged
RFO = refueling outage

Notes
1. Inspections were from hot-leg tube end through uppermost tube support on cold-leg end (i.e., no full-length inspections).

Table 3-89: Summer Causes of Tube Plugging

Cause of Tube Plugging/Outage		Year Pre-Op	1996 RFO 9	1997 RFO 10	1999 RFO 11	2000 RFO 12	2002 RFO 13	2003 RFO 14	Totals	Totals
Wear	AVB								0	0
	Tube Support								0	
Loose Part Wear	Confirmed, Periphery								0	0
	Confirmed, Interior								0	
	Not Confirmed, Periphery								0	
	Not Confirmed, Interior								0	
Obstruction Restriction	From PSI, No Progression								0	0
	Service-Induced								0	
Manufacturing/ Maintenance	Pre-Operation	3							3	8
	Other					5			5	
Inspection Issues	Probe Lodged								0	0
	Data Quality								0	
	Dent/Geometry								0	
	Permeability								0	
	Not Inspected								0	
Other	Top of Tubesheet								0	0
	Freespan								0	
	TSP								0	
	Other/Not Reported								0	
SCC	ID								0	0
	OD								0	
TOTALS		3	0	0	0	5			8	8
Notes:						1				

Notes
1. Tubes not fully expanded for full length of tubesheet.

Table 3-90: Summer: Tubes Plugged

STEAM GENERATOR A				
Tube	Location	RFO #	Characterization	Stabilized[1]
25-26		12	Tube not expanded in the tubesheet.	
25-31		12	Tube not expanded in the tubesheet.	
94-51		12	Tube not expanded in the tubesheet.	

STEAM GENERATOR B				
Tube	Location	RFO #	Characterization	Stabilized[1]
***		Pre-op	1 tube plugged prior to operation (specific tube not identified)	

STEAM GENERATOR C				
Tube	Location	RFO #	Characterization	Stabilized[1]
***		Pre-op	2 tubes were plugged prior to operation (specific tubes not identified)	
57-96		12	Tube not expanded in the tubesheet.	
99-100		12	Tube not expanded in the tubesheet.	

[1]An empty cell indicates that it was not reported whether the tube was stabilized or not.

4 SUMMARY

The following sections summarize salient points about the design and operating experience of steam generators with thermally treated Alloy 690 tubes.

4.1 Design Summary

As of December 2004, most of the steam generators in the United States with thermally treated Alloy 690 tubes were designed either by Westinghouse or Babcock and Wilcox International (BWI). Of the 30 units with thermally treated Alloy 690 steam generator tubes, Westinghouse designed the steam generators for 13 units (43 percent), and BWI designed the steam generators for 14 units (47 percent). The remaining three units (10 percent) have steam generators designed by either Framatome or ABB-CE (now a part of Westinghouse). The steam generators designed by Westinghouse contain 43 percent (204,306) of the 577,070 thermally treated Alloy 690 tubes, the steam generators designed by BWI contain 47 percent (317,936) of the tubes, and the remaining steam generators contain 10 percent (54,828) of the tubes.

The design of the steam generator tube supports differs among the vendors. For example, the tube supports are fabricated from Type 410 stainless steel in steam generators designed by BWI, Type 405 stainless steel in those designed by Westinghouse, and Type 409 stainless steel in those designed by ABB-CE. In addition, the tube supports are lattice grids in the steam generators designed by BWI and ABB-CE (except for the once-through steam generators where the tube supports are plates), and the tube supports are plates in Westinghouse-designed steam generators.

The 30 units with Alloy 690 as the tube material contain a total of 577,070 tubes. These tubes were fabricated either by Sandvik (in Sweden), Sumitomo (in Japan), or Valinox (in France). Of the 30 units with thermally treated Alloy 690 steam generator tubes, Sandvik has supplied the tubes for 15 units (50 percent), Sumitomo has supplied the tubes for 12 units (40 percent), and Valinox has supplied the tubes for 3 units (10 percent). Of the 577,070 thermally treated Alloy 690 tubes, Sandvik fabricated 251,950 (44 percent), Sumitomo fabricated 291,360 (50 percent), and Valinox fabricated 33,760 (6 percent). In the Westinghouse-designed steam generators, Sandvik supplied all of the steam generator tubes except for those used at Kewaunee. In the BWI-designed steam generators, Sumitomo supplied all of the steam generator tubes except for those used at Ginna and Millstone 2. Valinox supplied the Kewaunee, Ginna, and Millstone 2 steam generator tubes.

4.2 Operating Experience Summary

As depicted in Figure 2-2, 577,070 thermally treated Alloy 690 tubes were placed in service at 30 units between 1989 and 2004. Cumulatively, these 30 units have operated for approximately 173 calendar years and for an average of 6 calendar years each (as of December 2004). Of the 577,070 tubes in these units, only 333 tubes (0.06 percent) have been plugged. Table 4-1 summarizes the number and percentage of tubes plugged at the 30 units with thermally treated Alloy 690 tubes. Figures 4-1 and 4-2 depict the total number and percentage, respectively, of tubes plugged in units with thermally treated Alloy 690 tubes as a function of year. These figures were developed from the data provided in Tables 4-2 and 4-3.

Table 4-4 summarizes the number of tubes plugged as a function of the degradation mechanism. Figure 4-3 graphically depicts the information in this table. As this table and figure show, the dominant degradation mode (excluding manufacturing and maintenance reasons) of thermally treated Alloy 690 tubes is wear. Of the 333 tubes plugged, approximately 24 percent were plugged as a result of tube wear. Tube wear occurs as a result of contact between the tube and a support structure (e.g., an antivibration bar (AVB)) or a foreign object (e.g., a loose

-235-

part). Loose parts can be introduced during steam generator fabrication, during maintenance activities, or as a result of corrosion degradation of other components in the primary or secondary side of the steam generator (e.g., a split pin nut). The rate of tube wear from support structures is generally predictable and is readily managed. Wear from loose parts is usually unexpected and can be detected only by inspection (visual or eddy current), loose parts monitoring systems, or primary-to-secondary leakage. The wear in thermally treated tubes has occurred predominantly at the AVBs and near loose parts. A very limited number of tubes have been plugged for wear at the tube supports. Most of the tubes plugged for wear at a support structure were in BWI-designed steam generators.

Developed from data in Tables 4-5, 4-6, and 4-7, Figure 4-4 depicts the fraction of tubes plugged for a specific mechanism as a function of year. In this figure, tubes plugged before the steam generators went into commercial operation were treated as being plugged during the year the steam generator began commercial operation (in previous tables and figures in this report, these tubes were treated as a distinct group independent of the actual year/outage in which they were plugged).

4.2.1 Forced Outages

As of December 2004, the steam generator operating experience of units with thermally treated Alloy 690 has been favorable. These units account for approximately 43 percent of the currently operating pressurized-water reactors in the United States. A historical review identified only two unplanned outages as a result of steam generator issues in units with thermally treated Alloy 690 tubes (as of December 2004). Both of these outages resulted from primary-to-secondary leakage. During the preparation of this report in the first half of 2005, one unplanned outage attributed to primary-to-secondary leakage occurred (at Arkansas Nuclear One (ANO), Unit 2), one unscheduled inspection occurred during an outage at South Texas Project 1 because of loose parts found on the secondary side of the steam generator, and one unit experienced a chemical excursion resulting in the plant's having to correct the chemistry condition while shut down (Kewaunee). All of these outages are discussed below.

Only three units with thermally treated Alloy 690 tubes have experienced any significant primary-to-secondary leakage. In February 2004, Palo Verde 2 shut down approximately two months after replacing its steam generators as a result of an 11-gallon-per-day (gpd) primary-to-secondary leak. The location, shape, and size of the deformation in the leaking tube were consistent with damage that would occur if a screw (which was used during the crating process for shipment of the tube to the steam generator fabricator) penetrated completely through the packing material and came in contact with the tube. Nuclear Regulatory Commission (NRC) Information Notice 2004-16, "Tube Leakage Due to a Fabrication Flaw in a Replacement Steam Generator," gives additional details.

In May 2004, following a unit trip for an unrelated reason, the Harris plant investigated the source of a 5- to 10-gpd primary-to-secondary leak. The cause of the leak was a foreign object located on the secondary side of the steam generator. This object damaged three tubes, and the tubes were plugged. Additional details appear in NRC Information Notice 2004-17, "Loose Part Detection and Computerized Eddy Current Data Analysis in Steam Generators."

In March 2005, ANO 2 entered a refueling outage early as a result of a 30-gpd primary-to-secondary leak. The leak was caused by a foreign object.

In-situ pressure testing of the leaking tubes at all three units indicated that the tubes had adequate structural integrity and that the leakage integrity of the steam generator was not compromised (i.e., the accident-induced primary-to-secondary leakage was less than that assumed during the design/licensing of the facility).

In addition to these three primary-to-secondary leakage events, during the first half of 2005, one unit experienced a chemical excursion and another received an unscheduled inspection. In February 2005, approximately 1000 gallons of service water, which is drawn from Lake Michigan, entered the secondary side of the steam generators at Kewaunee during a plant shutdown/cooldown. The saturation temperature of the steam generators at the time of the introduction was approximately 310 °F. Both steam generators were drained and refilled in order to return the water chemistry to within normal specifications. During a refueling outage in March 2005 at South Texas Project 1, several hundred small wire fragments were identified in steam generator D. Because one tube exhibited wear resulting from these wire fragments, an eddy current inspection was performed in this steam generator. No eddy current inspections had been planned in any of the steam generators at South Texas Project 1 during this outage.

4.2.2 Tube Pulls

To characterize eddy current indications found during steam generator tube inservice inspections, portions of tubes are occasionally removed from the steam generators. As of December 2004, no portions of thermally treated Alloy 690 tubes have been removed for destructive examination.

4.2.3 Summary and Observations

Units with thermally treated Alloy 690 tubes have a variety of strategies for inspecting their steam generators. At several units, individual steam generator tubes have not been inspected for a period of 6 calendar years. In addition, many units have not performed tube inspections at every refueling outage. No instances have been reported in which a thermally treated Alloy 690 tube did not satisfy the criteria for structural integrity (e.g., three times the normal operating differential pressure). In addition, no instances have been reported in which a steam generator with thermally treated Alloy 690 tubing did not satisfy the accident-induced leakage performance criteria under these inspection strategies.

A review of the design and operating experience of steam generators with thermally treated Alloy 690 steam generator tubes leads to the following observations:

- No cracklike indications have been found in any of the thermally treated Alloy 690 tubes.

- Issues related to manufacturing/maintenance are the most common reason that tubes have been plugged. In fact, approximately one-third of the tubes plugged were at one unit (South Texas Project 1).

- Before 2004, there were no forced outages, chemical excursions, or unscheduled inspections at units with thermally treated Alloy 690 tubes. Since January 1, 2004, there have been five occurrences of forced outages, chemical excursions, or unscheduled inspections. The forced outages and unscheduled inspections all resulted from loose parts or foreign objects. These results highlight the need to limit the introduction of loose parts into the steam generator and the need to remain vigilant in inspecting the steam generators to ensure prompt identification of conditions adverse to quality.

- In several units with BWI-designed steam generators, there is a small population of tubes that are in close proximity. No tube wear has been associated with this condition (i.e., tube-to-tube proximity). This condition was expected to naturally correct itself after one or two cycles of operation in the vertical position; however, in at least one unit with BWI steam generators, additional tubes which had not been noticed to be in close proximity following fabrication were classified as being in close proximity following several cycles of operation.

- Three types of wear are postulated to occur at units with BWI-designed steam generators. Typical fan bar wear is a result of thermal hydraulic conditions and tube-to-support clearances which can vary because of manufacturing tolerances. Typical wear results in either uniform or tapered wear scars on the tube. Localized U-bend wear is a phenomenon "localized" to specific columns of tubes and possibly the adjacent column. It is theorized to be the result of arch-bar distortion instead of a more random manufacturing tolerance issue (which causes typical fan bar wear). The wear occurs as a result of some local distortion in the fan bar because of how it is attached at the U-bend superstructure which results in an increased tube-to-support gap. Localized U-bend wear has been observed at St. Lucie 1 and McGuire 1. Atypical U-bend wear refers to pitlike indications found at flat-bar supports. These indications are thought to be the result of asperities on the flat bars and are attributed to fabrication deficiencies. This mechanism has been observed in the steam generators at McGuire 1 and 2 and at St. Lucie 1.

- In one of the Byron 1 steam generators, 57 of the 671 tubes in row 1 on the hot-leg side of the tube bundle were found to be disengaged from the collector bar (the lowest fan bar). A flow-induced vibration analysis indicated that the tubes will remain fluid elastically stable, and there is no risk of high-cycle fatigue. The collector bar was most likely mispositioned during a fabrication repair that repositioned the fan bar support structures called J-tabs.

- During a unit shutdown/cooldown in February 2005 at Kewaunee, approximately 1000 gallons of service water, which is drawn from Lake Michigan, entered the secondary side of the steam generators. The steam generators were drained and refilled to reduce the level of chemical impurities in the steam generator.

- Following the first cycle of operation with replacement once-through steam generators, several thousand indications of wear at the tube support plate elevations were detected at Oconee 1 in 2005. At the time this report was being prepared, the root cause investigation was ongoing.

Far fewer tubes have been plugged in the steam generators with third-generation tube materials (i.e., thermally treated Alloy 690) than in earlier steam generators with comparable operating times. Improvements in the design and operation of the third-generation steam generators appear to have increased the corrosion resistance of the tubes, as evidenced by the general lack of significant amounts of corrosion degradation. The enhanced corrosion resistance is largely the result of the improved alloy and the thermal treatment process that has superseded the mill annealing process used in earlier steam generator designs.

The relatively good operating experience for units with thermally treated Alloy 690 steam generator tubes can be attributed to several factors in addition to the material selection and the heat treatment of the tubes. These include the hydraulic expansion of the tubes into the tubesheet, the design of the tube supports, and the stainless steel material used to fabricate the supports. The residual stress levels at the expansion transition in tubes hydraulically expanded into the tubesheet are lower than those observed in units whose tubes were expanded mechanically or explosively. Since crack growth rate and time to crack initiation depend, in part, on the stress level, lower stresses may result in lower crack growth rates and/or longer times before crack initiation.

Although the operating experience with thermally treated Alloy 690 tubes has been favorable to date, there is a continued need to monitor the tubes for the onset of tube degradation (including cracking) and to assure the structural and leakage integrity of the tubes during the intervals between inspections. Currently, wear attributed to loose parts appears to be the largest challenge for units with thermally treated Alloy 690 tubes (i.e., challenge in terms of determining the appropriate inspection interval).

Table 4-1: Total Number and Percentage of Tubes Plugged for All Models (12/2004)
(Part 1)

Unit	Number of Tubes Plugged[1]	Percent Plugged	Operating Time[2]
Arkansas Nuclear One 2	1	<0.01	4
Braidwood 1	30	0.11	6
Byron 1	1	<0.01	7
Calvert Cliffs 1	0	0.00	3
Calvert Cliffs 2	3	0.02	2
Catawba 1	26	0.10	8
Cook 1	4	0.03	4
Cook 2	16	0.11	16
Farley 1	0	0.00	5
Farley 2	0	0.00	4
Ginna	2	0.02	9
Harris	5	0.03	3
Indian Point 3	14	0.11	16
Kewaunee	0	0.00	3
McGuire 1	13	0.05	8
McGuire 2	11	0.04	7
Millstone 2	1	0.01	12
North Anna 1	2	0.02	12
North Anna 2	2	0.02	10
TOTALS	SEE NEXT PAGE		

[1]As of 12/31/2004
[2]Operating Time = calendar years of operation as of 12/31/2004

Table 4-1: Total Number and Percentage of Tubes Plugged for All Models (12/2004)
(Part 2)

Unit	Number of Tubes Plugged[1]	Percent Plugged	Operating Time[2]
Oconee 1	2	0.01	1
Oconee 2	5	0.02	1
Oconee 3	0	0.00	<1
Palo Verde 2	24	0.10	1
Point Beach 2	4	0.06	8
Prairie Island 1	0	0.00	<1
Sequoyah 1	31	0.16	2
South Texas Project 1	108	0.36	5
South Texas Project 2	6	0.02	2
St. Lucie 1	14	0.08	7
Summer	8	0.04	10
TOTALS	333	0.06	173

[1]As of 12/31/2004
[2]Operating Time = calendar years of operation as of 12/31/2004

Table 4-2: Plugging per Year (Part 1)

Year	ANO 2	Braidwood 1	Byron 1	Calvert Cliffs 1	Calvert Cliffs 2	Catawba 1	Cook 1	Cook 2
Pre-Op	1	3	1		3	19		1
1989								
1990								
1991								
1992								
1993								9
1994								
1995								
1996								5
1997								
1998								
1999								
2000		1						
2001							4	
2002						7		
2003		21						
2004		5						1
2005								

-241-

Table 4-2: Plugging per Year (Part 2)

Year	Farley 1	Farley 2	Ginna	Harris	Indian Point 3	Kewaunee	McGuire 1	McGuire 2
Pre-Op			2	2	2		10	2
1989								
1990								
1991								
1992								
1993								
1994								
1995								
1996								
1997								
1998							2	
1999								9
2000								
2001								
2002								
2003				3	12			
2004							1	
2005								

Table 4-2: Plugging per Year (Part 3)

Year	Millstone 2	North Anna 1	North Anna 2	Oconee 1	Oconee 2	Oconee 3	Palo Verde 2	Point Beach 2
Pre-Op	1			2	5		23	
1989								
1990								
1991								
1992								
1993								
1994								
1995								
1996								
1997		1						
1998								2
1999								2
2000			1					
2001			1					
2002								
2003							1	
2004		1						
2005								

Table 4-2: Plugging per Year (Part 4)

Year	Prairie Island 1	Sequoyah 1	South Texas 1	South Texas 2	St. Lucie 1	Summer
Pre-Op		20	108	6		3
1989						
1990						
1991						
1992						
1993						
1994						
1995						
1996						
1997						
1998						
1999					11	5
2000						
2001						
2002						
2003					3	
2004		11				
2005						

-244-

Table 4-3: Cumulative Plugging per Year (Part 1)

Year	ANO 2	Braidwood 1	Byron 1	Calvert Cliffs 1	Calvert Cliffs 2	Catawba 1	Cook 1	Cook 2
Pre-Op	1	3	1		3	19		1
1989								
1990								
1991								
1992								
1993								
1994								10
1995								
1996								
1997								15
1998								
1999								
2000		4						
2001								
2002						26	4	
2003		25						
2004		30						16
2005								

-245-

Table 4-3: Cumulative Plugging per Year (Part 2)

Year	Farley 1	Farley 2	Ginna	Harris	Indian Point 3	Kewaunee	McGuire 1	McGuire 2
Pre-Op			2	2	2		10	2
1989								
1990								
1991								
1992								
1993								
1994								
1995								
1996								
1997								
1998							12	
1999								11
2000								
2001								
2002								
2003				5	14			
2004							13	
2005								

Table 4-3: Cumulative Plugging per Year (Part 3)

Year	Millstone 2	North Anna 1	North Anna 2	Oconee 1	Oconee 2	Oconee 3	Palo Verde 2	Point Beach 2
Pre-Op	1			2	5		23	
1989								
1990								
1991								
1992								
1993								
1994								
1995								
1996								
1997								
1998		1						
1999								2
2000								4
2001			1					
2002			2					
2003								
2004		2					24	
2005								

Table 4-3: Cumulative Plugging per Year (Part 4)

Year	Prairie Island 1	Sequoyah 1	South Texas 1	South Texas 2	St. Lucie 1	Summer
Pre-Op		20	108	6		3
1989						
1990						
1991						
1992						
1993						
1994						
1995						
1996						
1997						
1998						
1999					11	
2000						8
2001						
2002						
2003						
2004		31			14	
2005						

Table 4-4: Number of Tubes Plugged as a Function of Mechanism, All Plants

	Cause of Tube Plugging	Tubes Plugged	Percentage of Plugs	Tubes Plugged	Percentage of Plugs
Wear	AVB	32	9.6%	33	9.9%
	Tube Support	1	0.3%		
Loose Part Wear	Confirmed, Periphery	34	10.2%	46	13.8%
	Confirmed, Interior	2	0.6%		
	Not Confirmed, Periphery	7	2.1%		
	Not Confirmed, Interior	3	0.9%		
Obstruction Restriction	From PSI, No Progression	0	0.0%	0	0.0%
	Service-Induced	0	0.0%		
Manufacturing/ Maintenance	Pre-Operation	214	64.3%	241	72.4%
	Other	27	8.1%		
Inspection Issues	Probe Lodged	0	0.0%	4	1.2%
	Data Quality	2	0.6%		
	Dent/Geometry	0	0.0%		
	Permeability	2	0.6%		
	Not Inspected	0	0.0%		
Other	Top of Tubesheet	0	0.0%	9	2.7%
	Freespan	5	1.5%		
	TSP	1	0.3%		
	Other/Not Reported	3	0.9%		
SCC	ID	0	0.0%	0	0.0%
	OD	0	0.0%		
	TOTALS	333	100.0%	333	100.0%

Total Tubes =   577070
Fraction Plugged =   0.06%

Table 4-5: Number of Tubes Plugged as a Function of Mechanism per Year (Detailed)

Causes of Tube Plugging/Year		1989	1990	1991	1992	1993	1994	1995	1996	1997	1998	1999	2000	2001	2002	2003	2004	2005	Totals	Totals
Wear	AVB										2	18	1				11		32	33
	Tube Support													1					1	
Loose Part Wear	Confirmed, Periphery									1						28	5		34	46
	Confirmed, Interior											2							2	
	Not Confirmed, Periphery									4							3		7	
	Not Confirmed, Interior																3		3	
Obstruction Restriction	From PSI, No Progression																		0	0
	Service-Induced																		0	
Manufacturing/ Maintenance	Pre-Operation	3				1	3		21	12	4		109	2	6	46	7		214	241
	Other						9					1	5			11	1		27	
Inspection Issues	Probe Lodged																		0	4
	Data Quality												2						2	
	Dent/Geometry																		0	
	Permeability														1	1			2	
	Not Inspected																		0	
Other	Top of Tubesheet																		0	9
	Freespan											1			4				5	
	TSP										1								1	
	Other/Not Reported																3		3	
SCC	ID																		0	0
	OD																		0	
TOTALS		3	0	0	0	1	12	0	21	17	7	22	117	3	11	86	33	0	333	333

Notes:

Table 4-6: Number of Tubes Plugged as a Function of Mechanism per Year (Summary)

Causes of Tube Plugging/Year	1989	1990	1991	1992	1993	1994	1995	1996	1997	1998	1999	2000	2001	2002	2003	2004	2005	Totals
Wear										2	18	1	1			11		33
Loose Part Wear									5		2				28	11		46
Obstruction Restriction																		0
Manufacturing/ Maintenance	3				1	12		21	12	4	1	114	2	6	57	8		241
Inspection Issues										1		2			1			4
Other											1			5		3		9
SCC																		0
	3	0	0	0	1	12	0	21	17	7	22	117	3	11	86	33	0	333

Notes:

Table 4-7: Fraction of Tubes Plugged as a Function of Mechanism per Year (Summary)

Causes of Tube Plugging/Year	1989	1990	1991	1992	1993	1994	1995	1996	1997	1998	1999	2000	2001	2002	2003	2004	2005	Totals
Wear										0.29	0.82	0.01	0.33			0.33		0.10
Loose Part Wear									0.29		0.09				0.33	0.33		0.14
Obstruction Restriction	1.00																	0.00
Manufacturing Maintenance					1.00	1.00		1.00	0.71	0.57	0.05	0.97	0.67	0.55	0.66	0.24		0.72
Inspection Issues												0.02		0.09	0.01			0.01
Other										0.14	0.05			0.36		0.09		0.03
SCC																		0.00
	1.00	0.00	0.00	0.00	1.00	1.00	0.00	1.00	1.00	1.00	1.00	1.00	1.00	1.00	1.00	1.00	0.00	1.00

Notes:

-252-

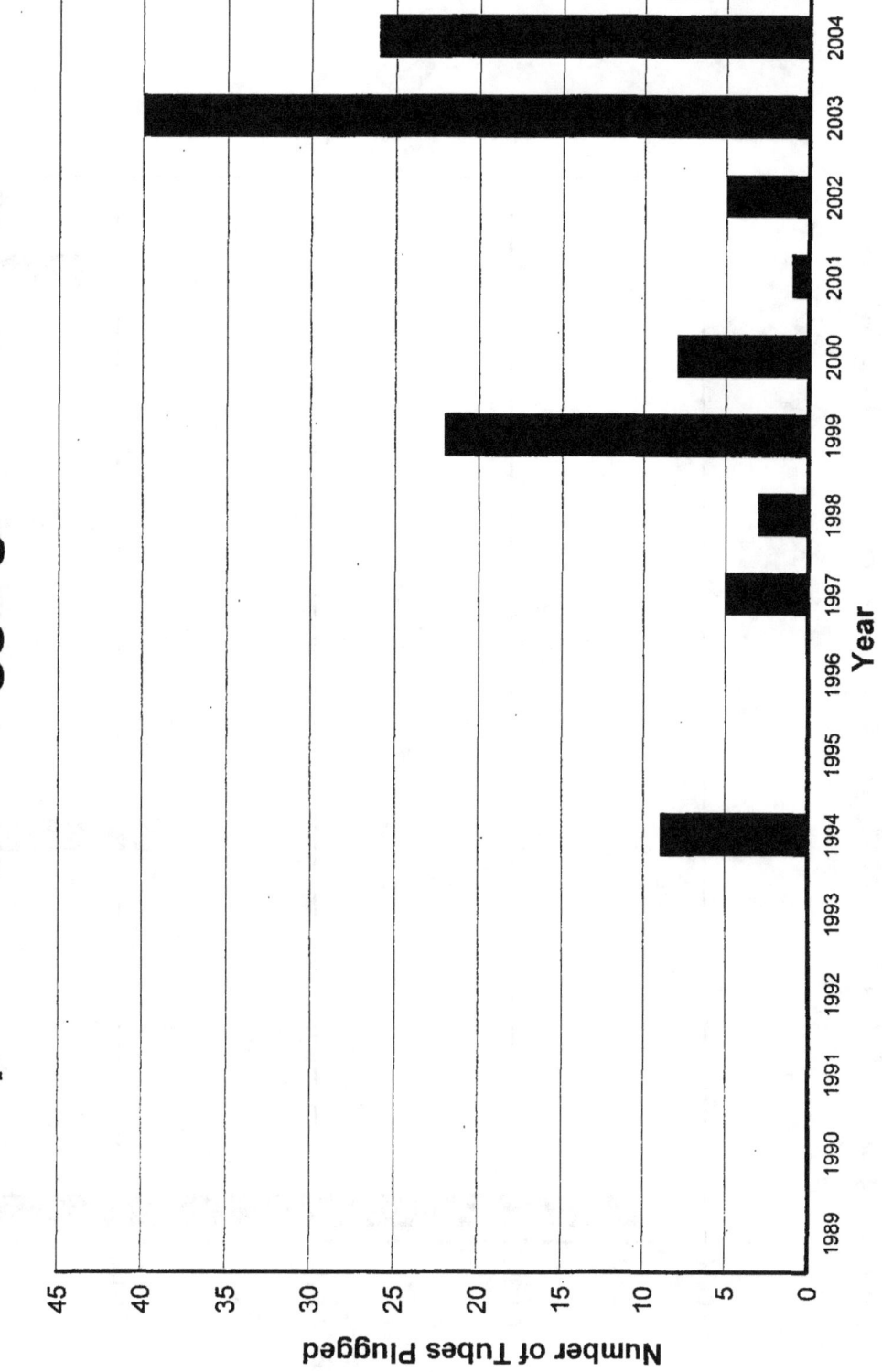

Figure 4-1: Number of Tubes Plugged per Year Preoperational Plugging Not Included

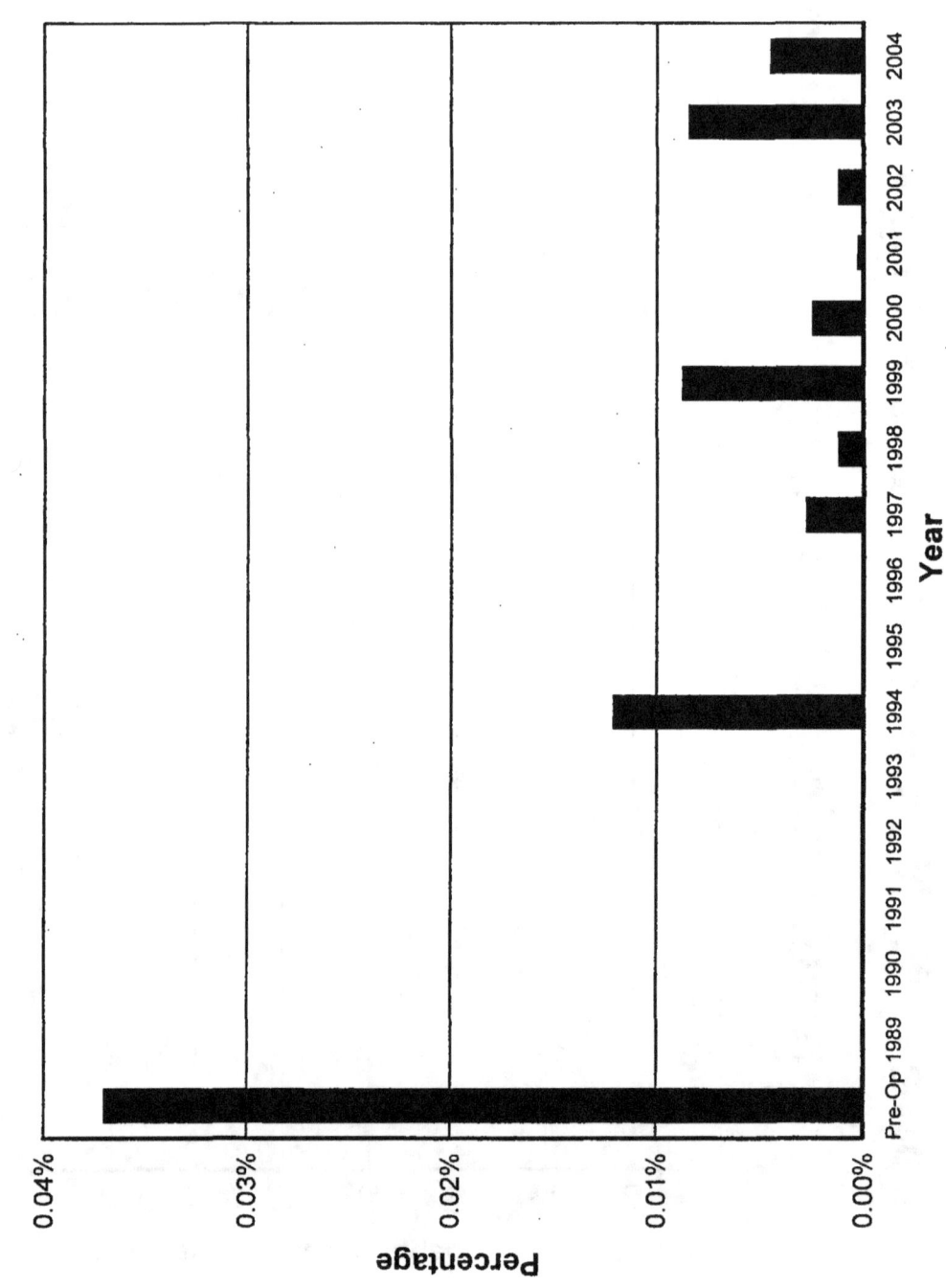

Figure 4-2: Percentage of Tubes Plugged per Year

Figure 4-3: Causes of Tube Plugging

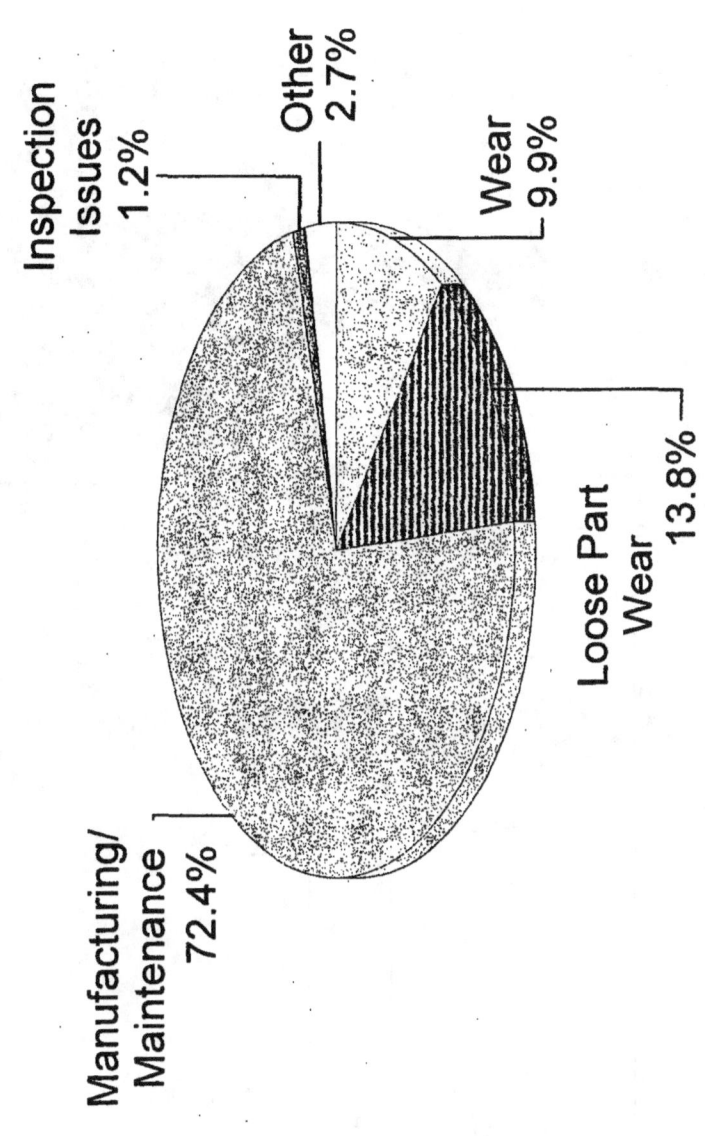

Inspection
Issues
1.2%

Other
2.7%

Wear
9.9%

Loose Part
Wear
13.8%

Manufacturing/
Maintenance
72.4%

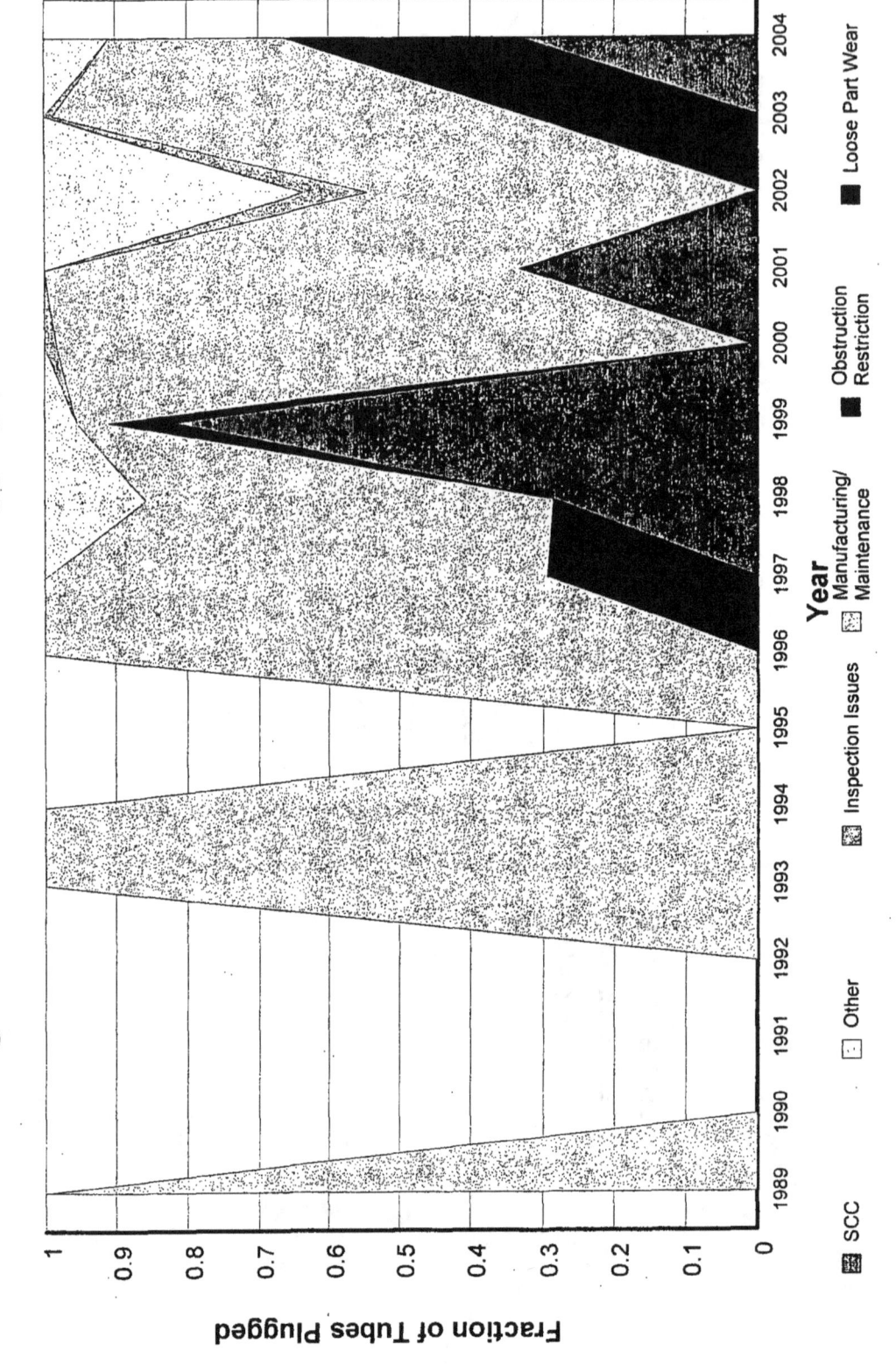

Figure 4-4: Causes of Tube Plugging per Year

APPENDIX A: ABBREVIATIONS

ABB	Asea Brown Boveri
ADI	absolute drift indication
ADS	absolute drift signal
ANO	Arkansas Nuclear One
AV	antivibration bar
AVB	antivibration bar
AVT	all volatile treatment
B&W	Babcock and Wilcox
BPC	cold-leg flow distribution baffle (baffle plate cold)
BPH	hot-leg flow distribution baffle (baffle plate hot)
BW	batwing
BWI	Babcock and Wilcox International
CBC	connector bar cold
CBH	connector bar hot
CDS	computerized data screening
CE	Combustion Engineering
CLP	confirmed loose part
ECT	eddy current testing
EFPM	effective full-power month
EFPY	effective full-power year
ENSA	Equipos Nucleares
EPRI	Electric Power Research Institute
F	fan support
FB	fan bar
FBC	cold-leg flow distribution baffle (flow baffle cold)
FBH	hot-leg flow distribution baffle (flow baffle hot)
FDB	flow distribution baffle
FDP	fan distribution plate
FOSAR	foreign object search and retrieval
fps	feet per second
Fr	Framatome
FS	freespan
ft^2	square feet
FURS	flat bar U-bend restraint system
gpd	gallons per day
gpm	gallons per minute
ID	inside diameter
LOCA	loss-of-coolant accident
LTE	lower tube end
LTS	lower tubesheet
MA	mill annealed
MAI	multiple axial indication
MBM	manufacturing burnishing mark
NDE	non-destructive examination
NQI	nonquantifiable indication
NRC	Nuclear Regulatory Commission
NUDOCS	Nuclear Documents System
OD	outside diameter
ODI	outside diameter indication
ODSCC	outside diameter stress-corrosion cracking
OTSG	once-through steam generator
PLP	possible loose part
ppb	parts per billion
ppm	parts per million

psi	pounds per square inch
PSI	preservice inspection
psig	pounds per square inch gauge
PWR	pressurized-water reactor
PWSCC	primary water stress-corrosion cracking
RFO	refueling outage
SAI	single axial indication
SCC	stress-corrosion cracking
SG	steam generator
TEC	tube end cold
TEH	tube end hot
TSC	tubesheet cold/cold-leg tubesheet
TSH	tubesheet hot/hot-leg tubesheet
TSP	tube support plate
TSTF	Technical Specification Task Force
TT	thermally treated
TW	through-wall
UT	ultrasonic testing
UTE	upper tube end
UTS	upper tubesheet
VS	vertical strap
W	Westinghouse

APPENDIX B: BIBLIOGRAPHY

General

Code of Federal Regulations

Title 10, "Energy," Part 50, "Domestic Licensing of Production and Utilization Facilities."

Title 10, "Energy," Part 100, "Reactor Site Criteria."

Electric Power Research Institute

EPRI NP-46655-SR, "Proceedings from the 1986 Electric Power Research Institute Workshop on Thermally Treated Alloy 690 Tubes for Steam Generators."

U.S. Nuclear Regulatory Commission

Bulletin 89-01, "Failure of Westinghouse Steam Generator Tube Mechanical Plugs," May 15, 1989.

Generic Letter 91-04, "Changes in Technical Specification Surveillance Intervals to Accommodate a 24-Month Fuel Cycle," April 2, 1991.

Generic Letter 95-03, "Circumferential Cracking of Steam Generator Tubes," April 28, 1995 Information Notice 2001-016, "Recent Foreign and Domestic Experience with Degradation of Steam Generator Tubes and Internals," October 31, 2001.

Information Notice 2002-21, "Axial Outside-Diameter Cracking Affecting Thermally Treated Alloy 600 Steam Generator Tubing," June 25, 2002.

Information Notice 2005-09, "Indications in Thermally Treated Alloy 600 Steam Generator Tubes and Tube-to-Tubesheet Welds," April 7, 2005.

NUREG-0966, "Safety Evaluation Report Related to the D2/D3 Steam Generator Design Modification."

NUREG-1014, "Safety Evaluation Report Related to the D4/D5/E Steam Generator Design Modification."

NUREG-1604, "Circumferential Cracking of Steam Generator Tubes," April 1997.

NUREG-1771, "U.S. Operating Experience with Thermally Treated Alloy 600 Steam Generator Tubes."

NUREG/CR-6365, "Steam Generator Tube Failures," April 1996.

Regulatory Guide 1.121, "Bases for Plugging Degraded PWR Steam Generator Tubes," August 1976.

ANO 2

Letter from W.D. Reckley, NRC, to Entergy Operations, Inc., dated August 25, 1998, "Summary of July 28, 1998, Meeting with Entergy Operations, Inc., Regarding the Replacement of Steam Generators at Arkansas Nuclear One, Unit 2 (TAC No. MA2396)." NUDOCS Accession No. 9808310003

Letter from C.R. Hutchinson, Entergy Operations, Inc., to the NRC, dated August 18, 1999, "Arkansas Nuclear One—Unit 2; Docket No. 50-368; License No. NPF-6; Proposed Technical Specification Change Request Concerning Steam Generator Inspection Requirements for the Replacement Steam Generators." NUDOCS Accession Nos. 9908250231 and 9908250232

Letter from C. Anderson, Entergy Operations, Inc., to the NRC, dated June 29, 2000, "Arkansas Nuclear One—Unit 2; Docket No. 50-368; License No. NPF-6; Response to Request for Additional Information Regarding the August 18, 1999, Steam Generator Inspection Requirements License Amendment Request." ADAMS Accession No. ML003728728

Letter from J.D. Vandergrift, Entergy Operations, Inc., to the NRC, dated July 19, 2000, "Arkansas Nuclear One—Unit 2; Docket No. 50-368; License No. NPF-6; Transmittal of Proprietary Document WCAP-15406, 'Regulatory Guide 1.121 Analysis for Arkansas Nuclear One Unit 2 Replacement Steam Generators.'" ADAMS Accession No. ML003734917

Letter from J.D. Vandergrift, Entergy Operations, Inc., to the NRC, dated August 9, 2000, "Arkansas Nuclear One—Unit 2; Docket No. 50-368; License No. NPF-6; Supplemental Information Regarding the Steam Generator Inspection License Amendment Request." ADAMS Accession No. ML003741772

Letter from J.D. Vandergrift, Entergy Operations, Inc., to the NRC, dated March 1, 2001, "Arkansas Nuclear One—Unit 2; Docket No. 50-368; License No. NPF-6; 2000 Annual Report of Steam Generator Tubing Inservice Inspections." ADAMS Accession No. ML010660066

Letter from J.D. Vandergrift, Entergy Operations, Inc., to the NRC, dated June 11, 2001, "Arkansas Nuclear One—Unit 2; Docket No. 50-368; License No. NPF-6; Post-Steam Generator Replacement Startup Testing Summary Report." ADAMS Accession No. ML011710429

Letter from S.R. Cotton, Entergy Operations, Inc., to the NRC, dated August 5, 2002, "Arkansas Nuclear One—Unit 2; Docket No. 50-368; License No. NPF-6; Power Uprate Startup Testing Report." ADAMS Accession No. ML022190570

Letter from C. Anderson, Entergy Operations, Inc., to the NRC, dated November 22, 2002, "Arkansas Nuclear One, Unit 2; Docket No. 50-368; Operating License Amendment Request to Modify Steam Generator Tube Inspection Frequency." ADAMS Accession No. ML023600429

Letter from C. Anderson, Entergy Operations, Inc., to the NRC, dated November 22, 2002, "Arkansas Nuclear One, Unit 2; Docket No. 50-368; Operating License Amendment Request to Modify Steam Generator Tube Inspection Frequency." ADAMS Accession No. ML023440166

Letter from S.R. Cotton, Entergy Operations, Inc., to the NRC, dated January 31, 2003, "Arkansas Nuclear One—Unit 2; Docket No. 50-368; License No. NPF-6; Power Uprate Startup Testing Report Supplement." ADAMS Accession No. ML030380382

Letter from S.R. Cotton, Entergy Operations, Inc., to the NRC, dated March 13, 2003, "Arkansas Nuclear One, Unit 2; Docket No. 50-368; Response to Request for Additional

Information on License Amendment for Steam Generator Tube Inspection Frequency." ADAMS Accession No. ML030780328

Letter from S.R. Cotton, Entergy Operations, Inc., to the NRC, dated March 26, 2003, "Arkansas Nuclear One, Unit 2; Docket No. 50-368; Response to Request for Additional Information Regarding the Power Uprate Startup Testing Report Supplement." ADAMS Accession No. ML030860239

Letter from S.R. Cotton, Entergy Operations, Inc., to the NRC, dated April 11, 2003, "Arkansas Nuclear One—Unit 2; Docket No. 50-368; License No. NPF-6; 2003 Annual Report of Steam Generator Tubing In-service Inspections." ADAMS Accession No. ML031080421

Letter from T.W. Alexion, NRC, to C.G. Anderson, Entergy Operations, Inc., dated May 28, 2003, "Arkansas Nuclear One—Unit 2, Issuance of Amendment Re: One-Time Change of Steam Generator Tube Inspection Frequency (TAC No. MB6808)." ADAMS Accession No. ML031490475

Letter from T.W. Alexion, NRC, to C.G. Anderson, Entergy Operations, Inc., dated July 1, 2003, "Arkansas Nuclear One, Unit No. 2—2003 Annual Reports of Steam Generator Tubing Inservice Inspections (TAC No. MB8452)." ADAMS Accession No. ML031820241

Letter from D.E. James, Entergy Operations, Inc., to the NRC, dated October 28, 2004, "Response to Generic Letter 2004-01, Requirements for Steam Generator Tube Inspections, Plant Name Arkansas Nuclear One, Units 1 and 2; Docket Nos. 50-313 and 50-368; License No. DPR-51 and NPF-6." ADAMS Accession No. ML043140261

Braidwood 1

Letter from T.J. Tulon, Commonwealth Edison Company, to the NRC, dated April 5, 2000, "Steam Generator Tube Inspection Report from Braidwood Unit 1 Refueling Outage Inspections." ADAMS Accession No. ML003701661

Letter from G.K. Schwartz, Commonwealth Edison Company, to the NRC, dated June 28, 2000, "Braidwood Station, Unit 1 Inservice Inspection Summary Report." ADAMS Accession No. ML003729023

Letter from R.M. Krich, Commonwealth Edison Company, to the NRC, dated July 31, 2000, "Steam Generator Laser Welded Sleeves." ADAMS Accession No. ML003738467

Letter from R.M. Krich, Exelon Corporation, to the NRC, dated February 9, 2001, "Request for Technical Specifications Change; Braidwood Station, Unit 1, Steam Generator Inspection Frequency Revision for the Fall 2001 Refueling Outage." ADAMS Accession No. ML010470080

Letter from J.D. von Suskil, Exelon Generation, to the NRC, dated March 27, 2001, "Steam Generator Tube Inspection Report from Braidwood Unit 1 Refueling Outage Inspections." ADAMS Accession No. ML010930262

Letter from R.M. Krich, Exelon Generation Company, LLC, to the NRC, dated May 18, 2001, "Response to Request for Additional Information for Technical Specifications Change to Revise Steam Generator Inspection Frequency for the Fall 2001 Refueling Outage for Braidwood Station, Unit 1." ADAMS Accession No. ML011440048

Letter from R.M. Krich, Exelon Generation, to the NRC, dated June 26, 2001, "Response to Request for Additional Information for Technical Specifications Change to Revise Steam

Generator Inspection Frequency for the Fall 2001 Refueling Outage for Braidwood Station, Unit 1." ADAMS Accession No. ML011840369

Letter from M. Chawla, NRC, to O.D. Kingsley, Exelon Generation Company, LLC, dated August 9, 2001, "Issuance of Amendments—Technical Specification Changes to Revise Steam Generator Inspection Frequency, Braidwood Station, Units 1 and 2 (TAC Nos. MB1226 and MB1227)." ADAMS Accession No. ML012040245

Letter from J.D. von Suskil, Exelon Generation Company, LLC, to the NRC, dated January 10, 2002, "Braidwood Station, Unit 1 Inservice Inspection Summary Report." ADAMS Accession No. ML020520667

Letter from J.D. von Suskil, Exelon Generation Company, LLC, to the NRC, dated May 7, 2003, "Tenth Refuel Outage, Steam Generator In-Service Inspection Report." ADAMS Accession No. ML031360617

Letter from M. Chawla, NRC, to J.L. Skolds, Exelon Nuclear, dated June 18, 2003, "Summary of Conference Call With Exelon Nuclear Regarding the 2003 Steam Generator Inspections at Braidwood Unit 1 (TAC No. MB8475)." ADAMS Accession No. ML031570110

Letter from M.J. Pacilio, Exelon Generation, to the NRC, dated July 29, 2003, "Braidwood Station, Unit 1 Tenth Refueling Outage, Steam Generator Inservice Inspection Summary Report." ADAMS Accession No. ML032190155

Letter from M.J. Pacilio, Exelon Generation Company, LLC, to the NRC, dated July 31, 2003, "Braidwood Station, Unit 1 Inservice Inspection Summary Report." ADAMS Accession No. ML032170925

Letter from K.A. Ainger, Exelon Generation, to the NRC, dated April 16, 2004, "Response to Request for Additional Information Regarding the Braidwood Station Unit 1 April 2003 Steam Generator Inspection." ADAMS Accession No. ML041140380

Letter from G.F. Dick, Jr., NRC, to C.M. Krane, Exelon Generation Company, LLC, dated July 27, 2004, "Evaluation of Steam Generator Inservice Inspection Summary Report, Braidwood, Unit 1 (TAC No. MC1894)." ADAMS Accession No. ML042020262

Letter from T.P. Joyce, Exelon Generation Company, LLC, to the NRC, dated October 29, 2004, "October 2004, Eleventh Refuel Outage Steam Generator Inservice Inspection Report." ADAMS Accession No. ML043090549

Letter from K.R. Jury, AmerGen Energy Company, LLC, to the NRC, dated October 29, 2004, "Response to NRC Generic Letter 2004-01, "Requirements for Steam Generator Tube Inspections." ADAMS Accession No. ML043060328

Letter from G.F. Dick, Jr., NRC, to C.M. Krane, Exelon Generation Company, LLC, dated December 21, 2004, "Braidwood Station, Unit 1—Summary of Conference Call Regarding Steam Generator Tube Inspections from Fall 2004 Outage (TAC No. MC4914)." ADAMS Accession No. ML043240231

Letter from K.J. Polson, Exelon Generation Company, LLC, to the NRC, dated January 19, 2005, "Braidwood Station, Unit 1 Inservice Inspection Summary Report." ADAMS Accession No. ML050250279

Letter from K.J. Polson, Exelon Generation Company, LLC, to the NRC, dated January 20, 2005, "Braidwood Station, Unit 1 Eleventh Refueling Outage Steam Generator Inservice Inspection Summary Report." ADAMS Accession No. ML050280208

<u>Byron 1</u>

Meeting Summary from G.F. Dick, Jr., NRC, "Summary of January 25, 1996, Steam Generator Replacement Meeting." NUDOCS Accession No. 9602130059

Letter from H.G. Stanley, Commonwealth Edison Company, to the NRC, dated March 20, 1998, "Braidwood Station Units 1 and 2; Byron Station Units 1 and 2; NRC Dockets Numbers: 50-456 and 50-457; NRC Dockets Numbers: 50-454 and 50-455; Commonwealth Edison Company (ComEd) response to NRC Generic Letter 97-06, 'Degradation of Steam Generator Internals' dated December 30, 1997." NUDOCS Accession No. 9803300446

Letter from W. Levis, Commonwealth Edison Company, to the NRC, dated April 22, 1999, "Steam Generator Tube Repairs Resulting from Byron Unit 1 Cycle 9 Refueling Outage Inservice Inspections." NUDOCS Accession No. 9905050232

Letter from W. Levis, Commonwealth Edison Company, to the NRC, dated July 9, 1999, "Steam Generator Inservice Inspection Summary Report." NUDOCS Accession Nos. 9907190037 and 9907190066

Letter from W. Levis, Commonwealth Edison Company, to the NRC, dated October 6, 2000, "Steam Generator Tube Repairs Resulting from Byron Unit 1, Cycle 10 Refueling Outage." ADAMS Accession No. ML003760517

Letter from W. Levis, Commonwealth Edison Company, to the NRC, dated January 4, 2001, "Byron Station Unit 1 Cycle 10 Steam Generator Eddy Current Examination, 90-Day Summary Report." ADAMS Accession No. ML010230109

Letter from R.M. Krich, Exelon Corporation, to the NRC, dated February 9, 2001, "Request for Technical Specifications Change; Braidwood Station, Unit 1, Steam Generator Inspection Frequency Revision for the Fall 2001 Refueling Outage." ADAMS Accession No. ML010470080

Letter from R.P. Lopriore, Exelon Generation Company, LLC, to the NRC, dated March 28, 2002, "Steam Generator Tube Repairs Resulting from Byron Station, Unit 1, Cycle 11 Refueling Outage." ADAMS Accession No. ML020940528

Letter from R.P. Lopriore, Exelon Generation Company, LLC, to the NRC, dated June 13, 2002, "Byron Station Unit 1 Steam Generator Inservice Inspection Summary Report." ADAMS Accession No. ML021720555

Letter from G.F. Dick, Jr., NRC, to J.L. Skolds, Exelon Generation Company, LLC, dated December 12, 2003, "Evaluation of Steam Generator Inservice Inspection Summary Report, Byron, Unit 1 (TAC No. MC0962)." ADAMS Accession No. ML033420151

Letter from S.E. Kuczynski, Exelon Generation Company, LLC, to the NRC, dated January 12, 2004, "Byron Station Unit 1 90-Day Inservice Inspection Report for Interval 2, Period 3, Outage 1 (B1R12)." ADAMS Accession No. ML040210017

Letter from K.R. Jury, AmerGen Energy Company, LLC, to the NRC, dated October 29, 2004, "Response to NRC Generic Letter 2004-01, 'Requirements for Steam Generator Tube Inspections.'" ADAMS Accession No. ML043060328

Letter from S.E. Kuczynski, Exelon Generation Company, LLC, to the NRC, dated March 28, 2005, "Steam Generator Tube Repairs Completed during the Byron Station, Unit 1, Refueling Outage 13." ADAMS Accession No. ML050950421

Letter from G.F. Dick, Jr., NRC, to C.M. Crane, Exelon Generation Company, LLC, dated May 25, 2005, "Byron Station, Unit 1—Summary of Conference Telephone Call Regarding Steam Generator Inspections from the Spring 2005 Outage (TAC No. MC6415)." ADAMS Accession No. ML051400413

Letter from S.E. Kuczynski, Exelon Generation Company, LLC, to the NRC, dated June 3, 2005, "Byron Station Unit 1 Steam Generator Inservice Inspection Summary Report." ADAMS Accession No. ML051600185

Calvert Cliffs 1

Meeting Summary from A.W. Dromerick, NRC, "Summary of November 1, 2000, Meeting Regarding Calvert Cliffs Nuclear Power Plant, Unit Nos. 1 and 2 Re: Steam Generator Replacement Project (TAC Nos. MB0061 and MB0062)." ADAMS Accession Nos. ML003773346 and ML003774728

Letter from C.H. Cruse, Constellation Nuclear, to the NRC, dated December 20, 2000, "Calvert Cliffs Nuclear Power Plant; Unit Nos. 1 & 2; Docket Nos. 50-317 & 50-318; License Amendment Request: Revision to the Technical Specifications to Support Steam Generator Replacement." ADAMS Accession No. ML003780962

Presentation to the NRC on April 25, 2001, "Calvert Cliffs Nuclear Power Plant Steam Generator Replacement." ADAMS Accession No. ML011430032

Letter from C.H. Cruse, Constellation Nuclear, to the NRC, dated July 12, 2001, "Calvert Cliffs Nuclear Power Plant; Unit Nos. 1 & 2; Docket Nos. 50-317 & 50-318; Response to NRC Request for Additional Information Regarding License Amendment Request to Support Steam Generator Replacement." ADAMS Accession No. ML011970135

Letter from K.J. Nietmann, Constellation Energy Group, to the NRC, dated February 28, 2003, "Calvert Cliffs Nuclear Power Plant; Unit Nos. 1 & 2; Docket Nos. 50-317 & 50-318; 2002 Steam Generator Tube Inspection Results Report." ADAMS Accession No. ML030650343

Letter from K.J. Nietmann, Constellation Energy Group, to the NRC, dated February 26, 2004, "Calvert Cliffs Nuclear Power Plant; Unit Nos. 1 & 2; Docket Nos. 50-317 & 50-318; 2003 Steam Generator Tube Inspection Results Report." ADAMS Accession No. ML040620475

Letter from K.J. Nietmann, Constellation Energy Group, to the NRC, dated May 6, 2004, "Calvert Cliffs Nuclear Power Plant; Unit No. 1; Docket No. 50-317; Report of Steam Generator Tube Plugging." ADAMS Accession No. ML041330142

Letter from G. Vanderheyden, Constellation Energy, to the NRC, dated October 25, 2004, "Calvert Cliffs Nuclear Power Plant; Unit Nos. 1 & 2; Docket Nos. 50-317 & 50-318; Response to NRC Generic Letter 2004-01, 'Requirements for Steam Generator Tube Inspections.'" ADAMS Accession No. ML043030450

Letter from K.J. Nietmann, Constellation Energy, to the NRC, dated February 25, 2005, "Calvert Cliffs Nuclear Power Plant; Unit Nos. 1 & 2; Docket Nos. 50-317 & 50-318; 2004 Steam Generator Tube Inspection Results Report." ADAMS Accession No. ML050610714

Letter from G. Vanderheyden, Constellation Energy, to the NRC, dated April 28, 2005, "Calvert Cliffs Nuclear Power Plant; Unit No. 1; Docket No. 50-317; Response to Request for Additional Information Regarding Steam Generator Inservice Inspection Summary Report (TAC No. MC6320)." ADAMS Accession No. ML051250065

Calvert Cliffs 2

Letter from K.J. Nietmann, Constellation Energy Group, to the NRC, dated February 26, 2004, "Calvert Cliffs Nuclear Power Plant; Unit Nos. 1 & 2; Docket Nos. 50-317 & 50-318; 2003 Steam Generator Tube Inspection Results Report." ADAMS Accession No. ML040620475

Letter from G. Vanderheyden, Constellation Energy, to the NRC, dated October 25, 2004, "Calvert Cliffs Nuclear Power Plant; Unit Nos. 1 & 2; Docket Nos. 50-317 & 50-318; Response to NRC Generic Letter 2004-01, 'Requirements for Steam Generator Tube Inspections.'" ADAMS Accession No. ML043030450

Letter from K.J. Nietmann, Constellation Energy, to the NRC, dated February 25, 2005, "Calvert Cliffs Nuclear Power Plant; Unit Nos. 1 & 2; Docket Nos. 50-317 & 50-318; 2004 Steam Generator Tube Inspection Results Report." ADAMS Accession No. ML050610714

Letter from G. Vanderheyden, Constellation Energy, to the NRC, dated April 28, 2005, "Calvert Cliffs Nuclear Power Plant; Unit No. 1; Docket No. 50-317; Response to Request for Additional Information Regarding Steam Generator Inservice Inspection Summary Report (TAC No. MC6320)." ADAMS Accession No. ML051250065

Catawba 1

Letter from G.R. Peterson, Duke Power Company, to the NRC, dated January 13, 1998, "Catawba Nuclear Station, Unit 1; Docket No. 50-413; Special Report—Steam Generator Tube Plugging." NUDOCS Accession No. 9801220291

Letter from M.S. Tuckman, Duke Power Company, to the NRC, dated March 26, 1998, "Duke Energy Corporation; Oconee Nuclear Station—Units 1, 2, and 3; Docket Nos. 50-269, 50-270, and 50-287; McGuire Nuclear Station—Units 1 and 2; Docket Nos. 50-369 and 50-370; Catawba Nuclear Station—Units 1 and 2; Docket Nos. 50-413, 50-414; Response to Generic Letter 97-06." NUDOCS Accession No. 9803310294

Letter from G.R. Peterson, Duke Power, to the NRC, dated September 11, 2000, "Duke Energy Corporation; Catawba Nuclear Station; Docket Nos. 50-413, 50-414; Steam Generator Tube Inspection Summary Reports." ADAMS Accession Nos. ML003751637 and ML003751594

Letter from G.R. Peterson, Duke Power, to the NRC, dated November 14, 2000, "Duke Energy Corporation; Catawba Nuclear Station, Unit 1; Docket No. 50-413; Steam Generator Tube Inspection Report." ADAMS Accession No. ML003769896

Letter from G.R. Peterson, Duke Power, to the NRC, dated February 19, 2001, "Duke Energy Corporation; Catawba Nuclear Station, Unit 1; Docket Number 50-413; Steam Generator Outage Summary Report for End of Cycle 12 Refueling Outage." ADAMS Accession No. ML010580184

Letter from G.R. Peterson, Duke Power, to the NRC, dated June 26, 2002, "Catawba Nuclear Station, Unit 1; Docket No. 50-413; Steam Generator Tube Inspection Report; 1 End of Core (EOC) 13." ADAMS Accession No. ML021920017

Letter from D.M. Jamil, Duke Power, to the NRC, dated December 23, 2003, "Catawba Nuclear Station, Unit 2; Docket No. 50-414; Steam Generator Tube Inspection Report; 1 End of Core (EOC) 14." ADAMS Accession No. ML040020318

Letter from D.M. Jamil, Duke Power, to the NRC, dated January 7, 2004, "Catawba Nuclear Station, Unit 1; Docket No. 50-413; Errata: Steam Generator Tube Inspection Report; 1 End of Core (EOC) 14." ADAMS Accession No. ML040160602

Letter from D.M. Jamil, Duke Power, to the NRC, dated May 4, 2004, "Duke Energy Corporation; Catawba Nuclear Station, Unit 1; Docket Number 50-413; Steam Generator Outage Summary Report for End of Cycle 14 Refueling Outage; Reply to Request for Additional Information (TAC Number MC1703)." ADAMS Accession No. ML041420128

Letter from S.E. Peters, NRC, to G.R. Peterson, Duke Energy Corporation, dated July 1, 2004, "Catawba Nuclear Station, Unit 1 Re: Steam Generator Tube Inspection Report for the End of Core 14 Refueling Outage (TAC No. MC1703)." ADAMS Accession No. ML041900327

Letter from W.R. McCollum, Jr., Duke Power, to the NRC, dated October 28, 2004, "Duke Energy Corporation; Oconee Nuclear Station, Units 1, 2, & 3; Docket Nos. 50-269, 50-270, 50-287; McGuire Nuclear Station, Units 1 & 2; Docket Nos. 50-369 and 50-370; Catawba Nuclear Station, Units 1 & 2; Docket Nos. 50-413, 50-414; Response to NRC Generic Letter 2004-01, Requirements for Steam Generator Tube Inspections." ADAMS Accession No. ML043090390

Letter from L.N. Olshan, NRC, to D.M. Jamil, Duke Energy Corporation, dated January 13, 2005, "Catawba Nuclear Station, Units 1 and 2 Re: Issuance of Amendments (TAC Nos. MB7842 and MB7843)."

Cook 1

Letter from S.A. Greenlee, Indiana Michigan Power Company, to the NRC, dated March 22, 2001, "Donald C. Cook Nuclear Plant Unit 1; NIS 1 Report for Inservice Inspection (ISI) Activities." ADAMS Accession No. ML010860229

Letter from S.A. Greenlee, Indiana Michigan Power Company, to the NRC, dated February 26, 2002, "Donald C. Cook Nuclear Plant Units 1 and 2; 2001 Annual Operating Report." ADAMS Accession No. ML020600132

Letter from S.A. Greenlee, Indiana Michigan Power Company, to the NRC, dated June 3, 2002, "Donald C. Cook Nuclear Plant Unit 1; Steam Generator Tube Inservice Inspection Report." ADAMS Accession No. ML021560396

Letter from S.A. Greenlee, Indiana Michigan Power Company, to the NRC, dated September 6, 2002, "Donald C. Cook Nuclear Plant Unit 1; NIS-1 Report for Inservice Inspection (ISI) Activities." ADAMS Accession No. ML022530146

Letter from S.A. Greenlee, Indiana Michigan Power Company, to the NRC, dated February 28, 2003, "Donald C. Cook Nuclear Plant Units 1 and 2; 2002 Annual Operating Report." ADAMS Accession No. ML030700530

Letter from J.A. Zwolinski, Indiana Michigan Power Company, to the NRC, dated September 15, 2003, "Donald C. Cook Nuclear Plant Units 1 and 2; Steam Generator Inspection—Request for Additional Information (TAC Nos. MB8121 and MB8122)." ADAMS Accession No. ML032671024

Letter from J.A. Zwolinski, Indiana Michigan Power Company, to the NRC, dated November 14, 2003, "Donald C. Cook Nuclear Plant Unit 1; Steam Generator Tube Inservice Inspection Report." ADAMS Accession No. ML033290423

Letter from J.A. Zwolinski, Indiana Michigan Power Company, to the NRC, dated February 19, 2004, "Donald C. Cook Nuclear Plant Unit 1; Inservice Inspection Summary Report." ADAMS Accession No. ML040560060

Letter from J.A. Zwolinski, Indiana Michigan Power Company, to the NRC, dated March 1, 2004, "Donald C. Cook Nuclear Plant Units 1 and 2; 2003 Annual Operating Report." ADAMS Accession No. ML040700910

Letter from J.F. Stang, NRC, to M.K. Nazar, American Electric Power, dated March 31, 2004, "Donald C. Cook Nuclear Plant, Units 1 and 2—Summary of Steam Generator Inspection Reports from January 2002 and May 2002 (TAC Nos. MB8121 and MB8122)." ADAMS Accession No. ML040850589

Letter from J.G. Lamb, NRC, to M.K. Nazar, American Electric Power, dated June 15, 2004, "Donald C. Cook Nuclear Plant, Unit 1—Summary of the Nuclear Regulatory Commission Staff's Review of Steam Generator Tube Inspection Summary Report for the Fall 2003 Outage (TAC No. MC2992)." ADAMS Accession No. ML041590049

Letter from J.N. Jensen, Indiana Michigan Power Company, to the NRC, dated October 27, 2004, "Donald C. Cook Nuclear Plant Units 1 and 2; Nuclear Regulatory Commission Generic Letter 2004-01; Requirements for Steam Generator Tube Inspections." ADAMS Accession No. ML043080352

Letter from J.N. Jensen, Indiana Michigan Power, to the NRC, dated February 25, 2005, "Donald C. Cook Nuclear Plant Units 1 and 2; 2004 Annual Operating Report." ADAMS Accession No. ML050630237

Cook 2

Letter from W.G. Smith, Jr., Indiana Michigan Power Company, to the NRC, dated February 28, 1989, "Donald C. Cook Nuclear Plant Annual Operating Report: 1988," NUDOCS Accession No. 8903100329

Letter from M.P. Alexich, Indiana Michigan Power Company, to the NRC, dated June 20, 1989, "Response to NRC Bulletin 89-01; Failure of Westinghouse Steam Generator Tube Mechanical Plugs." NUDOCS Accession No. 8906290111

Letter from M.P. Alexich, Indiana Michigan Power Company, to the NRC, dated February 28, 1990, "1989 Annual Operating Report." NUDOCS Accession No. 9003080293

Letter from M.P. Alexich, Indiana Michigan Power Company, to the NRC, dated February 28, 1991, "Annual Operating Report." NUDOCS Accession No. 9103060045

Letter from M.P. Alexich, Indiana Michigan Power Company, to the NRC, dated March 25, 1991, "Annual Operating Report Corrected Pages." NUDOCS Accession No. 9103280264

Letter from E.E. Fitzpatrick, Indiana Michigan Power Company, to the NRC, dated July 29, 1991, "Response to NRC Bulletin 89-01 Supplement 2: Failure of Westinghouse Steam Generator Tube Mechanical Plugs." NUDOCS Accession No. 9108020118

Letter from E.E. Fitzpatrick, Indiana Michigan Power Company, to the NRC, dated February 26, 1992, "Annual Operating Report." NUDOCS Accession No. 9203020147

Letter from E.E. Fitzpatrick, Indiana Michigan Power Company, to the NRC, dated February 26, 1993, "Annual Operating Report." NUDOCS Accession No. 9303030283

Letter from E.E. Fitzpatrick, Indiana Michigan Power Company, to the NRC, dated February 24, 1994, "Annual Operating Report." NUDOCS Accession No. 9403040328

Letter from E.E. Fitzpatrick, Indiana Michigan Power Company, to the NRC, dated February 24, 1995, "Donald C. Cook Nuclear Plant Units 1 and 2; 1994 Annual Operating Report." NUDOCS Accession No. 9503030159

Letter from L.S. Gibson, Indiana Michigan Power Company, to the NRC, dated March 3, 1995, "ISI NIS-1 Report." NUDOCS Accession No. 9503130325

Letter from E.E. Fitzpatrick, Indiana Michigan Power Company, to the NRC, dated March 17, 1995, "Donald C. Cook Nuclear Plant Units 1 and 2; Interim Report on Westinghouse Alloy 600 Steam Generator Mechanical Plugs." NUDOCS Accession No. 9503230165

Letter from E.E. Fitzpatrick, Indiana Michigan Power Company, to the NRC, dated February 28, 1996, "Donald C. Cook Nuclear Plant Units 1 and 2; 1995 Annual Operating Report." NUDOCS Accession No. 9603110426

Letter from E.E. Fitzpatrick, Indiana Michigan Power Company, to the NRC, dated February 28, 1997, "Donald C. Cook Nuclear Plant Units 1 and 2; 1996 Annual Operating Report." NUDOCS Accession No. 9703100078

Letter from E.E. Fitzpatrick, Indiana Michigan Power Company, to the NRC, dated November 20, 1997, "Donald C. Cook Nuclear Plant Unit 2; Steam Generator 15 Day Inspection Report." NUDOCS Accession No. 9711260075

Letter from E.E. Fitzpatrick, Indiana Michigan Power Company, to the NRC, dated April 14, 1998, "Donald C. Cook Nuclear Plant Units 1 and 2; Response to NRC Generic Letter (GL) 97-06: Degradation of Steam Generator Internals." NUDOCS Accession No. 9804200484

Letter from J.R. Sampson, Indiana Michigan Power Company, to the NRC, dated May 14, 1998, "Donald C. Cook Nuclear Plant Units 1 and 2; 1997 Annual Operating Report." NUDOCS Accession No. 9805220409

Letter from R.P. Powers, Indiana Michigan Power Company, to the NRC, dated November 6, 1999, "Donald C. Cook Nuclear Plant Units 1 and 2; Retransmittal of the 1998 Annual Operating Report." ADAMS Accession Nos. ML993190401 and ML993190404

Letter from A.C. Bakken III, Indiana Michigan Power Company, to the NRC, dated March 13, 2000, "Donald C. Cook Nuclear Plant Units 1 and 2; 1999 Annual Operating Report." ADAMS Accession No. ML003694275

Letter from W.J. Kropp, Indiana Michigan Power Company, to the NRC, dated September 19, 2000, "Donald C. Cook Nuclear Plant Unit 2; NIS-1 Report for Inservice Inspection (ISI) Activities." ADAMS Accession No. ML003752153

Letter from R.P. Powers, Indiana Michigan Power Company, to the NRC, dated September 30, 2000, "Donald C. Cook Nuclear Plant Unit 2; License Amendment Request; Steam Generator Tube Surveillance Interval Extension." ADAMS Accession No. ML003756902

Letter from M.W. Rencheck, Indiana Michigan Power Company, to the NRC, dated November 22, 2000, "Donald C. Cook Nuclear Plant Unit 2; Response to Request for Additional Information on Steam Generator Tube Surveillance Interval Extension (TAC No. MB0156)." ADAMS Accession No. ML003771977

Letter from M.W. Rencheck, Indiana Michigan Power Company, to the NRC, dated December 20, 2000, "Donald C. Cook Nuclear Plant Unit 2; Supplement to License Amendment Request; Steam Generator Tube Surveillance Interval Extension (TAC No. MB0156)." ADAMS Accession No. ML003780797

Letter from J.F. Stang, NRC to R.P. Powers, Indiana Michigan Power Company, dated January 30, 2001, "Donald C. Cook Nuclear Plant, Unit 2—Issuance of Amendment (TAC No. MB0156)." ADAMS Accession No. ML010320218

Letter from S.A. Greenlee, Indiana Michigan Power Company, to the NRC, dated February 19, 2002, "Donald C. Cook Nuclear Plant Unit 2; Steam Generator Tube Inservice Inspection Report." ADAMS Accession No. ML020520524

Letter from S.A. Greenlee, Indiana Michigan Power Company, to the NRC, dated February 26, 2002, "Donald C. Cook Nuclear Plant Units 1 and 2; 2001 Annual Operating Report." ADAMS Accession No. ML020600132

Letter from S.A. Greenlee, Indiana Michigan Power Company, to the NRC, dated May 15, 2002, "Donald C. Cook Nuclear Plant Unit 2; NIS-1 Report for Inservice Inspection (ISI) Activities." ADAMS Accession No. ML021370050

Letter from S.A. Greenlee, Indiana Michigan Power Company, to the NRC, dated September 6, 2002, "Donald C. Cook Nuclear Plant Unit 1; NIS-1 Report for Inservice Inspection (ISI) Activities." ADAMS Accession No. ML022530146

Letter from S.A. Greenlee, Indiana Michigan Power Company, to the NRC, dated February 28, 2003, "Donald C. Cook Nuclear Plant Units 1 and 2; 2002 Annual Operating Report." ADAMS Accession No. ML030700530

Letter from J.A. Zwolinski, Indiana Michigan Power Company, to the NRC, dated September 15, 2003, "Donald C. Cook Nuclear Plant Units 1 and 2; Steam Generator Inspection—Request for Additional Information (TAC Nos. MB8121 and MB8122)." ADAMS Accession No. ML032671024

Letter from J.A. Zwolinski, Indiana Michigan Power Company, to the NRC, dated September 18, 2003, "Donald C. Cook Nuclear Plant Unit 2; Inservice Inspection Summary Report." ADAMS Accession No. ML032670622

Letter from J.A. Zwolinski, Indiana Michigan Power Company, to the NRC, dated March 1, 2004, "Donald C. Cook Nuclear Plant Units 1 and 2; 2003 Annual Operating Report." ADAMS Accession No. ML040700910

Letter from J.F. Stang, NRC, to M.K. Nazar, American Electric Power, dated March 31, 2004, "Donald C. Cook Nuclear Plant, Units 1 and 2—Summary of Steam Generator Inspection Reports from January 2002 and May 2002 (TAC Nos. MB8121 and MB8122)." ADAMS Accession No. ML040850589

Letter from J.N. Jensen, Indiana Michigan Power Company, to the NRC, dated October 27, 2004, "Donald C. Cook Nuclear Plant Units 1 and 2; Nuclear Regulatory Commission Generic Letter 2004-01; Requirements for Steam Generator Tube Inspections." ADAMS Accession No. ML043080352

Letter from J.N. Jensen, Indiana Michigan Power Company, to the NRC, dated October 28, 2004, "Donald C. Cook Nuclear Plant Unit 2; Steam Generator Tube Inservice Inspection Report." ADAMS Accession No. ML043090491

Letter from D.P. Fadel, American Electric Power, to the NRC, dated January 31, 2005, "Donald C. Cook Nuclear Plant Unit 2; Inservice Inspection Summary Report." ADAMS Accession No. ML050390357

Letter from J.N. Jensen, Indiana Michigan Power, to the NRC, dated February 25, 2005, "Donald C. Cook Nuclear Plant Units 1 and 2; 2004 Annual Operating Report." ADAMS Accession No. ML050630237

Farley 1

Letter from D. Morey, Southern Nuclear Operating Company, Inc., to the NRC, dated October 22, 2001, "Joseph M. Farley Nuclear Plant—Unit 1; Steam Generator Plugging 15-Day Report." ADAMS Accession No. ML020020385

Inservice Inspection Report, dated January 18, 2002. ADAMS Accession No. ML020300072

Letter from D. Morey, Southern Nuclear Operating Company, Inc., to the NRC, dated March 4, 2002, "Farley Nuclear Plant, Unit 1; Request for Technical Specifications Change; Steam Generator Inspection Frequency Revision for the Spring 2003 Refueling Outage." ADAMS Accession No. ML020650389

Letter from D. Morey, Southern Nuclear Operating Company, to the NRC, dated July 11, 2002, "Joseph M. Farley Nuclear Plant, Unit 1; Request for Additional Information Related to Technical Specifications Change; Steam Generator Inspection Frequency Revision for the Spring 2003 Refueling Outage." ADAMS Accession No. ML021960109

Letter from F. Rinaldi, NRC, to D.N. Morey, Southern Nuclear Operating Company, Inc., dated September 20, 2002, "Joseph M. Farley Nuclear Plant, Unit 1 RE: Issuance of Amendment (TAC No. MB4310)." ADAMS Accession No. ML022340746

Letter from F. Rinaldi, NRC to J.B. Beasley, Jr., Southern Nuclear Operating Company, Inc., dated April 18, 2003, "Joseph M. Farley Nuclear Plant, Units 1 and 2 RE: Review of Steam Generator Tube Inservice Inspection Report for Refueling Outage 1R17 (Fall 2001) (TAC No. MB6476)." ADAMS Accession No. ML031110259

Letter from J.B. Beasley, Jr., Southern Nuclear Operating Company, Inc., to the NRC, dated July 25, 2003, "Joseph M. Farley Nuclear Plant; Inservice Inspection Summary Report." ADAMS Accession No. ML032110126

Letter from J.B. Beasley, Jr., Southern Nuclear Operating Company, Inc., to the NRC, dated September 12, 2003, "Joseph M. Farley Nuclear Plant; Application for License Renewal." ADAMS Accession Nos. ML032721353 and ML032721360

Letter from L.M. Stinson, Southern Nuclear Operating Company, Inc., to the NRC, dated October 25, 2004, "Joseph M. Farley Nuclear Plant; Vogtle Electric Generating Plant; Response to NRC Generic Letter 2004-01, 'Requirements for Steam Generator Tube Inspections.'" ADAMS Accession No. ML043010265

Letter from L.M. Stinson, Southern Nuclear Operating Company, Inc., to the NRC, dated January 7, 2005, "Joseph M. Farley Nuclear Plant; Vogtle Electric Generating Plant; Supplemental Response to NRC Generic Letter 2004-01, 'Requirements for Steam Generator Tube Inspections.'" ADAMS Accession No. ML050110224

Farley 2

Letter from J.B. Beasley, Jr., Southern Nuclear Operating Company, Inc., to the NRC, dated October 8, 2002, "Joseph M. Farley Nuclear Plant—Unit 2; Steam Generator Plugging 15-Day Report." ADAMS Accession No. ML022840014

Letter from J.B. Beasley, Jr., Southern Nuclear Operating Company, Inc., to the NRC, dated January 21, 2003, "Joseph M. Farley Nuclear Plant; Inservice Inspection Summary Report." ADAMS Accession No. ML022840014

Letter from J.B. Beasley, Jr., Southern Nuclear Operating Company, Inc., to the NRC, dated February 11, 2003, "Joseph M. Farley Nuclear Plant, Unit 2; Request for Technical Specifications Change; Steam Generator Inspection Frequency Revision for the Spring 2004 Refueling Outage." ADAMS Accession No. ML030520106

Letter from F. Rinaldi, NRC, to J.B. Beasley, Jr., Southern Nuclear Operating Company, Inc., dated July 14, 2003, "Joseph M. Farley Nuclear Plant, Unit 2 RE: Revising the Steam Generator Inspection Frequency (TAC No. MB7938)." ADAMS Accession No. ML031980003

Letter from J.B. Beasley, Jr., Southern Nuclear Operating Company, Inc., to the NRC dated September 12, 2003, "Joseph M. Farley Nuclear Plant; Application for License Renewal." ADAMS Accession Nos. ML032721353 and ML032721360

Letter from L.M. Stinson, Southern Nuclear Operating Company, Inc., to the NRC, dated October 25, 2004, "Joseph M. Farley Nuclear Plant; Vogtle Electric Generating Plant; Response to NRC Generic Letter 2004-01, 'Requirements for Steam Generator Tube Inspections.'" ADAMS Accession No. ML043010265

Letter from L.M. Stinson, Southern Nuclear Operating Company, Inc., to the NRC, dated December 17, 2004, "Joseph M. Farley Nuclear Plant—Unit 2; Response to Request for Additional Information, Regarding Fall 2002 Steam Generator Inspection." ADAMS Accession No. ML043570226

Letter from L.M. Stinson, Southern Nuclear Operating Company, Inc., to the NRC, dated January 7, 2005, "Joseph M. Farley Nuclear Plant; Vogtle Electric Generating Plant; Supplemental Response to NRC Generic Letter 2004-01, 'Requirements for Steam Generator Tube Inspections.'" ADAMS Accession No. ML050110224

Ginna

Letter from R.C. Mecredy, Rochester Gas and Electric Corporation, to the NRC, dated March 30, 1998, "Response to NRC Generic Letter 97-06, dated December 30, 1997; Subject: Degradation of Steam Generator Internals; R.E. Ginna Nuclear Power Plant; Docket No. 50-244." NUDOCS Accession Nos. 9804090312 and 9804130153

Letter from R.C. Mecredy, Rochester Gas and Electric Corporation, to the NRC, dated April 6, 1998, "Transmittal of Inservice Inspection Report for the Third Interval (1990–1999), Third Period, First Outage (1997); R.E. Ginna Nuclear Power Plant; Docket No. 50/244." NUDOCS Accession No. 9804160108

Letter from R.C. Mecredy, Rochester Gas and Electric Corporation, to the NRC, dated July 21, 1999, "Transmittal of Inservice Inspection Report for the Third Interval (1990–1999), Third Period, Second Outage (1999); R.E. Ginna Nuclear Power Plant; Docket No. 50-244." NUDOCS Accession No. 9908030082

Letter from R.C. Mecredy, Rochester Gas and Electric Corporation, to the NRC, dated February 23, 2001, "Transmittal of Inservice Inspection Report for the Fourth Interval (2000–2009), First Period, Fourth Outage (2000)—IWE/IWL; R.E. Ginna Nuclear Power Plant; Docket No. 50-244." ADAMS Accession No. ML010660279

Letter from R.C. Mecredy, Rochester Gas and Electric Corporation, to the NRC, dated July 9, 2002, "Transmittal of Inservice Inspection Report for the Fourth Interval (2000–2009), First

Period, Second Outage (2002)—ISI and First Interval (1997–2008), Second Period, First Outage (2002)—IWE/IWL; R.E. Ginna Nuclear Power Plant; Docket No. 50-244." ADAMS Accession No. ML021960257

Letter from R.C. Mecredy, Rochester Gas and Electric Corporation, to the NRC, dated October 17, 2002, "Supplementary Information Associated with the 2002 Steam Generator Inservice Inspection; Rochester Gas and Electric Corporation; R.E. Ginna Nuclear Power Plant; Docket No. 50-244." ADAMS Accession No. ML022960389

Letter from R.C. Mecredy, Rochester Gas and Electric Corporation, to the NRC, dated May 29, 2003, "Response to Request for Additional Information Associated with the 2002 Steam Generator Inservice Inspection Report; Rochester Gas and Electric Corporation; R.E. Ginna Nuclear Power Plant; Docket No. 50-244." ADAMS Accession No. ML031560413

Letter from R.L. Clark, NRC, to R.C. Mecredy, Rochester Gas and Electric Corporation, dated September 25, 2003, "Summary of the U.S. Nuclear Regulatory Commission (NRC) Staff's Review of the R.E. Ginna Nuclear Power Plant (Ginna) Steam Generator Tube Inspection Report Dated July 2002 (TAC No. MB6467)." ADAMS Accession No. ML032681130

Letter from R.C. Mecredy, Rochester Gas and Electric Corporation, to the NRC, dated January 13, 2004, "Transmittal of Inservice Inspection Report for the Fourth Interval (2000–2009), Second Period, First Outage (2003)—ISI and First Interval (1997–2008), Second Period, Second Outage (2003)—IWE/IWL; R.E. Ginna Nuclear Power Plant; Docket No. 50-244." ADAMS Accession No. ML040220185

Letter from J.A. Widay, Constellation Energy, to the NRC, dated October 27, 2004, "Sixty (60) Day Response to Generic Letter (GL) 2004-01; Requirements for Steam Generator Tube Inspections; R.E. Ginna Nuclear Power Plant; Docket No. 50-244." ADAMS Accession No. ML043090480

Harris

Letter from R.J. Field, Carolina Power and Light Company, to the NRC, dated April 1, 2002, "Shearon Harris Nuclear Power Plant; Docket 50-400/License No. NPF-63; Inservice Inspection Summary Report." ADAMS Accession No. ML020990567

Letter from J.R. Caves, Carolina Power and Light Company, to the NRC, dated May 15, 2003, "Shearon Harris Nuclear Power Plant; Docket 50-400/License No. NPF-63; 15 Day Special Report—Steam Generator Tubes Plugged." ADAMS Accession No. ML031420075

Letter from J.R. Caves, Progress Energy Carolinas, Inc., to the NRC, dated August 13, 2003, "Shearon Harris Nuclear Power Plant Unit 1; Docket 50-400/License No. NPF-63; Inservice Inspection Summary Report." ADAMS Accession No. ML032680868

Letter from J. Scarola, Progress Energy Carolinas, Inc., to the NRC, dated December 8, 2003, "Shearon Harris Nuclear Power Plant, Unit No. 1; Docket 50-400/License No. NPF-63; Request for License Amendment; Technical Specification 4.4.5.3a." ADAMS Accession No. ML033530429

Letter from J. Scarola, Progress Energy Carolinas, Inc., to the NRC, dated April 15, 2004, "Shearon Harris Nuclear Power Plant, Unit No. 1; Docket 50-400/License No. NPF-63; Response to Request for Additional Information Concerning the License Amendment Request for a One-Time Increase in Steam Generator Inspection Interval (TAC No. MC1633)." ADAMS Accession No. ML041120371

Letter from J.R. Caves, Progress Energy Carolinas, Inc., to the NRC, dated May 7, 2004, "Shearon Harris Nuclear Power Plant, Unit No. 1; Docket 50-400/License No. NPF-63; One-Year Special Report; Steam Generator Tube Inservice Inspection Results." ADAMS Accession No. ML041320496

Letter from B.C. McCabe, Progress Energy Carolinas, Inc., to the NRC, dated May 27, 2004, "Shearon Harris Nuclear Power Plant, Unit No. 1; Docket 50-400/License No. NPF-63; 15-Day Special Report; Steam Generator Tube Plugging." ADAMS Accession No. ML041560343

Letter from C.P. Patel, NRC, to J. Scarola, Carolina Power and Light Company, dated August 17, 2004, "Review of the 2003 Shearon Harris Steam Generator Tube Inspection Results (TAC Number MC3820)." ADAMS Accession No. ML042360545

Letter from T.C. Morton, Progress Energy Carolinas, Inc., to the NRC, dated October 28, 2004, "Shearon Harris Nuclear Power Plant, Unit No. 1; Docket 50-400/License No. NPF-63; 60-Day Response to NRC Generic Letter 2004-01, 'Requirements for Steam Generator Tube Inspections.'" ADAMS Accession No. ML043090488

Letter from C.P. Patel, NRC to J. Scarola, Carolina Power and Light Company, dated January 12, 2005, "Summary of Conference Call with Progress Energy Regarding the Spring 2004 Steam Generator Tube Inspections at Shearon Harris (TAC No. MC3133)." ADAMS Accession No. ML043440363

Letter from D.H. Corlett, Progress Energy Carolinas, Inc., to the NRC, dated February 15, 2005, "Shearon Harris Nuclear Power Plant Unit 1; Docket 50-400/License No. NPF-63; Inservice Inspection Summary Report." ADAMS Accession No. ML050600144

Letter from D.H. Corlett, Progress Energy Carolinas, Inc., to the NRC, dated August 1, 2005, "Shearon Harris Nuclear Power Plant, Unit No. 1; Docket 50-400/License No. NPF-63; One-Year Special Report; Steam Generator Tube Inservice Inspection Results." ADAMS Accession No. ML052210457

Indian Point 3

"Summary of Meeting Held on July 30, 1987 to Discuss Replacement of Indian Point 3 Steam Generators," dated August 11, 1987. NUDOCS Accession No. 8708190408

Letter from J.C. Brons, New York Power Authority, to the NRC, dated September 15, 1989, "Indian Point 3 Nuclear Power Plant; Docket No. 50-286; Preservice Eddy Current Inspection Summary Report." NUDOCS Accession No. 8909220195

Letter from W.A. Josiger, New York Power Authority, to the NRC, dated May 18, 1989, "Indian Point 3 Nuclear Power Plant; NRC Bulletin No. 89-01: 'Failure of Westinghouse Steam Generator Tube Mechanical Plugs.'" NUDOCS Accession No. 8906050057

Letter from J.C. Brons, New York Power Authority, to the NRC, dated October 12, 1989, "Indian Point 3 Nuclear Power Plant; Docket No. 50-286; Inservice Inspection Report for Cycle 6/7; Refueling Outage and Preservice Examination Report for Replacement Steam Generators." NUDOCS Accession No. 8910240290

Letter from J.E. Russell, New York Power Authority, to the NRC, dated December 14, 1990, NUDOCS Accession No. 9012270284

Letter from J.E. Russell, New York Power Authority, to the NRC, dated July 16, 1991, "Indian Point 3 Nuclear Power Plant; NRC Bulletin 89-01, Supplement 2, Failure of Westinghouse Steam Generator Tube Mechanical Plugs." NUDOCS Accession No. 9107260068

Letter from J.E. Russell, New York Power Authority, to the NRC, dated September 16, 1992, NUDOCS Accession No. 9209240286

Letter from J.E. Russell, New York Power Authority, to the NRC, dated March 14, 1996, "Indian Point 3 Nuclear Power Plant; Docket No. 50-286; License No. DPR-64; Proposed Technical Specification Change Regarding Steam Generator Tube Inspection Interval." NUDOCS Accession No. 9603200079

Letter from G.F. Wunder, NRC, to W.J. Cahill, Power Authority of the State of New York, dated June 19, 1996, "Issuance of Amendment for Indian Point Nuclear Generating Unit No. 3 (TAC No. M94980)." NUDOCS Accession No. 9606280221

Letter from R.J. Barrett, New York Power Authority, to the NRC, dated April 15, 1997, "Indian Point 3 Nuclear Power Plant; Docket No. 50-286; License No. DPR-64; Clarification of Amendment 166 to Indian Point 3 Technical Specifications; Regarding One-Time Deferral of Steam Generator Tube Inspection." NUDOCS Accession No. 9704210225

Letter from J. Knubel, New York Power Authority, to the NRC, dated December 10, 1997, "Indian Point 3 Nuclear Power Plant; Docket No. 50-286; Inservice Inspection Report for Cycle 9/10 Refueling Outage." NUDOCS Accession No. 9712220213

Letter from R.J. Barrett, New York Power Authority, to the NRC, dated December 19, 1997, "Indian Point 3 Nuclear Power Plant; Docket No. 50-286; License No. DPR-64; Steam Generator Inservice Inspection Results." NUDOCS Accession No. 9712310066

Letter from R.J. Barrett, New York Power Authority, to the NRC, dated February 23, 1998, "Indian Point 3 Nuclear Power Plant; Docket No. 50-286; Response to NRC Generic Letter 97-05: Steam Generator Tube Inspection Techniques." NUDOCS Accession No. 9803030287

Letter from R.J. Barrett, New York Power Authority, to the NRC, dated March 10, 1998, "Indian Point 3 Nuclear Power Plant; Docket No. 50-286; Response to NRC Generic Letter 97-06: Degradation of Steam Generator Internals." NUDOCS Accession No. 9803200015

Letter from J. Knubel, New York Power Authority, to the NRC, dated March 26, 1998, "Indian Point 3 Nuclear Power Plant; Docket No. 50-286; Change to Technical Specification Basis Regarding Steam Generator Tube Plugging Limit." NUDOCS Accession No. 9803310138

Letter from R.J. Barrett, New York Power Authority, to the NRC, dated December 17, 1999, "Indian Point 3 Nuclear Power Plant; Docket No. 50-286; License No. DPR-64; Steam Generator Inservice Inspection Results." ADAMS Accession No. ML003670350

Letter from R.J. Barrett, New York Power Authority, to the NRC, dated January 11, 2000, "Indian Point 3 Nuclear Power Plant; Docket No. 50-286; License No. DPR-64; Inservice Inspection Report for Cycle 10/11 Refueling Outage." ADAMS Accession No. ML003679018

Letter from F.R. Dacimo, Entergy Nuclear Northeast, to the NRC, dated April 25, 2003, "Indian Point 3 Nuclear Power Plant; Docket No. 50-286; License No. DPR-64; Steam Generator Tube Inspection—Technical Specification 5.6.8 Report on Tubes Plugged." ADAMS Accession No. ML031200250

Letter from F.R. Dacimo, Entergy Nuclear Northeast, to the NRC, dated August 19, 2003, "Steam Generator Tube Inspection—Revision to Technical Specification 5.6.8 Report on Tubes Plugged." ADAMS Accession No. ML032330262

Letter from F.R. Dacimo, Entergy Nuclear Northeast, to the NRC, dated December 8, 2003, "Steam Generator Tube Inservice Examination Results for 2003 Refueling Outage 12 (R12)." ADAMS Accession No. ML033450339

Letter from M.R. Kansler, Entergy Nuclear Northeast, to the NRC, dated May 25, 2004, "Indian Point Nuclear Generating Unit No. 3; Docket 50-286; Request for Additional Information Regarding Steam Generator Tube Inspection Reports for the 2003 Outage, Indian Point Nuclear Generating Unit No. 3 (TAC No. MC1912)." ADAMS Accession No. ML041560463

Letter from P.D. Milano, NRC, to M. Kansler, Entergy Nuclear Operations, Inc., dated August 18, 2004, "Indian Point Nuclear Generating Unit No. 3—Evaluation of Steam Generator Tube Inspection Results for 2003 (TAC No. MC1912)." ADAMS Accession No. ML042300253

Letter from M.R. Kansler, Entergy Nuclear Northeast, to the NRC, dated October 27, 2004, "Indian Point Nuclear Generating Unit No. 2; Docket 50-247; Indian Point Nuclear Generating Unit No. 3; Docket 50-286; 60-Day Response to Generic Letter 2004-01, 'Requirements for Steam Generator Tube Inspection.'" ADAMS Accession No. ML043070599

Kewaunee

Letter from J.A. Grobe, NRC, to M. Reddemann, Nuclear Management Company, LLC, dated February 26, 2001, "Summary of the February 21, 2001 Kewaunee Public Meeting." ADAMS Accession Nos. ML010570188 and ML010540279

Letter from M.E. Reddemann, Nuclear Management Company, LLC, to the NRC, dated May 25, 2001, "Docket 50-305; Operating License DPR-43; Kewaunee Nuclear Power Plant; Proposed Amendment 177 to Kewaunee Nuclear Power Plant Technical Specification 4.2." ADAMS Accession No. ML011500373

Letter from M.E. Reddemann, Nuclear Management Company, LLC, to the NRC, dated August 17, 2001, "Docket 50-305; Operating License DPR-43; Kewaunee Nuclear Power Plant; Response to Nuclear Regulatory Commission Request for Additional Information Regarding Proposed Amendment 177 to Remove Steam Generator Alternate Repair Criteria from Kewaunee Nuclear Power Plant Technical Specification 4.2." ADAMS Accession No. ML012330117

Letter from J.G. Lamb, NRC, to M. Reddemann, Nuclear Management Company, LLC, dated September 20, 2001, "Kewaunee Nuclear Power Plant—Issuance of Amendment (TAC No. MB2047)." ADAMS Accession No. ML012470175

Letter from J.G. Lamb, NRC, to M. Reddemann, Nuclear Management Company, LLC, dated October 11, 2001, "Kewaunee Nuclear Power Plant—Correction to Issuance of Amendment 158 (TAC No. MB2047)." ADAMS Accession No. ML012820330

Letter from T. Coutu, Nuclear Management Company, LLC, to the NRC, dated February 22, 2002, "Docket 50-305; Operating License; Kewaunee Nuclear Power Plant; 2001 Inservice Inspection (ISI) Summary Report." ADAMS Accession Nos. ML020720559 and ML020720547

Letter from T. Coutu, Nuclear Management Company, LLC, to the NRC, dated February 28, 2002, "Docket 50-305; Operating License DPR-43; Kewaunee Nuclear Power Plant; 2001 Annual Operating Report." ADAMS Accession No. ML020700410

Letter from T. Coutu, Nuclear Management Company, LLC, to the NRC, dated December 19, 2002, "Docket 50-305; Operating License DPR-43; Kewaunee Nuclear Power Plant; License Amendment Request 192 to the Kewaunee Nuclear Power Plant Technical Specifications,

'Changes to Steam Generator Inspection Reporting Criteria.'" ADAMS Accession No. ML023650500

Letter from T. Coutu, Nuclear Management Company, LLC, to the NRC, dated February 27, 2003, "Kewaunee Nuclear Power Plant, Docket 50-305; License No. DPR-43; 2002 Annual Operating Report." ADAMS Accession No. ML030650954

Letter from T. Coutu, Nuclear Management Company, LLC, to the NRC, dated July 25, 2003, "Kewaunee Nuclear Power Plant; Docket 50-305; License No. DPR-43; Response to Request for Additional Information Related to License Amendment Request 192 to the Kewaunee Nuclear Power Plant Technical Specifications." ADAMS Accession No. ML032170612

Letter from T. Coutu, Nuclear Management Company, LLC, to the NRC, dated August 6, 2003, "Kewaunee Nuclear Power Plant; Docket 50-305; License No. DPR-43; 2003 Inservice Inspection (ISI) Summary Report." ADAMS Accession No. ML032250165

Letter from J.G. Lamb, NRC, to T. Coutu, Nuclear Management Company, LLC, dated August 15, 2003, "Kewaunee Nuclear Power Plant—Summary of the Staff's Review of the Steam Generator Tube Preservice Inspection Report from June 2001 (TAC No. MB6941)." ADAMS Accession No. ML032230154

Letter from T. Coutu, Nuclear Management Company, LLC, to the NRC, dated October 8, 2003, "Kewaunee Nuclear Power Plant; Docket 50-305; License No. DPR-43; License Amendment Request 199, 'Steam Generator Eddy Current Inspection Frequency Extension,' to the Kewaunee Nuclear Power Plant Technical Specifications." ADAMS Accession No. ML032901121

Letter from A.C. McMurtray, NRC, to T. Coutu, Nuclear Management Company, LLC, dated November 20, 2003, "Kewaunee Nuclear Power Plant—Issuance of Amendment (TAC No. MB6993)." ADAMS Accession No. ML032940169

Letter from T. Coutu, Nuclear Management Company, LLC, to the NRC, dated December 15, 2003, "Kewaunee Nuclear Power Plant; Docket 50-305; License No. DPR-43; Responses to NRC Clarification Questions Regarding License Amendment Request 195, Stretch Power Uprate for Kewaunee Nuclear Power Plant." ADAMS Accession No. ML033570513

Letter from T. Coutu, Nuclear Management Company, LLC, to the NRC, dated February 27, 2004, "Kewaunee Nuclear Power Plant, Docket 50-305; License No. DPR-43; 2003 Annual Operating Report." ADAMS Accession No. ML040650370

Letter from T. Coutu, Nuclear Management Company, LLC, to the NRC, dated February 27, 2004, "Kewaunee Nuclear Power Plant; Docket 50-305; License No. DPR-43; Response to Request for Additional Information Related to License Amendment Request 199 to the Kewaunee Nuclear Power Plant Technical Specifications." ADAMS Accession No. ML040720434

Letter from T. Coutu, Nuclear Management Company, LLC, to the NRC, dated May 3, 2004, "Kewaunee Nuclear Power Plant; Docket 50-305; License No. DPR-43; Response to Request for Additional Information Related to License Amendment Request 199 'Steam Generator Eddy Current Inspection Frequency Extension' to the Kewaunee Nuclear Power Plant Technical Specifications." ADAMS Accession No. ML041320470

Letter from C.F. Lyon, NRC, to T. Coutu, Nuclear Management Company, LLC, dated June 18, 2004, "Kewaunee Nuclear Power Plant—Issuance of Amendment RE: Steam Generator Eddy Current Inspection Frequency Extension (TAC Nos. MC1049 and MC0927)." ADAMS Accession No. ML041470119

Letter from T. Coutu, Nuclear Management Company, LLC, to the NRC, dated October 28, 2004, "Kewaunee Nuclear Power Plant; Docket 50-305; License No. DPR-43; Generic Letter 2004-01: Requirements for Steam Generator Tube Inspections 60-Day Response." ADAMS Accession No. ML043140252

Memorandum from L. Raghavan, NRC, to A. Mohseni, NRC, dated March 18, 2005, "Kewaunee Nuclear Power Plant—Summary of Telephone Conference RE: Steam Generator Water Chemistry Excursion (TAC No. M40695)." ADAMS Accession No. ML050750545

McGuire 1

Letter from M.S. Tuckman, Duke Power Company, to the NRC, dated September 30, 1994, "McGuire Nuclear Station; Docket Nos. 50-369, 50-370; Replacement Steam Generator Proposed Tech Spec Amendment." NUDOCS Accession No. 9410070181

Letter from H.B. Barron, Duke Energy Corporation, to the NRC, dated March 16, 2000, "McGuire Nuclear Station; Docket Nos. 50-369, 50-370; Steam Generator Tube Inspection Summary Reports." ADAMS Accession Nos. ML003693923, ML003694073, ML003693925, and ML003693934

Letter from M.S. Tuckman, Duke Power Company, to the NRC, dated February 24, 1998, "Duke Energy Corporation; Oconee Nuclear Station—Units 1, 2, and 3; Docket Nos. 50-269, 50-270, and 50-287; McGuire Nuclear Station—Units 1 and 2; Docket Nos. 50-369 and 50-370; Catawba Nuclear Station—Units 1 and 2; Docket Nos. 50-413, 50-414; Response to Generic Letter 97-05." NUDOCS Accession No. 9803040131

Letter from M.S. Tuckman, Duke Power Company, to the NRC, dated March 26, 1998, "Duke Energy Corporation; Oconee Nuclear Station—Units 1, 2, and 3; Docket Nos. 50-269, 50-270, and 50-287; McGuire Nuclear Station—Units 1 and 2; Docket Nos. 50-369 and 50-370; Catawba Nuclear Station—Units 1 and 2; Docket Nos. 50-413, 50-414; Response to Generic Letter 97-06." NUDOCS Accession No. 9803310294

Letter from H.B. Barron, Duke Energy Corporation, to the NRC, dated April 10, 2001, "McGuire Nuclear Station; Docket Nos. 50-369; Unit 1 Steam Generators; EOC-14 Refueling Outage." ADAMS Accession No. ML011070648

Letter from H.B. Barron, Duke Energy Corporation, to the NRC, dated June 18, 2001, "McGuire Nuclear Station—Unit 1; Docket No. 50-369; Steam Generator Inservice Inspection Report." ADAMS Accession No. ML011770253

Letter from H.B. Barron, Duke Energy Corporation, to the NRC, dated June 26, 2001, "McGuire Nuclear Station Units 1 and 2; Docket Nos. 50-369, 50-370; Third Ten-Year Interval; Steam Generator Inservice Inspection Plan (ISI) for Unit 1; Relief Request 01-005 for Units 1 and 2; Relief Request 01-008 for Units 1 and 2." ADAMS Accession No. ML011870155

Letter from D.M. Jamil, Duke Power, to the NRC, dated October 10, 2002, "McGuire Nuclear Station, Unit 1; Docket No. 50-369; Steam Generator Tube Inspection Report for End of Core (EOC) 15." ADAMS Accession No. ML022960340

Letter from G.R. Peterson, Duke Power, to the NRC, dated April 29, 2004, "McGuire Nuclear Station; Docket No. 50-369; Unit 1 Steam Generators; EOC-16 Refueling Outage." ADAMS Accession No. ML041270476

Letter from G.R. Peterson, Duke Power, to the NRC, dated June 22, 2004, "McGuire Nuclear Station; Docket Nos. 50-369; Steam Generator In-Service Inspection Report; Unit 1, End of Cycle (EOC) 16," ADAMS Accession No. ML041820197

Letter from W.R. McCollum, Jr., Duke Power, to the NRC, dated October 28, 2004, "Duke Energy Corporation; Oconee Nuclear Station, Units 1, 2, & 3; Docket Nos. 50-269, 50-270, 50-287; McGuire Nuclear Station, Units 1 & 2; Docket Nos. 50-369 and 50-370; Catawba Nuclear Station, Units 1 & 2; Docket Nos. 50-413, 50-414; Response to NRC Generic Letter 2004-01, Requirements for Steam Generator Tube Inspections." ADAMS Accession No. ML043090390

Letter from G.R. Peterson, Duke Power, to the NRC, dated January 12, 2005, "McGuire Nuclear Station, Unit 1; Docket No. 50-369; NRC Request for Additional Information; End of Cycle 16 Steam Generator Inservice Inspection Summary Report." ADAMS Accession No. ML050190290

Letter from J.J. Shea, NRC, to G.R. Peterson, Duke Energy Corporation, dated March 4, 2005, "McGuire Nuclear Station, Unit 1—Review of EOC-16 Steam Generator Tube Inservice Inspection Summary Reports (TAC No. MC4275)." ADAMS Accession No. ML050610218

McGuire 2

Letter from M.S. Tuckman, Duke Power Company, to the NRC, dated September 30, 1994, "McGuire Nuclear Station; Docket Nos. 50-369, 50-370; Replacement Steam Generator Proposed Tech Spec Amendment." NUDOCS Accession No. 9410070181

Letter from H.B. Barron, Duke Energy Corporation, to the NRC, dated March 16, 2000, "McGuire Nuclear Station; Docket Nos. 50-369, 50-370; Steam Generator Tube Inspection Summary Reports." ADAMS Accession Nos. ML003693923, ML003693988, and ML003693995

Letter from M.S. Tuckman, Duke Power Company, to the NRC, dated February 24, 1998, "Duke Energy Corporation; Oconee Nuclear Station—Units 1, 2, and 3; Docket Nos. 50-269, 50-270, and 50-287; McGuire Nuclear Station—Units 1 and 2; Docket Nos. 50-369 and 50-370; Catawba Nuclear Station—Units 1 and 2; Docket Nos. 50-413, 50-414; Response to Generic Letter 97-05." NUDOCS Accession No. 9803040131

Letter from M.S. Tuckman, Duke Power Company, to the NRC, dated March 26, 1998, "Duke Energy Corporation; Oconee Nuclear Station—Units 1, 2, and 3; Docket Nos. 50-269, 50-270, and 50-287; McGuire Nuclear Station—Units 1 and 2; Docket Nos. 50-369 and 50-370; Catawba Nuclear Station—Units 1 and 2; Docket Nos. 50-413, 50-414; Response to Generic Letter 97-06." NUDOCS Accession No. 9803310294

Letter from H.B. Barron, Duke Energy Corporation, to the NRC, dated October 5, 2000, "McGuire Nuclear Station; Docket Nos. 50-370; Unit 2 Steam Generators, EOC-13 Refueling Outage." ADAMS Accession No. ML003758969

Letter from H.B. Barron, Duke Energy Corporation, to the NRC, dated December 14, 2000, "McGuire Nuclear Station, Unit 2; Docket No. 50-370; Steam Generator Tube Inspection Summary Report; September 2000, EOC-13 Refueling Outage." ADAMS Accession No. ML003779260

Letter from H.B. Barron, Duke Energy Corporation, to the NRC, dated March 20, 2002, "McGuire Nuclear Station; Docket Nos. 50-370; Unit 2 Steam Generators; EOC-14 Refueling Outage." ADAMS Accession No. ML020870351

Letter from H.B. Barron, Duke Energy Corporation, to the NRC, dated May 30, 2002, "McGuire Nuclear Station; Docket Nos. 50-370; Steam Generator In-Service Inspection Summary Report; Unit 2, End of Cycle (EOC) 14." ADAMS Accession No. ML021640320

Letter from D.M. Jamil, Duke Power, to the NRC, dated March 11, 2003, "McGuire Nuclear Station, Unit 2; Docket Nos. 50-370; Steam Generator Tube Inspections; Response to Request for Additional Information." ADAMS Accession No. ML030840642

Letter from R.E. Martin, NRC, to D. Jamil, Duke Energy Corporation, dated June 23, 2003, "McGuire Nuclear Station, Unit 2 RE: Summary of NRC's Review of McGuire 2 Steam Generator Tube Inservice Inspection Report for Their Spring 2002 Outage." ADAMS Accession No. ML031740393

Letter from W.R. McCollum, Jr., Duke Power, to the NRC, dated October 28, 2004, "Duke Energy Corporation; Oconee Nuclear Station, Units 1, 2, & 3; Docket Nos. 50-269, 50-270, 50-287; McGuire Nuclear Station, Units 1 & 2; Docket Nos. 50-369 and 50-370; Catawba Nuclear Station, Units 1 & 2; Docket Nos. 50-413, 50-414; Response to NRC Generic Letter 2004-01, Requirements for Steam Generator Tube Inspections." ADAMS Accession No. ML043090390

Millstone 2

"Summary of Meeting of January 23, 1991, with Representatives of Northeast Utilities Concerning the Program for the Replacement of Steam Generators for Millstone 2 in 1992," dated February 11, 1991. NUDOCS Accession No. 9102180104

Letter from E.A. DeBarba, Northeast Utilities System, to the NRC, dated February 28, 1995, "Millstone Nuclear Power Station, Unit Nos. 1, 2, and 3; Annual Report." NUDOCS Accession No. 9503100447

Letter from E.A. DeBarba, Northeast Utilities System, to the NRC, dated October 24, 1995, "Millstone Nuclear Power Station, Unit 2; Proposed Revision to Technical Specifications; Steam Generator Surveillance Requirement Extension." NUDOCS Accession No. 9510260302

Letter from F.R. Dacimo, Northeast Utilities System, to the NRC, dated March 7, 1996, "Millstone Nuclear Power Station, Unit No. 2; Additional Information Regarding Proposed Revision to Technical Specifications to Extend Steam Generator Surveillance Requirement." NUDOCS Accession No. 9603190181

Letter from M.L. Bowling, Jr., Northeast Utilities System, to the NRC, dated February 4, 1997, "Millstone Nuclear Power Station, Unit No. 2; Retraction of Proposed Revision to Technical Specifications Regarding One-Time Extension to Surveillance Requirement for Steam Generator Tubes." NUDOCS Accession No. 9702100474

Letter from M.L. Bowling, Jr., Northeast Nuclear Energy, to the NRC, dated July 11, 1997, "Millstone Nuclear Power Station, Unit No. 2; Steam Generator Tube Plugging." NUDOCS Accession No. 9707150264

Letter from M.L. Bowling, Jr., Northeast Nuclear Energy, to the NRC, dated October 23, 1997, "Millstone Nuclear Power Station Unit No. 2; Steam Generator Tube Inservice Inspection Report." NUDOCS Accession No. 9710290050

Letter from M.L. Bowling, Jr., Northeast Nuclear Energy, to the NRC, dated February 26, 1998, "Millstone Nuclear Power Station, Unit No. 2; Annual Report." NUDOCS Accession No. 9803040419

Letter from M.L. Bowling, Jr., Northeast Nuclear Energy, to the NRC, dated March 10, 1998, "Millstone Nuclear Power Station, Unit No. 2; Response to Generic Letter 97-05, Steam Generator Tube Inspection Techniques." NUDOCS Accession No. 9803190306

Letter from M.L. Bowling, Jr., Northeast Nuclear Energy, to the NRC, dated March 26, 1998, "Millstone Nuclear Power Station, Unit No. 2; Response to Generic Letter 97-06, Degradation of Steam Generator Internals." NUDOCS Accession No. 9804070392

Letter from W.M. Dean, NRC, to M. L. Bowling, Jr., Northeast Nuclear Energy Company, dated August 20, 1998, "Millstone Nuclear Power Station, Units 2 and 3—Closeout of Generic Letter 97-05 (TAC Nos. MA0473 and MA0474)." NUDOCS Accession No. 9809250008

Letter from R.B. Eaton, NRC, to R. P. Necci, Northeast Nuclear Energy Company, dated October 31, 1999, "Millstone Nuclear Power Station, Unit No. 2, RE: Closeout of Generic Letter 97-06, 'Degradation of Steam Generator Internals' (TAC No. MA0924)." ADAMS Accession No. ML993200078

Letter from S.E. Scace, Northeast Nuclear Energy, to the NRC, dated February 22, 2000, "Millstone Nuclear Power Station, Unit Nos. 2 and 3; Annual Report." ADAMS Accession No. ML003687006

Letter from C.J. Schwarz, Northeast Nuclear Energy, to the NRC, dated May 22, 2000, "Millstone Nuclear Power Station, Unit No. 2; Steam Generator Tube Plugging." ADAMS Accession No. ML003719737

Letter from R.G. Lizotte, Northeast Nuclear Energy, to the NRC, dated February 28, 2001, "Millstone Nuclear Power Station, Unit Nos. 2 and 3; Annual Report." ADAMS Accession No. ML010640097

Letter from J.A. Price, Dominion Nuclear Connecticut, Inc., to the NRC, dated February 28, 2002, "Millstone Nuclear Power Station, Unit Nos. 1, 2, and 3; Annual Report." ADAMS Accession No. ML020710669

Letter from C.J. Schwarz, Dominion Nuclear Connecticut, Inc., to the NRC, dated March 12, 2002, "Millstone Nuclear Power Station, Unit No. 2; Steam Generator Tube Plugging." ADAMS Accession No. ML020870655

Letter from J.A. Price, Dominion Nuclear Connecticut, Inc., to the NRC, dated February 28, 2003, "Millstone Power Station, Unit Nos. 1, 2, and 3; Annual Report." ADAMS Accession No. ML030700380

Letter from R.B. Ennis, NRC, to D. A. Christian, Dominion Nuclear Connecticut, Inc., dated September 10, 2003, "Millstone Power Station, Unit No. 2—Review of Steam Generator Tube Inspection Reports for the 2002 Outage." ADAMS Accession No. ML032390710

Letter from J.A. Price, Dominion Nuclear Connecticut, Inc., to the NRC, dated November 5, 2003, "Millstone Power Station, Unit No. 2; Special Report—Steam Generator Tube Inservice Examination." ADAMS Accession No. ML033240373

Letter from J.A. Price, Dominion Nuclear Connecticut, Inc., to the NRC, dated February 26, 2004, "Dominion Nuclear Connecticut, Inc.; Millstone Power Station Units 2 and 3; Technical Specifications Annual Report." ADAMS Accession No. ML040690874

Letter from L.N. Hartz, Dominion Nuclear Connecticut, Inc., to the NRC, dated September 23, 2004, "Dominion Nuclear Connecticut, Inc.; Millstone Power Station Unit 2; Response to Request for Additional Information Regarding Steam Generator Tube Inspection Summary for Fall 2003 Outage." ADAMS Accession No. ML042670416

Letter from V. Nerses, NRC, to D. A. Christian, Dominion Nuclear Connecticut, Inc., dated November 9, 2004, "Review of Steam Generator Tube Inservice Inspection Report for the 2003

Refueling Outage at Millstone Power Station, Unit No. 2 (TAC No. MC2525)." ADAMS Accession No. ML043070352

Letter from W.R. Matthews, Dominion Nuclear Connecticut, Inc., Virginia Electric and Power Company, to the NRC, dated October 29, 2004, "Virginia Electric and Power Company (Dominion); Dominion Nuclear Connecticut, Inc. (DNC); North Anna Power Station Units 1 and 2; Surry Power Station Units 1 and 2; Millstone Power Station Units 2 and 3; Sixty Day Response to NRC Generic Letter 2004-01; Requirements for Steam Generator Tube Inspections." ADAMS Accession No. ML043060099

North Anna 1

Letter from W.L. Stewart, Virginia Electric and Power Company, to the NRC, dated July 2, 1993, "Virginia Electric and Power Company; North Anna Power Station Units 1 and 2; Proposed Technical Specifications Change; Steam Generator Inspection Scope Reduction." NUDOCS Accession No. 9307150044

Letter from M.L. Bowling, Virginia and Electric Power Company, to the NRC, dated February 23, 1994, "Virginia Electric and Power Company; North Anna Power Station Units 1 and 2; 1993 Annual Steam Generator Inservice Inspection Report." NUDOCS Accession No. 9403040255

Letter from W.L. Stewart, Virginia Electric and Power Company, to the NRC, dated March 1, 1994, "Virginia Electric and Power Company; North Anna Power Station Units 1 and 2; Withdrawal of Proposed License Amendment." NUDOCS Accession No. 9403090173

Letter from M.L. Bowling, Virginia Electric and Power Company, to the NRC, dated October 11, 1994, "Virginia Electric and Power Company; North Anna Power Station Unit 1; Steam Generator Tube Plugging Report." NUDOCS Accession No. 9410190251

Letter from J.P. O'Hanlon, Virginia Electric and Power Company, to the NRC, dated November 10, 1994, "Virginia Electric and Power Company; North Anna Power Station Units 1 and 2; Reduced NRC Reviews and Inspections; Cost Beneficial Licensing Action." NUDOCS Accession No. 9411210268

Letter from M.L. Bowling, Virginia Electric and Power Company, to the NRC, dated February 22, 1995, "Virginia Electric and Power Company; North Anna Power Station Units 1 and 2; 1994 Annual Steam Generator Inservice Inspection Report." NUDOCS Accession No. 9502280283

Letter from M.L. Bowling, Virginia Electric and Power Company, to the NRC, dated February 21, 1996, "Virginia Electric and Power Company; North Anna Power Station Units 1 and 2; 1995 Annual Steam Generator Inservice Inspection Report." NUDOCS Accession No. 9602260342

Letter from M.L. Bowling, Virginia Electric and Power Company, to the NRC, dated March 26, 1996, "Virginia Electric and Power Company; North Anna Power Station Unit 1; Steam Generator Tube Plugging Report." NUDOCS Accession No. 9603290091

Letter from S.P. Sarver, Virginia Electric and Power Company, to the NRC, dated January 29, 1997, "Virginia Electric and Power Company; North Anna Power Station Units 1 and 2; Annual Steam Generator Tube Inservice Inspection Summary Report." NUDOCS Accession No. 9702050424

Letter from S.P. Sarver, Virginia Electric and Power Company, to the NRC, dated June 9, 1997, "Virginia Electric and Power Company; North Anna Power Station Unit 1; Steam Generator Tube Inspection Report." NUDOCS Accession No. 9706170129

Licensee Event Report, dated June 10, 1997, "Missed Surveillance on First Inservice Inspection of Steam Generator Tubes." NUDOCS Accession No. 9706180407

Letter from J.H. McCarthy, Virginia Electric and Power Company, to the NRC, dated February 4, 1998, "Virginia Electric and Power Company; North Anna Power Station Units 1 and 2; Annual Steam Generator Tube Inservice Inspection Summary Report." NUDOCS Accession No. 9802120274

Letter from J.P. O'Hanlon, Virginia Electric and Power Company, to the NRC, dated March 17, 1998, "Virginia Electric and Power Company; Surry Power Station Units 1 and 2; North Anna Power Station Units 1 and 2; NRC Generic Letter (GL) 97-05: Steam Generator Tube Inspection Techniques." NUDOCS Accession No. 9803230262

Letter from J.P. O'Hanlon, Virginia Electric and Power Company, to the NRC, dated March 30, 1998, "Virginia Electric and Power Company; Surry Power Station Units 1 and 2; North Anna Power Station Units 1 and 2; NRC Generic Letter (GL) 97-06: Degradation of Steam Generator Internals." NUDOCS Accession No. 9804080029

Letter from J.H. McCarthy, Virginia Electric and Power Company, to the NRC, dated October 5, 1998, "Virginia Electric and Power Company; North Anna Power Station Unit 1; Steam Generator Tube Inspection Report." NUDOCS Accession No. 9810130122

Letter from J.H. McCarthy, Virginia Electric and Power Company, to the NRC, dated February 11, 1999, "Virginia Electric and Power Company; North Anna Power Station Units 1 and 2; Annual Steam Generator Inservice Inspection Summary Report." NUDOCS Accession No. 9902220321

Letter from D.A. Christian, Virginia Electric and Power Company, to the NRC, dated May 3, 1999, "Virginia Electric and Power Company; North Anna Power Station Units 1 and 2; Technical Specifications Changes; Primary-to-Secondary Leakage Rate Provisions and Detection System Operability Requirements." NUDOCS Accession No. 9905100156

Letter from D.A. Christian, Virginia Electric and Power Company, to the NRC, dated February 28, 2000, "Virginia Electric and Power Company; North Anna Power Station Units 1 and 2; Annual Steam Generator Inservice Inspection Summary Report." ADAMS Accession No. ML003692133

Letter from J.H. McCarthy, Virginia Electric and Power Company, to the NRC, dated April 3, 2000, "Virginia Electric and Power Company; North Anna Power Station Unit 1; Steam Generator Tube Inspection Report." ADAMS Accession No. ML003701796

Letter from S.P. Sarver, Virginia Electric and Power Company, to the NRC, dated February 26, 2001, "Virginia Electric and Power Company; North Anna Power Station Units 1 and 2; Annual Steam Generator Inservice Inspection Summary Report." ADAMS Accession No. ML010650296

Letter from S.P. Sarver, Virginia Electric and Power Company, to the NRC, dated October 4, 2001, "Virginia Electric and Power Company (Dominion); North Anna Power Station Unit 1; Steam Generator Tube Inspection Report." ADAMS Accession No. ML012850443

Letter from S.P. Sarver, Virginia Electric and Power Company, to the NRC, dated February 28, 2002, "Virginia Electric and Power Company (Dominion); North Anna Power Station Units 1, and 2; Annual Steam Generator Inservice Inspection Summary Report." ADAMS Accession No. ML020710697

Letter from S.P. Sarver, Virginia Electric and Power Company, to the NRC, dated October 16, 2002, "Virginia Electric and Power Company; North Anna Power Station Units 1 and 2; Request for Additional Information; Steam Generator Inservice Inspection Report." ADAMS Accession No. ML022960530

Letter from C.L. Funderburk, Virginia Electric and Power Company, to the NRC, dated February 26, 2003, "Virginia Electric and Power Company (Dominion); North Anna Power Station Units 1 and 2; Annual Steam Generator Inservice Inspection Summary Report." ADAMS Accession No. ML030690373

Letter from S.R. Monarque, NRC, to D.A. Christian, Virginia Electric and Power Company, dated March 17, 2003, "North Anna Power Station, Units 1 and 2—Review of Steam Generator Tube Inservice Inspection Reports for the 2001 Refueling Outages." ADAMS Accession No. ML030760028

Letter from C.L. Funderburk, Virginia Electric and Power Company, to the NRC, dated January 7, 2004, "Virginia Electric and Power Company (Dominion); North Anna Power Station Units 1 and 2; Annual Steam Generator Inservice Inspection Summary Report." ADAMS Accession No. ML040140749

Letter from C.L. Funderburk, Dominion Resources Services, Inc., to the NRC, dated October 4, 2004, "Virginia Electric and Power Company; North Anna Power Station Unit 1; Steam Generator Tube Plugging Report." ADAMS Accession No. ML042780547

Letter from W.R. Matthews, Dominion Nuclear Connecticut, Inc., Virginia Electric and Power Company, to the NRC, dated October 29, 2004, "Virginia Electric and Power Company (Dominion); Dominion Nuclear Connecticut, Inc. (DNC); North Anna Power Station Units 1 and 2; Surry Power Station Units 1 and 2; Millstone Power Station Units 2 and 3; Sixty Day Response to NRC Generic Letter 2004-01; Requirements for Steam Generator Tube Inspections." ADAMS Accession No. ML043060099

Letter from C.L. Funderburk, Dominion Resources Services, Inc., to the NRC, dated February 21, 2005, "Virginia Electric and Power Company (Dominion); North Anna Power Station Units 1 and 2; Annual Steam Generator Tube Inspection Report." ADAMS Accession No. ML050530049

Letter from S. Monarque, NRC, to D. A. Christian, Virginia Electric and Power Company, dated May 23, 2005, "North Anna Power Station, Unit 1—Review of Steam Generator Tube Inservice Inspection Report for the Fall 2004 Refueling Outage (TAC No. MC6162)." ADAMS Accession No. ML051470175

North Anna 2

Letter from W.L. Stewart, Virginia Electric and Power Company, to the NRC, dated July 2, 1993, "Virginia Electric and Power Company; North Anna Power Station Units 1 and 2; Proposed Technical Specifications Change; Steam Generator Inspection Scope Reduction." NUDOCS Accession No. 9307150044

Letter from M.L. Bowling, Virginia and Electric Power Company, to the NRC, dated February 23, 1994, "Virginia Electric and Power Company; North Anna Power Station Units 1 and 2; 1993 Annual Steam Generator Inservice Inspection Report." NUDOCS Accession No. 9403040255

Letter from W.L. Stewart, Virginia Electric and Power Company, to the NRC, dated March 1, 1994, "Virginia Electric and Power Company; North Anna Power Station Units 1 and 2; Withdrawal of Proposed License Amendment." NUDOCS Accession No. 9403090173

Letter from J.P. O'Hanlon, Virginia Electric and Power Company, to the NRC, dated November 10, 1994, "Virginia Electric and Power Company; North Anna Power Station Units 1 and 2; Reduced NRC Reviews and Inspections; Cost Beneficial Licensing Action." NUDOCS Accession No. 9411210268

Letter from M.L. Bowling, Virginia Electric and Power Company, to the NRC, dated February 22, 1995, "Virginia Electric and Power Company; North Anna Power Station Units 1 and 2; 1994 Annual Steam Generator Inservice Inspection Report." NUDOCS Accession No. 9502280283

Letter from J.P. O'Hanlon, Virginia Electric and Power Company, to the NRC, dated October 17, 1995, "Virginia Electric and Power Company; North Anna Power Station Unit 2; Proposed Technical Specifications Change; Steam Generator Inspection Scope Reduction." NUDOCS Accession No. 9510240269

Letter from J.P. O'Hanlon, Virginia Electric and Power Company, to the NRC, dated February 19, 1996, "Virginia Electric and Power Company; North Anna Power Station Unit 2; Withdrawal of Proposed License Amendment; Steam Generator Inspection Scope Reduction." NUDOCS Accession No. 9602220387

Letter from M.L. Bowling, Virginia Electric and Power Company, to the NRC, dated February 21, 1996, "Virginia Electric and Power Company; North Anna Power Station Units 1 and 2; 1995 Annual Steam Generator Inservice Inspection Report." NUDOCS Accession No. 9602260342

Letter from R.F. Saunders, Virginia Electric and Power Company, to the NRC, dated October 15, 1996, "Virginia Electric and Power Company; North Anna Power Station Unit 2; Steam Generator Tube Inspection Report." NUDOCS Accession No. 9610230201

Letter from S.P. Sarver, Virginia Electric and Power Company, to the NRC, dated January 29, 1997, "Virginia Electric and Power Company; North Anna Power Station Units 1 and 2; Annual Steam Generator Tube Inservice Inspection Summary Report." NUDOCS Accession No. 9702050424

Letter from J.H. McCarthy, Virginia Electric and Power Company, to the NRC, dated February 4, 1998, "Virginia Electric and Power Company; North Anna Power Station Units 1 and 2; Annual Steam Generator Tube Inservice Inspection Summary Report." NUDOCS Accession No. 9802120274

Letter from J.P. O'Hanlon, Virginia Electric and Power Company, to the NRC, dated March 17, 1998, "Virginia Electric and Power Company; Surry Power Station Units 1 and 2; North Anna Power Station Units 1 and 2; NRC Generic Letter (GL) 97-05: Steam Generator Tube Inspection Techniques." NUDOCS Accession No. 9803230262

Letter from J.P. O'Hanlon, Virginia Electric and Power Company, to the NRC, dated March 30, 1998, "Virginia Electric and Power Company; Surry Power Station Units 1 and 2; North Anna Power Station Units 1 and 2; NRC Generic Letter (GL) 97-06: Degradation of Steam Generator Internals." NUDOCS Accession No. 9804080029

Letter from J.H. McCarthy, Virginia Electric and Power Company, to the NRC, dated April 13, 1998, "Virginia Electric and Power Company; North Anna Power Station Unit 2; Steam Generator Tube Inspection Report." NUDOCS Accession No. 9804210129

Letter from J.H. McCarthy, Virginia Electric and Power Company, to the NRC, dated February 11, 1999, "Virginia Electric and Power Company; North Anna Power Station Units 1 and 2; Annual Steam Generator Inservice Inspection Summary Report." NUDOCS Accession No. 9902220321

Letter from D.A. Christian, Virginia Electric and Power Company, to the NRC, dated May 3, 1999, "Virginia Electric and Power Company; North Anna Power Station Units 1 and 2; Technical Specifications Changes; Primary-to-Secondary Leakage Rate Provisions and Detection System Operability Requirements." NUDOCS Accession No. 9905100156

Letter from D.A. Christian, Virginia Electric and Power Company, to the NRC, dated February 28, 2000, "Virginia Electric and Power Company; North Anna Power Station Units 1 and 2; Annual Steam Generator Inservice Inspection Summary Report." ADAMS Accession No. ML003692133

Letter from S.P. Sarver, Virginia Electric and Power Company, to the NRC, dated February 26, 2001, "Virginia Electric and Power Company; North Anna Power Station Units 1 and 2; Annual Steam Generator Inservice Inspection Summary Report." ADAMS Accession No. ML010650296

Letter from S.P. Sarver, Virginia Electric and Power Company, to the NRC, dated April 2, 2001, "Virginia Electric and Power Company; North Anna Power Station Unit 2; Steam Generator Tube Plugging Report." ADAMS Accession No. ML010990142

Letter from S.P. Sarver, Virginia Electric and Power Company, to the NRC, dated February 28, 2002, "Virginia Electric and Power Company (Dominion); North Anna Power Station Units 1, and 2; Annual Steam Generator Inservice Inspection Summary Report." ADAMS Accession No. ML020710697

Letter from E.S. Grecheck, Virginia Electric and Power Company, to the NRC, dated October 9, 2002, "Virginia Electric and Power Company; North Anna Power Station Unit 2; Steam Generator Tube Plugging Report." ADAMS Accession No. ML022960254

Letter from S.P. Sarver, Virginia Electric and Power Company, to the NRC, dated October 16, 2002, "Virginia Electric and Power Company; North Anna Power Station Units 1 and 2; Request for Additional Information; Steam Generator Inservice Inspection Report." ADAMS Accession No. ML022960530

Letter from C.L. Funderburk, Virginia Electric and Power Company, to the NRC, dated February 26, 2003, "Virginia Electric and Power Company (Dominion); North Anna Power Station Units 1 and 2; Annual Steam Generator Inservice Inspection Summary Report." ADAMS Accession No. ML030690373

Letter from S.R. Monarque, NRC, to D.A. Christian, Virginia Electric and Power Company, dated March 17, 2003, "North Anna Power Station, Units 1 and 2—Review of Steam Generator Tube Inservice Inspection Reports for the 2001 Refueling Outages." ADAMS Accession No. ML030760028

Letter from S.R. Monarque, NRC, to D.A. Christian, Virginia Electric and Power Company, dated May 21, 2003, "North Anna Power Station, Unit 2—Review of Annual Steam Generator Tube Inservice Inspection Report for the 2002 Refueling Outage." ADAMS Accession No. ML031410748

Letter from C.L. Funderburk, Virginia Electric and Power Company, to the NRC, dated January 7, 2004, "Virginia Electric and Power Company (Dominion); North Anna Power Station

Units 1 and 2; Annual Steam Generator Inservice Inspection Summary Report." ADAMS Accession No. ML040140749

Letter from W.R. Matthews, Dominion Nuclear Connecticut, Inc., Virginia Electric and Power Company, to the NRC, dated October 29, 2004, "Virginia Electric and Power Company (Dominion); Dominion Nuclear Connecticut, Inc. (DNC); North Anna Power Station Units 1 and 2; Surry Power Station Units 1 and 2; Millstone Power Station Units 2 and 3; Sixty Day Response to NRC Generic Letter 2004-01; Requirements for Steam Generator Tube Inspections." ADAMS Accession No. ML043060099

Letter from C.L. Funderburk, Dominion Resources Services, Inc., to the NRC, dated February 21, 2005, "Virginia Electric and Power Company (Dominion); North Anna Power Station Units 1 and 2; Annual Steam Generator Tube Inspection Report." ADAMS Accession No. ML050530049

Oconee 1

"Summary of the October 28, 1999, Meeting on Steam Generator Replacement Project (TAC Nos. MA6354, MA6355, and MA6356)," dated November 5, 1999. ADAMS Accession No. ML993210043

Letter from R.A. Jones, Duke Power, to the NRC, dated February 19, 2003, "Duke Energy Corporation; Oconee Nuclear Station Units 1, 2, and 3; Docket Nos. 50-269, 50-270, and 50-287; Proposed Amendment to the Renewed Facility Operating License and Technical Specifications for Steam Generator Replacement (TSCR 2002-01)." ADAMS Accession No. ML030560879

Letter from L.N. Olshan, NRC, to R.A. Jones, Duke Energy Corporation, dated September 4, 2003, "Oconee Nuclear Station, Units 1, 2 and 3 RE: Issuance of Amendments (TAC Nos. MB7737, MB7738, and MB7739)." ADAMS Accession Nos. ML032520601 and ML032540688

Letter from R.A. Jones, Duke Power, to the NRC, dated January 29, 2004, "Duke Energy Corporation; Oconee Nuclear Station, Unit 1; Docket No. 50-269; Steam Generator Inservice Inspection and Repairs; Steam Generator Tube 30-Day and 3-Month Reports." ADAMS Accession No. ML040370544

Letter from R.A. Jones, Duke Power, to the NRC, dated April 9, 2004, "Oconee Nuclear Station; Docket No. 50-269; Unit 1 EOC-21 Refueling Outage; Steam Generator Inservice Inspection; Steam Generator Three Month Report." ADAMS Accession No. ML041270237

Presentation by J. Batton, K. Davis, M. Klatt, and M. Addario, "Oconee Nuclear Station Replacement OTSG Pre-Service Inspection Using X-Probe™," 23rd EPRI Steam Generator NDE Conference, Chicago, Illinois, July 12–14, 2004.

Letter from W.R. McCollum, Jr., Duke Power, to the NRC, dated October 28, 2004, "Duke Energy Corporation; Oconee Nuclear Station, Units 1, 2, & 3; Docket Nos. 50-269, 50-270, 50-287; McGuire Nuclear Station, Units 1 & 2; Docket Nos. 50-369 and 50-370; Catawba Nuclear Station, Units 1 & 2; Docket Nos. 50-413, 50-414; Response to NRC Generic Letter 2004-01, Requirements for Steam Generator Tube Inspections." ADAMS Accession No. ML043090390

Letter from R.A. Jones, Duke Power, to the NRC, dated June 13, 2005, "Duke Energy Corporation; Oconee Nuclear Station, Unit 1; Docket No. 50-269; Steam Generator Inservice Inspection 1EOC22; Steam Generator Tube Plugging and Repair 30-Day Report." ADAMS Accession No. ML051730463

"Summary of Conference Calls with Oconee, Unit 1 Regarding Their 2005 Steam Generator Tube Inspections," dated June 29, 2005. ADAMS Accession No. ML051780116

"Summary of June 22, 2005, Meeting to Discuss Results of Spring 2005 Steam Generator Inspection at Oconee Nuclear Station, Unit 1 (TAC No. MC7234)," dated July 26, 2005. ADAMS Accession No. ML051940482

Letter from R.A. Jones, Duke Power, to the NRC, dated August 2, 2005, "Duke Energy Corporation; Oconee Nuclear Station, Unit 1; Docket No. 50-269; 1EOC22 Refueling Outage, April 2005; Steam Generator Inservice Inspection; Steam Generator 3-Month Report." ADAMS Accession No. ML052230208

Oconee 2

"Summary of the October 28, 1999, Meeting on Steam Generator Replacement Project (TAC Nos. MA6354, MA6355, and MA6356)," dated November 5, 1999. ADAMS Accession No. ML993210043

Letter from R.A. Jones, Duke Power, to the NRC, dated February 19, 2003, "Duke Energy Corporation; Oconee Nuclear Station Units 1, 2, and 3; Docket Nos. 50-269, 50-270, and 50-287; Proposed Amendment to the Renewed Facility Operating License and Technical Specifications for Steam Generator Replacement (TSCR 2002-01)." ADAMS Accession No. ML030560879

Letter from L.N. Olshan, NRC, to R.A. Jones, Duke Energy Corporation, dated September 4, 2003, "Oconee Nuclear Station, Units 1, 2 and 3 RE: Issuance of Amendments (TAC Nos. MB7737, MB7738, and MB7739)." ADAMS Accession Nos. ML032520601 and ML032540688

Letter from R.A. Jones, Duke Power, to the NRC, dated July 6, 2004, "Duke Energy Corporation; Oconee Nuclear Station, Unit 2; Docket No. 50-270; Steam Generator Inservice Inspection; Steam Generator Tube Plugging and Repair 30-Day Report." ADAMS Accession No. ML041950183

Presentation by J. Batton, K. Davis, M. Klatt, and M. Addario, "Oconee Nuclear Station Replacement OTSG Pre-Service Inspection Using X-Probe™," 23rd EPRI Steam Generator NDE Conference, Chicago, Illinois, July 12–14, 2004.

Letter from R.A. Jones, Duke Power, to the NRC, dated September 22, 2004, "Oconee Nuclear Station; Docket No. 50-270; Unit 2 EOC-21 Refueling Outage; Steam Generator Inservice Inspection; Steam Generator Three Month Report." ADAMS Accession No. ML042740471

Letter from W.R. McCollum, Jr., Duke Power, to the NRC, dated October 28, 2004, "Duke Energy Corporation; Oconee Nuclear Station, Units 1, 2, & 3; Docket Nos. 50-269, 50-270, 50-287; McGuire Nuclear Station, Units 1 & 2; Docket Nos. 50-369 and 50-370; Catawba Nuclear Station, Units 1 & 2; Docket Nos. 50-413, 50-414; Response to NRC Generic Letter 2004-01, Requirements for Steam Generator Tube Inspections." ADAMS Accession No. ML043090390

Oconee 3

"Summary of the October 28, 1999, Meeting on Steam Generator Replacement Project (TAC Nos. MA6354, MA6355, and MA6356)," dated November 5, 1999. ADAMS Accession No. ML993210043

Letter from R.A. Jones, Duke Power, to the NRC, dated February 19, 2003, "Duke Energy Corporation; Oconee Nuclear Station Units 1, 2, and 3; Docket Nos. 50-269, 50-270, and 50-287; Proposed Amendment to the Renewed Facility Operating License and Technical

Specifications for Steam Generator Replacement (TSCR 2002-01)." ADAMS Accession No. ML030560879

Letter from L.N. Olshan, NRC, to R.A. Jones, Duke Energy Corporation, dated September 4, 2003, "Oconee Nuclear Station, Units 1, 2 and 3 RE: Issuance of Amendments (TAC Nos. MB7737, MB7738, and MB7739)." ADAMS Accession Nos. ML032520601 and ML032540688

Presentation by J. Batton, K. Davis, M. Klatt, and M. Addario, "Oconee Nuclear Station Replacement OTSG Pre-Service Inspection Using X-Probe™," 23rd EPRI Steam Generator NDE Conference, Chicago, Illinois, July 12–14, 2004.

Letter from W.R. McCollum, Jr., Duke Power, to the NRC, dated October 28, 2004, "Duke Energy Corporation; Oconee Nuclear Station, Units 1, 2, & 3; Docket Nos. 50-269, 50-270, 50-287; McGuire Nuclear Station, Units 1 & 2; Docket Nos. 50-369 and 50-370; Catawba Nuclear Station, Units 1 & 2; Docket Nos. 50-413, 50-414; Response to NRC Generic Letter 2004-01, Requirements for Steam Generator Tube Inspections." ADAMS Accession No. ML043090390

Letter from R.A. Jones, Duke Power, to the NRC, dated May 10, 2005, "Oconee Nuclear Station; Docket No. 50-287; Unit 3 EOC-21 Refueling Outage; Steam Generator Inservice Inspection and Repairs; Steam Generator Tube 30-Day and Three-Month Reports." ADAMS Accession No. ML051370377

Palo Verde 2

Letter from D. Mauldin, Arizona Public Service Company, to the NRC, dated August 29, 2002, "Palo Verde Nuclear Generating Station (PVNGS) Unit 2, Docket No. STN 50-529; Response to Request for Additional Information Regarding Steam Generator Replacement and Power Uprate License Amendment Request." ADAMS Accession No. ML022470278

Letter from D.D. Chamberlain, NRC, to G.R. Overbeck, Arizona Public Service Company, dated May 4, 2004, "Palo Verde Nuclear Generator Station, Unit 2–NRC Special Inspection Report 05000529/2004-009." ADAMS Accession No. ML041260002

Letter from D.M. Smith, Arizona Public Service Company, to the NRC, dated June 1, 2004, "Palo Verde Nuclear Generating Station (PVNGS) Unit 2; Docket No. STN 50-529; License No. NPF-51; Licensee Event Report 2004-001-00." ADAMS Accession No. ML041600565

Letter from D. Mauldin, Arizona Public Service Company, to the NRC, dated October 28, 2004, "Palo Verde Nuclear Generating Station (PVNGS) Units 1, 2, and 3; Docket Nos. STN 50-528, STN 50-529, and STN 50-530; 60-Day Response to NRC Generic Letter 2004-01, 'Requirements for Steam Generator Tube Inspections.'" ADAMS Accession No. ML043090485

Point Beach 2

Letter from B. Link, Wisconsin Electric Power Company, to the NRC, dated September 26, 1996, "Dockets 50-266 and 50-301; Supplement to Technical Specifications Change Requests 188 and 189; Point Beach Nuclear Plant, Units 1 and 2." NUDOCS Accession No. 9610020093

Letter from D.F. Johnson, Wisconsin Electric Power Company, to the NRC, dated February 27, 1997, "Dockets 50-266 and 50-301; Annual Results and Data Report—1996 Point Beach Nuclear Plant, Units 1 and 2." NUDOCS Accession No. 9703050035

Letter from D.F. Johnson, Wisconsin Electric Power Company, to the NRC, dated December 11, 1997, "Docket 50-301; Filing of Owner's Inservice Inspection Summary Report; Point Beach Nuclear Plant, Unit 2." NUDOCS Accession No. 9712170473

Letter from D.F. Johnson, Wisconsin Electric Power Company, to the NRC, dated February 27, 1998, "Dockets 50-266 and 50-301; 1997 Annual Results and Data Report; Point Beach Nuclear Plant, Units 1 and 2." NUDOCS Accession No. 9803040445

Letter from D.F. Johnson, Wisconsin Electric Power Company, to the NRC, dated March 17, 1998, "Dockets 50-266 and 50-301; Response to Generic Letter 97-05; Steam Generator Tube Inspection Techniques; Point Beach Nuclear Plant, Units 1 and 2." NUDOCS Accession No. 9803240368

Letter from D.F. Johnson, Wisconsin Electric Power Company, to the NRC, dated March 30, 1998, "Dockets 50-266 and 50-301; Response to Generic Letter 97-06; Degradation of Steam Generator Internals; Point Beach Nuclear Plants, Units 1 and 2." NUDOCS Accession No. 9804080150

Letter from V.A. Kaminskas, Wisconsin Electric Power Company, to the NRC, dated October 14, 1998, "Dockets 50-266 and 50-301; Clarification of Response to Generic Letter 97-06; Degradation of Steam Generator Internals; Point Beach Nuclear Plants, Units 1 and 2." NUDOCS Accession No. 9810210004

Letter from V.A. Kaminskas, Wisconsin Electric Power Company, to the NRC, dated February 8, 1999, "Docket 50-301; Steam Generator Tube Plugging; Point Beach Nuclear Plant, Unit 2." NUDOCS Accession No. 9902180285

Letter from V.A. Kaminskas, Wisconsin Electric Power Company, to the NRC, dated February 25, 1999, "Dockets 50-266 and 50-301; 1998 Annual Results and Data Report; Point Beach Nuclear Plant, Units 1 and 2." NUDOCS Accession No. 9903100033

Letter from A.J. Cayia, Wisconsin Electric Power Company, to the NRC, dated February 28, 2000, "Dockets 50-266 and 50-301; 1999 Annual Results and Data Report; Point Beach Nuclear Plant, Units 1 and 2." ADAMS Accession No. ML003689804

Letter from D.E. Cole, Nuclear Management Company, LLC, to the NRC, dated November 9, 2000, "Docket 50-301; Steam Generator Tube Plugging; Point Beach Nuclear Plant, Unit 2." ADAMS Accession No. ML003769283

Letter from A.J. Cayia, Nuclear Management Company, LLC, to the NRC, dated February 28, 2001, "Dockets 50-266 and 50-301; 2000 Annual Results and Data Report; Point Beach Nuclear Plant, Units 1 and 2." ADAMS Accession No. ML010670316

Letter from A.J. Cayia, Nuclear Management Company, LLC, to the NRC, dated November 7, 2003, "Point Beach Nuclear Plant Unit 2; Docket 50-301; License No. DPR 27; Summary of Fall 2003 Unit 2 (U2R26) Steam Generator Eddy Current Examinations." ADAMS Accession No. ML040060249

Letter from D.L. Koehl, Nuclear Management Company, LLC, to the NRC, dated October 13, 2004, "Point Beach Nuclear Plant Unit 2; Docket 50-301; License No. DPR 27; Response to Request for Additional Information; Unit 2 Fall 2003 (U2R26) Steam Generator Eddy Current Examinations (TAC No. MC2070)." ADAMS Accession No. ML042990532

Letter from D.L. Koehl, Nuclear Management Company, LLC, to the NRC, dated October 29, 2004, "Point Beach Nuclear Plant, Units 1 and 2; Dockets 50-266 and 50-301; License Nos. DPR-24 and DPR 27; 60-Day Response Generic Letter 2004-01, 'Requirements for Steam Generator Tube Inspections.'" ADAMS Accession No. ML043100523

Letter from H. K. Chernoff, NRC, to D. Koehl, Nuclear Management Company, LLC, dated April 27, 2005, "Review of Point Beach Nuclear Plant, Unit 2, Steam Generator Tube Inspection Reports for 2003 Refueling Outage (TAC No. MC2070)." ADAMS Accession No. ML050460097

Prairie Island 1

Letter from T. Kim, NRC, to Nuclear Management Company, LLC, dated April 4, 2002, "Summary of March 19, 2002, Meeting Regarding the Steam Generator Replacement Program (TAC No. MB4607)." ADAMS Accession Nos. ML020980319 and ML020940042

Letter from J.G. Lamb, NRC, to Nuclear Management Company, LLC, dated April 1, 2003, "Summary of Meeting Between the Nuclear Regulatory Commission Staff and Nuclear Management Company, LLC, Held on March 12, 2003, Regarding the Steam Generator Replacement Program (TAC No. MB4607)." ADAMS Accession Nos. ML030910082 and ML030980830

Letter from D. Hills, NRC, to J. Solymossy, Nuclear Management Company, LLC, dated September 10, 2003, "Public Meeting to Discuss the Prairie Island Unit 1 Steam Generator Replacement Project." ADAMS Accession Nos. ML032530189 and ML032530383

Letter from J.M. Solymossy, Nuclear Management Company, LLC, to the NRC, dated October 29, 2004, "Prairie Island Nuclear Generating Plant Units 1 and 2; Dockets 50-282 and 50-306; License Nos. DPR-42 and DPR-60; 60-Day Response to Generic Letter 2004-01, 'Requirements for Steam Generator Tube Inspections.'" ADAMS Accession No. ML043090553

Letter from J.M. Solymossy, Nuclear Management Company, LLC, to the NRC, dated February 22, 2005, "Prairie Island Nuclear Generating Plant; Docket No. 50-282; License No. DPR-42; Unit 1 Inservice Inspection Summary Report, Interval 3, Period 3, Refueling Outage Dates: 9-10-2004 to 11-23-2004 Fuel Cycle 22: 12-7-2002 to 11-23-2004." ADAMS Accession No. ML0050910245

Letter from T.J. Palmisano, Nuclear Management Company, LLC, to the NRC, dated August 22, 2005, "Prairie Island Nuclear Generating Plant Units 1 and 2; Dockets 50-282 and 50-306; License Nos. DPR-42 and DPR-60; Response to Request for Additional Information Regarding the 'Unit 1 Inservice Inspection Summary Report, Interval 3, Period 3, Refueling Outage Dates: 9-10-2004 to 11-23-2004 Fuel Cycle 22: 12-7-2002 to 11-23-2004.'" ADAMS Accession No. ML052340676

Sequoyah 1

Letter from P. Salas, Tennessee Valley Authority, to the NRC, dated March 29, 2002, "Sequoyah Nuclear Plant (SQN)—Unit 1—Technical Specification (TS) Change No. 02-02, Steam Generator (SG) Alternate Repair Criteria (ARC) Deletion and SG Inspection Interval Revision." ADAMS Accession No. ML021020324

Letter from P. Salas, Tennessee Valley Authority, to the NRC, dated October 10, 2002, "Sequoyah Nuclear Plant (SQN)—Unit 1—Technical Specification (TS) Change No. 02-02, Steam Generator (SG) Alternate Repair Criteria (ARC) Deletion and SG Inspection Interval Revision—Partial Withdrawal." ADAMS Accession No. ML022960514

Letter from P. Salas, Tennessee Valley Authority, to the NRC, dated September 11, 2003, "Sequoyah Nuclear Plant (SQN)—Unit 1 Cycle 12 (U1C12) 90-Day Inservice Inspection (ISI) Summary Report." ADAMS Accession No. ML032660885

Letter from P. Salas, Tennessee Valley Authority, to the NRC, dated March 15, 2004, "Sequoyah Nuclear Plant (SQN)—Unit 1—Response to Request for Information Regarding

Inservice Inspection (ISI) Summary Report (TAC No. MC0940)." ADAMS Accession No. ML040760432

Letter from R.J. Pascarelli, NRC, to Tennessee Valley Authority, dated September 10, 2004, "Summary of June 17, 2003, Telephone Conference with the Tennessee Valley Authority Regarding the Preservice Steam Generator Tube Inspections at Sequoyah Unit 1 (TAC No. MC3377)." ADAMS Accession No. ML042570286

Letter from R.J. Pascarelli, NRC, to Tennessee Valley Authority, dated September 10, 2004, "Summary of May 20, 2004, Telephone Conference with Tennessee Valley Authority Regarding Potential Fabrication Damage to the Steam Generator Tubes at Sequoyah, Unit 1 (TAC No. MC3377)." ADAMS Accession No. ML042570229

Letter from P.L. Pace, Tennessee Valley Authority, to the NRC, dated October 29, 2004, "Sequoyah Nuclear Plant (SQN) and Watts Bar Nuclear Plant (WBN)—60 Day Response to Generic Letter (GL) 2004-01, 'Requirements for Steam Generator (SG) Tube Inspection,' dated August 30, 2004." ADAMS Accession No. ML043080413

Letter from P.L. Pace, Tennessee Valley Authority, to the NRC, dated February 15, 2005, "Sequoyah Nuclear Plant (SQN)—Unit 1 Cycle 13 (U1C13) 90-Day Inservice Inspection (ISI) Summary Report." ADAMS Accession No. ML050550413

South Texas Project 1

Presentation from South Texas Project Electric Generating Station to the NRC, September 16, 1996, "Unit 1 Steam Generator Replacement Project."

Letter from T.J. Jordan, STP Nuclear Operating Company, to the NRC, dated September 25, 2000, "South Texas Project Units 1 and 2; Docket Nos. STN 50-498 and STN 50-499; Steam Generator Tube Inservice Inspection Plan for the Second Inspection Interval at STP Units 1 and 2." ADAMS Accession No. ML003756607

Letter from T.J. Jordan, STP Nuclear Operating Company, to the NRC, dated November 6, 2001, "South Texas Project Unit 1; Docket No. STN 50-498; Special Report on Steam Generator Tube Plugging." ADAMS Accession No. ML020110160

Letter from T.J. Jordan, STP Nuclear Operating Company, to the NRC, dated January 22, 2002, "South Texas Project Unit 1; Docket No. STN 50-498; Special Report—1RE10 Refueling Outage; Inservice Inspection Results for Steam Generator Tubing." ADAMS Accession No. ML020390361

Letter from J.J. Sheppard, STP Nuclear Operating Company, to the NRC, dated January 28, 2002, "South Texas Project Units 1 and 2; Docket No. STN 50-498 and STN 50-499; License Amendment Request—Proposed Amendment to Technical Specification 4.4.5.3a." ADAMS Accession No. ML020310018

Letter from J.J. Sheppard, STP Nuclear Operating Company, to the NRC, dated May 23, 2002, "South Texas Project Units 1 and 2; Docket Nos. STN 50-498 and STN 50-499; License Amendment Request—Proposed Amendment to Operating License and Technical Specifications Regarding Steam Generators." ADAMS Accession No. ML021540264

Letter from T.J. Jordan, STP Nuclear Operating Company, to the NRC, dated June 20, 2002, "South Texas Project Units 1 and 2; Docket No. STN 50-498 and STN 50-499; License Amendment Request—Revised Proposed Amendment to Technical Specification 4.4.5.3a." ADAMS Accession No. ML021780019

Letter from T.J. Jordan, STP Nuclear Operating Company, to the NRC, dated July 3, 2002, "South Texas Project Unit 1; Docket No. STN 50-498; Response to Request for Additional Information." ADAMS Accession No. ML021910231

Letter from S.M. Head, STP Nuclear Operating Company, to the NRC, dated July 30, 2002, "South Texas Project Unit 1; Docket No. STN 50-498; Submittal of Information Previously Sent by Facsimile." ADAMS Accession No. ML022210287

Letter from J.L. Minns, NRC, to W.T. Cottle, STP Nuclear Operating Company, dated July 31, 2002, "South Texas Project, Unit 1—Issuance of Amendment on Steam Generator Surveillance Requirements (TAC No. MB3963)." ADAMS Accession No. ML022040265

Letter from D.H. Jaffe, NRC, to J.J. Sheppard, STP Nuclear Operating Company, dated August 21, 2003, "South Texas Project, Unit 1—NRC Staff's Review of the South Texas Project Electric Generating Station, Unit 1 Steam Generator Tube Inspection Report from the Fall 2001 Outage (TAC No. MB6541)." ADAMS Accession No. ML032340061

Letter from T.J. Jordan, STP Nuclear Operating Company, to the NRC, dated October 16, 2003, "South Texas Project Unit 1; Docket No. STN 50-498; License Amendment Request—Proposed Amendment to Technical Specification 4.4.5.3a." ADAMS Accession No. ML032930285

Letter from T.J. Jordan, STP Nuclear Operating Company, to the NRC, dated March 3, 2004, "South Texas Project Unit 1; Docket No. STN 50-498; Response to Request for Additional Information Regarding Proposed Amendment to Technical Specification 4.4.5.3a." ADAMS Accession No. ML040700529

Letter from M. Webb, NRC, to J.J. Sheppard, STP Nuclear Operating Company, dated June 8, 2004, "South Texas Project, Unit 1—Issuance of Amendment RE: One-Time Extension to Steam Generator Inservice Inspection Frequency (TAC No. MC1046)." ADAMS Accession No. ML041610073

Letter from T.J. Jordan, STP Nuclear Operating Company, to the NRC, dated June 21, 2004, "South Texas Project Units 1 and 2; Docket No. STN 50-498 and STN 50-499; License Amendment Request—Proposed Amendment to Technical Specification 4.4.5.3a." ADAMS Accession No. ML041750147

Letter from T.J. Jordan, STP Nuclear Operating Company, to the NRC, dated August 12, 2004, "South Texas Project Units 1 and 2; Docket No. STN 50-498 and STN 50-499; License Amendment Request—Proposed Amendment to Technical Specifications for Steam Generators." ADAMS Accession No. ML042310447

Letter from D.H. Jaffe, NRC, to J.J. Sheppard, STP Nuclear Operating Company, dated August 30, 2004, "South Texas Project, Units 1 and 2 RE: Withdrawal of an Amendment Request (TAC No. MC3613 and MC3614)." ADAMS Accession No. ML042450203

Letter from T.J. Jordan, STP Nuclear Operating Company, to the NRC, dated October 5, 2004, "South Texas Project Units 1 and 2; Docket No. STN 50-498 and STN 50-499; License Amendment Request—Revision to Proposed Amendment to Technical Specifications for Steam Generators." ADAMS Accession No. ML042860035

Letter from G.L. Parkey, STP Nuclear Operating Company, to the NRC, dated October 11, 2004, "South Texas Project Units 1 and 2; Docket No. STN 50-498 and STN 50-499; License Amendment Request—Correction to Proposed Amendment to Technical Specifications for Steam Generators." ADAMS Accession No. ML042890342

Letter from T.J. Jordan, STP Nuclear Operating Company, to the NRC, dated October 26, 2004, "South Texas Project Units 1 and 2; Docket No. STN 50-498 and STN 50-499; Response to NRC Generic Letter 2004-01." ADAMS Accession No. ML043070351

Letter from D.H. Jaffe, NRC, to J.J. Sheppard, STP Nuclear Operating Company, dated November 24, 2004, "South Texas Project, Units 1 and 2—Issuance of Amendments RE: Steam Generator Tube Integrity (TAC Nos. MC4048 and MC4049)." ADAMS Accession Nos. ML043290311 and ML043370370

Letter from T.J. Jordan, STP Nuclear Operating Company, to the NRC, dated March 3, 2005, "South Texas Project Units 1 and 2; Docket Nos. STN 50-498 and STN 50-499; Revised Steam Generator Tube Inservice Inspection Plan for the Second Inspection Interval at South Texas Project Units 1 and 2." ADAMS Accession No. ML050690200

Letter from D.H. Jaffe, NRC, to J.J. Sheppard, STP Nuclear Operating Company, dated May 27, 2005, "South Texas Project, Unit 1—RE: Discussions Concerning Foreign Objects Found in Steam Generators." ADAMS Accession No. ML051510378

South Texas Project 2

Letter from T.J. Jordan, STP Nuclear Operating Company, to the NRC, dated September 25, 2000, "South Texas Project Units 1 and 2; Docket Nos. STN 50-498 and STN 50-499; Steam Generator Tube Inservice Inspection Plan for the Second Inspection Interval at STP Units 1 and 2." ADAMS Accession No. ML003756607

Letter from J.J. Sheppard, STP Nuclear Operating Company, to the NRC, dated January 28, 2002, "South Texas Project Units 1 and 2; Docket No. STN 50-498 and STN 50-499; License Amendment Request—Proposed Amendment to Technical Specification 4.4.5.3a." ADAMS Accession No. ML020310018

Letter from J.J. Sheppard, STP Nuclear Operating Company, to the NRC, dated May 23, 2002, "South Texas Project Units 1 and 2; Docket Nos. STN 50-498 and STN 50-499; License Amendment Request—Proposed Amendment to Operating License and Technical Specifications Regarding Steam Generators." ADAMS Accession No. ML021540264

Letter from T.J. Jordan, STP Nuclear Operating Company, to the NRC, dated June 20, 2002, "South Texas Project Units 1 and 2; Docket No. STN 50-498 and STN 50-499; License Amendment Request—Revised Proposed Amendment to Technical Specification 4.4.5.3a." ADAMS Accession No. ML021780019

Letter from T.J. Jordan, STP Nuclear Operating Company, to the NRC, dated January 7, 2003, "South Texas Project Units 1 and 2; Docket Nos. STN 50-498 and STN 50-499; Corrections to Proposed Amendment to Operating License and Technical Specifications Regarding Steam Generators." ADAMS Accession No. ML030210106

Letter from A.P. Kent, STP Nuclear Operating Company, to the NRC, dated March 5, 2003, "South Texas Project Unit 2; Docket No. STN 50-499; Preservice Inspection Summary Report for Tubing in Replacement Steam Generators—2RE09." ADAMS Accession No. ML030710429

Letter from M. Thadani, NRC, to J.J. Sheppard, STP Nuclear Operating Company, dated July 21, 2003, "South Texas Project, Units 1 and 2—Issuance of Amendments to Technical Specifications and Unit 2 Operating License Regarding Steam Generators (TAC Nos. MB5158 and MB5159)." ADAMS Accession No. ML032130048

Letter from J.W. Crenshaw, STP Nuclear Operating Company, to the NRC, dated May 4, 2004, "South Texas Project Unit 2; Docket No. STN 50-499; Special Report on Steam Generator Tube Plugging." ADAMS Accession No. ML041450308

Letter from J.W. Crenshaw, STP Nuclear Operating Company, to the NRC, dated June 16, 2004, "South Texas Project Unit 2; Docket No. STN 50-499; Special Report—2RE10 Refueling Outage; Inservice Inspection Results for Steam Generator Tubing." ADAMS Accession No. ML041730355

Letter from T.J. Jordan, STP Nuclear Operating Company, to the NRC, dated June 21, 2004, "South Texas Project Units 1 and 2; Docket No. STN 50-498 and STN 50-499; License Amendment Request—Proposed Amendment to Technical Specification 4.4.5.3a." ADAMS Accession No. ML041750147

Letter from T.J. Jordan, STP Nuclear Operating Company, to the NRC, dated August 12, 2004, "South Texas Project Units 1 and 2; Docket No. STN 50-498 and STN 50-499; License Amendment Request—Proposed Amendment to Technical Specifications for Steam Generators." ADAMS Accession No. ML042310447

Letter from D.H. Jaffe, NRC, to J.J. Sheppard, STP Nuclear Operating Company, dated August 30, 2004, "South Texas Project, Units 1 and 2 RE: Withdrawal of an Amendment Request (TAC No. MC3613 and MC3614)." ADAMS Accession No. ML042450203

Letter from T.J. Jordan, STP Nuclear Operating Company, to the NRC, dated October 5, 2004, "South Texas Project Units 1 and 2; Docket No. STN 50-498 and STN 50-499; License Amendment Request—Revision to Proposed Amendment to Technical Specifications for Steam Generators." ADAMS Accession No. ML042860035

Letter from G.L. Parkey, STP Nuclear Operating Company, to the NRC, dated October 11, 2004, "South Texas Project Units 1 and 2; Docket No. STN 50-498 and STN 50-499; License Amendment Request—Correction to Proposed Amendment to Technical Specifications for Steam Generators." ADAMS Accession No. ML042890342

Letter from T.J. Jordan, STP Nuclear Operating Company, to the NRC, dated October 26, 2004, "South Texas Project Units 1 and 2; Docket No. STN 50-498 and STN 50-499; Response to NRC Generic Letter 2004-01." ADAMS Accession No. ML043070351

Letter from J.W. Crenshaw, STP Nuclear Operating Company, to the NRC, dated November 11, 2004, "South Texas Project Unit 2; Docket No. STN 50-499; Response to Request for Additional Information Regarding 2RE10 Steam Generator Inspection." ADAMS Accession No. ML043230294

Letter from D.H. Jaffe, NRC, to J.J. Sheppard, STP Nuclear Operating Company, dated November 24, 2004, "South Texas Project, Units 1 and 2—Issuance of Amendments RE: Steam Generator Tube Integrity (TAC Nos. MC4048 and MC4049)." ADAMS Accession Nos. ML043290311 and ML043370370

Letter from D.H. Jaffe, NRC, to J.J. Sheppard, STP Nuclear Operating Company, dated December 9, 2004, "South Texas Project, Unit 2—RE: Steam Generator Tube Inspection Reports for the Spring 2004 Outage (TAC No. MC3648)." ADAMS Accession No. ML043440065

Letter from T.J. Jordan, STP Nuclear Operating Company, to the NRC, dated March 3, 2005, "South Texas Project Units 1 and 2; Docket Nos. STN 50-498 and STN 50-499; Revised Steam Generator Tube Inservice Inspection Plan for the Second Inspection Interval at South Texas Project Units 1 and 2." ADAMS Accession No. ML050690200

St. Lucie 1

Letter from J.A. Stall, Florida Power & Light Company, to the NRC, dated May 28, 1997, "St. Lucie Unit 1; Docket No. 50-335; Steam Generator Eddy Current Testing." NUDOCS Accession No. 9706040289

Letter from L.A. Wiens, NRC, to T.F. Plunkett, Florida Power and Light Company, dated July 7, 1997, "Replacement Steam Generator Eddy Current Inspection for St. Lucie Unit 1 (TAC No. M98979)." NUDOCS Accession No. 9707100293

Letter from R.S. Kundalkar, Florida Power & Light Company, to the NRC, dated March 11, 1998, "St. Lucie Units 1 and 2; Docket Nos. 50-335 and 50-389; Generic Letter 97-05—Response." NUDOCS Accession No. 9803190069

Letter from J.A. Stall, Florida Power & Light Company, to the NRC, dated March 30, 1998, "St. Lucie Units 1 and 2; Docket Nos. 50-335 and 50-389; Generic Letter 97-06." NUDOCS Accession No. 9804070424

Letter from J.A. Stall, Florida Power & Light Company, to the NRC, dated June 23, 1998, "St. Lucie Units 1 and 2; Docket Nos. 50-335 and 50-389; Supplemental Response; Generic Letter 97-06." NUDOCS Accession No. 9806290310

Letter from J.A. Stall, Florida Power & Light Company, to the NRC, dated October 1, 1999, "St. Lucie Unit 1; Docket No. 50-335; Special Report; Steam Generator Tube Plugging Report." NUDOCS Accession No. 9910080125

Letter from R.S. Kundalkar, Florida Power & Light Company, to the NRC, dated January 28, 2000, "St. Lucie Unit 1; Docket No. 50-335; 1999 Steam Generator Tube Inservice Inspection Special Report." ADAMS Accession No. ML003684169

Letter from R.S. Kundalkar, Florida Power & Light Company, to the NRC, dated April 20, 2001, "St. Lucie Unit 1; Docket No. 50-335; Special Report; Steam Generator Tube Plugging Report." ADAMS Accession No. ML011140179

Letter from D.E. Jernigan, Florida Power & Light Company, to the NRC, dated August 20, 2001, "St. Lucie Unit 1; Docket No. 50-335; Refueling Outage SL1-17; Steam Generator Tube Inservice Inspection Special Report." ADAMS Accession No. ML012390098

Letter from W. Jefferson, Jr., Florida Power & Light Company, to the NRC, dated April 13, 2004, "St. Lucie Unit 1; Docket No. 50-335; Special Report; Steam Generator Tube Plugging Report." ADAMS Accession No. ML041130244

Letter from J.A. Stall, Florida Power & Light Company, to the NRC, dated October 29, 2004, "NRC Generic Letter 2004-01; Requirements for Steam Generator Tube Inspections." ADAMS Accession No. ML043070353

Letter from W. Jefferson, Jr., Florida Power & Light Company, to the NRC, dated April 8, 2005, "St. Lucie Unit 1; Docket No. 50-335; Refueling Outage SL1-19; Steam Generator Tube Inservice Inspection Special Report." ADAMS Accession No. ML051090258

Summer

Letter from J.L. Skolds, South Carolina Electric & Gas Company, to the NRC, dated March 11, 1994, "Virgil C. Summer Nuclear Station; Docket No. 50/395; Operating License No. NPF-12; Steam Generator Replacement Technical Specification Change Request (REM 6000-10, TSP 930019)." NUDOCS Accession Nos. 9403170301 and 9209160206

Letter from G.J. Taylor, South Carolina Electric & Gas Company, to the NRC, dated May 13, 1996, "Virgil C. Summer Nuclear Station (VCSNS); Docket No. 50/395; Operating License No. NPF-12; Special Report (SPR 960004)." NUDOCS Accession No. 9605170213

Letter from G.J. Taylor, South Carolina Electric & Gas Company, to the NRC, dated July 10, 1996, "Virgil C. Summer Nuclear Station; Docket No. 50/395; Operating License No. NPF-12; Special Report (SPR 960004), Revision 1." NUDOCS Accession No. 9607160001

Letter from G.J. Taylor, South Carolina Electric & Gas Company, to the NRC, dated November 3, 1997, "Virgil C. Summer Nuclear Station (VCSNS); Docket No. 50/395; Operating License No. NPF-12; Special Report (SPR 970001)." NUDOCS Accession No. 9711070068

Letter from G.J. Taylor, South Carolina Electric & Gas Company, to the NRC, dated February 2, 1998, "Virgil C. Summer Nuclear Station; Docket No. 50/395; Operating License No. NPF-12; Tenth Refueling Inservice Inspection Report." NUDOCS Accession No. 9805130081

Letter from G.J. Taylor, South Carolina Electric & Gas Company, to the NRC, dated March 13, 1998, "Virgil C. Summer Nuclear Station (VCSNS); Docket No. 50/395; Operating License No. NPF-12; Response to Generic Letter 97-05; 'Steam Generator Tube Inspection Techniques.'" NUDOCS Accession No. 9803230224

Letter from G.J. Taylor, South Carolina Electric & Gas Company, to the NRC, dated March 19, 1998, "Virgil C. Summer Nuclear Station (VCSNS); Docket No. 50/395; Operating License No. NPF-12; Response to Generic Letter 97-06; 'Degradation of Steam Generator Internals.'" NUDOCS Accession No. 9803230491

Letter from G.J. Taylor, South Carolina Electric & Gas Company, to the NRC, dated February 1, 1999, "Virgil C. Summer Nuclear Station (VCSNS); Docket No. 50/395; Operating License No. NPF-12; Supplemental Response to Generic Letter 97-06 'Degradation of Steam Generator Internals.'" NUDOCS Accession No. 9902040309

Letter from G.J. Taylor, South Carolina Electric & Gas Company, to the NRC, dated April 29, 1999, "Virgil C. Summer Nuclear Station; Docket No. 50/395; Operating License No. NPF-12; Special Report (SPR 1999-003)." NUDOCS Accession No. 9905050083

Letter from S.A. Byrne, South Carolina Electric & Gas Company, to the NRC, dated May 31, 2000, "Virgil C. Summer Nuclear Station (VCSNS); Docket No. 50/395; Operating License No. NPF-12; Supplemental Response to Generic Letter 97-06 'Degradation of Steam Generator Internals.'" ADAMS Accession No. ML003721068

Letter from S.A. Byrne, South Carolina Electric & Gas Company, to the NRC, dated November 8, 2000, "Virgil C. Summer Nuclear Station; Docket No. 50/395; Operating License No. NPF-12; Special Report (SPR 2000-005)." ADAMS Accession No. ML003769321

Letter from S.A. Byrne, South Carolina Electric & Gas Company, to the NRC, dated January 14, 2003, "Virgil C. Summer Nuclear Station (VCSNS); Docket No. 50/395; Operating License No. NPF-12; License Amendment Request—LAR 02-2767; Steam Generators—One-Time Exclusion of Inspection Frequency." ADAMS Accession No. ML030170174

Letter from S.A. Byrne, South Carolina Electric & Gas Company, to the NRC, dated July 1, 2003, "Virgil C. Summer Nuclear Station (VCSNS); Docket No. 50/395; Operating License No. NPF-12; Response to Request for Additional Information; License Amendment Request LAR-02-2767; Steam Generator Inspection Frequency." ADAMS Accession No. ML031840466

Letter from S.A. Byrne, South Carolina Electric & Gas Company, to the NRC, dated August 20, 2003, "Virgil C. Summer Nuclear Station (VCSNS); Docket No. 50/395; Operating License No.

NPF-12; Revision to License Amendment Request—LAR 02-2767; Steam Generators—One Time Exclusion of Inspection Frequency." ADAMS Accession No. ML032380053

Letter from K.R. Cotton, NRC, to S.A. Byrne, South Carolina Electric & Gas Company, dated October 29, 2003, "Virgil C. Summer Nuclear Station, Unit No. 1—Issuance of Amendment RE: One-Time Extension of the Steam Generator Inspection Frequency (TAC No. MB7312)." ADAMS Accession No. ML033020450

Letter from K.R. Cotton, NRC, to S.A. Byrne, South Carolina Electric & Gas Company, dated December 15, 2003, "Virgil C. Summer Nuclear Station, Unit No. 1—Correction to Amendment No. 165 Regarding Request for a One-Time Extension of the Steam Generator Inspection Frequency (TAC No. MB7312)." ADAMS Accession No. ML033490314

Letter from J.B. Archie, South Carolina Electric & Gas Company, to the NRC, dated October 27, 2004, "Virgil C. Summer Nuclear Station (VCSNS); Docket No. 50/395; Operating License No. NPF-12; 60-Day Response to NRC Generic Letter 2004-01; Requirements for Steam Generator Tube Inspections." ADAMS Accession No. ML043060420

NRC FORM 335 (9-2004) NRCMD 3.7	U.S. NUCLEAR REGULATORY COMMISSION	1. REPORT NUMBER (Assigned by NRC, Add Vol., Supp., Rev., and Addendum Numbers, if any.)
	BIBLIOGRAPHIC DATA SHEET *(See instructions on the reverse)*	NUREG-1841

2. TITLE AND SUBTITLE

U.S. Operating Experience With Thermally Treated Alloy 690 Steam Generator Tubes

3. DATE REPORT PUBLISHED	
MONTH	YEAR
August	2007

4. FIN OR GRANT NUMBER

N/A

5. AUTHOR(S)

Kenneth J. Karwoski, Gregory L. Makar, and Matthew G. Yoder

6. TYPE OF REPORT

Technical

7. PERIOD COVERED *(Inclusive Dates)*

01/1989 to 12/2004

8. PERFORMING ORGANIZATION - NAME AND ADDRESS *(If NRC, provide Division, Office or Region, U.S. Nuclear Regulatory Commission, and mailing address; if contractor, provide name and mailing address.)*

Division of Component Integrity
Office of Nuclear Reactor Regulation
U.S. Nuclear Regulatory Commission
Washington, DC 20555-0001

9. SPONSORING ORGANIZATION - NAME AND ADDRESS *(If NRC, type "Same as above"; if contractor, provide NRC Division, Office or Region, U.S. Nuclear Regulatory Commission, and mailing address.)*

Same as above

10. SUPPLEMENTARY NOTES

11. ABSTRACT *(200 words or less)*

This report documents the background and performance of thermally treated Alloy 690 steam generator tubing in U.S. commercial pressurized water reactors (PWRs). This material has been used extensively for replacement steam generators beginning in 1989, and as of December 31, 2004 it was being used in 30 units, or about 43% of the operating PWRs in the U.S. Of the 577,070 thermally treated Alloy 690 tubes placed in service, only 333 tubes (0.06%) have been plugged after approximately 173 calendar years of operation. The majority of these tubes (65%) were plugged prior to placing the steam generators in service. The dominant inservice degradation mode, responsible for about 24% of the plugged tubes, has been wear caused by a support structure or loose part. No corrosion or cracking had been detected as of the time this report was prepared. The superior performance experienced to date with thermally treated Alloy 690 tubes compared to earlier tube materials is attributed to the alloy chemistry (principally the higher chromium content), the corrosion-resistant microstructure developed by the combination of alloy chemistry and thermal processing, and design improvements in replacement steam generators.

12. KEY WORDS/DESCRIPTORS *(List words or phrases that will assist researchers in locating the report.)*

PWR
steam generator (SG)
steam generator tube
steam generator inspection
replacement steam generator
Alloy 690
thermally treated
eddy current testing
nondestructive evaluation

13. AVAILABILITY STATEMENT

unlimited

14. SECURITY CLASSIFICATION

(This Page)

unclassified

(This Report)

unclassified

15. NUMBER OF PAGES

16. PRICE

NRC FORM 335 (9-2004)

PRINTED ON RECYCLED PAPER